Light and Death

Light and Death

Figuration in Spenser, Kepler, Donne, Milton

Judith H. Anderson

FORDHAM UNIVERSITY PRESS

New York 2017

Fordham University Press also publishes its books in a variety of electronic formats. Some content that appears in print may not be available in electronic books.

Visit us online at www.fordhampress.com.

Library of Congress Cataloging-in-Publication Data

Names: Anderson, Judith H., author.
Title: Light and death : figuration in Spenser, Kepler, Donne, Milton / Judith H. Anderson.
Description: New York : Fordham University Press, 2017. | Includes bibliographical references and index.
Identifiers: LCCN 2016015035 | ISBN 9780823272778 (hardback)
Subjects: LCSH: English literature—Early modern, 1500–1700—History and criticism. | Death in literature. | Metaphor in literature. | Analogy in literature. | Allegory. | BISAC: LITERARY CRITICISM / Renaissance. | SCIENCE / Philosophy & Social Aspects.
Classification: LCC PR428.D4 A53 2017 | DDC 820.9/3548—dc23
LC record available at https://lccn.loc.gov/2016015035

Printed in the United States of America

19 18 17 5 4 3 2 1

First edition

in memory of Helen Tartar

CONTENTS

Light and Death

Introduction

Issues of Death, Light, and Analogy

Introducing this book, the word *issue*, derived from Latin *exire*, "to go out" or "go forth," embraces a range of meanings, among them "outflows," "questions," or "problems," which in turn suggest "results," "departures," "developments," or even "extensions."[1] English *issues* is itself a historical extension of *exire*, one issuing from this verb over time. For readers of early modern texts, *issues* and its implied Latin root hold a haunting memory of Donne's play on it in *Deaths Duell*, famously his own funeral sermon, which examines the issues from, in, and through death—Latin *à* (*ab*), *in*, and *per* death—and charges them with lingering emotional content.[2] By using this word, I want to encompass my thematic concern with death, light, and figuration and, as this book progresses, specifically with analogical expressions of figuration. These concerns variously overlap, now two of them in a given chapter, now all three. Remaining central, they generate associated and extended interests: the relation of literature and mathematics, the methodology of thought and argument, and the processes of narrative, discovery, and interpretation. Creativity, optics, rhetoric, and language, which inhere in my central themes and their offshoots, or outflows, recurrently come into focus as well.

This book itself began as an intellectual process, and I have wanted to reflect this process in my presentation. The process, as realized, is at once personal and professional—indeed, experienced scholarship. Over the course of the volume, it aims for something broadly human, historical, and inclusive, employing mathematical science, logical semantics, rhetoric, grammar, and major poems inclusively rather than oppositely. The body of my introduction, which follows, includes first the experience of reading a short poetic excerpt and a narrative of the project's beginning with the engagement of death, darkness, privation—evil. Coincidentally, though not just by chance, this is what the last chapter begins with, too—Milton's hell. Narrative and argument are both ways of organizing knowledge and experience, and the one can coexist with the other or lead into it. The texts I examine are cases in point, as is my introduction to them.

In Spenser's posthumously published Mutabilitie Cantos, Death is a personified figure in a moving pageant, last in line after Life. Both figures immediately follow on and appear to follow from various markers of time in the previous stanza: Day, Night, and the Hours, who are the "faire daughters of high Ioue, / And timely Night," daughters of the highest of hierarchs and darkness. Such figures of time usher in the pageant's concluding stanza:

> And after all came *Life*, and lastly *Death*;
> *Death* with most grim and grisly visage seene,
> Yet is he nought but parting of the breath;
> Ne ought to see, but like a shade to weene,
> Vnbodied, vnsoul'd, vnheard, vnseene.[3]

Death, the final figure in the pageant, is characterized or, rather, literalized as "nought but parting of the breath." This line pretty much says what can be said about physical death and, perhaps, everything about death that is real. Continuing, however, the stanza describes Death as being "like a shade" lacking substance—a phantasm that is "Vnbodied, vnsoul'd, vnheard, vnseene." Breath may stop, but imagination lingers on nothingness. Just before the line defining death as merely the cessation of breath, however, Death is said to have a "grim and grisly visage," that is, a "forbidding and ghastly appearance." A *visage* is the face that confronts us, what is perceived, or first seen; one of its cognates is *visor*, the front piece of a helmet—a covering, a concealment, a disguise.[4] Physical death, the last breath, comes after this visage, as if in neutralizing response to it, yet the figuration of Death does not stop here. Instead, nothingness itself motivates figuration anew. Figuration, an expression of *poiesis*, "making," has itself become a charged *issue*, at once an outcome and a concern in this stanza, and for this reason the stanza leads into my project: *Light and Death: Figuration in Spenser, Kepler, Donne, Milton*.

This book began with another arresting figure, "the body of this Death" in Paul's Epistle to the Romans (7:24). Happenstance, an invitation to write about bodily disease, occasioned my attention to Paul's memorable phrase in relation to sin, body, and death in Spenser, Donne, and Milton. The resulting article, now with a modified title, has become Chapter 1 in this volume. A second invitation, this time to write about Spenser's Mutabilitie Cantos, established my interest in early modern thinking about death. The resulting essay, now revised in Chapter 2, treats "Mutability and Mortality in *The Faerie Queene*." Mutability is another arresting figure. Her Cantos begin with her relation to the Fall, more exactly to sin and death, and end with her skepticism about men's "minds (which they immortall call)" and thus about the immortality of the intellective soul (VII.vii.19). Within this frame, the chapter examines the relation of sin and death elsewhere in Spenser's poetry, especially in Books I and II of *The Faerie Queene*. My third engagement with death, again my response to an invitation, is, in modified form and retitled, now Chapter 3: "Satanic Ethos: Evil, Death, and Individuality in *Paradise Lost*."[5] This chapter locates the origin of evil in Satan's envy of the Son when God exalts him and promises that under the Son's reign all shall "abide / United as one individual Soule" (V.609–10). God's promise, which conveys more than a single meaning, introduces a central conflict in the poem between union and autonomy, loyalty and revolt, the threat of absorption and the standalone, single being. Satan's envious response to the Son is essentially negative—at once evil and deathly—for, as noted in Milton's *Christian Doctrine*, "all evil, and whatever is seen to lead to ruin, is encompassed summarily in Scripture under the sign of *death*."[6] Whereas Satan first responds to the Son with envy, he subsequently responds with pride to the promise of unity that inheres in God's plan. A realized character Satan may be, but the shifting affinity of his figure with prosopopeia, an imagined person, not to mention its cousin personification, is hardly in question.[7]

Having written three pieces that variously engage death and the figuration of death in the early modern period, the third of these solely on Milton, I looked at numerous books about attitudes to death and burial practices over the centuries from the classical period to our own.[8] Two curious, mundane facts stood out: a recent reprinting of Philippe Ariès's *Hour of Our Death*, a comprehensive history of this anthropologically fascinating subject, is housed in the distant medical school of my university, and in the actual funerals I have recently attended, elegiac poetry has been present as a conspicuous feature, whether complementing or replacing biblical readings in poetry or prose. I should not make too much of this coincidence, but since my writing about death, then by contrast about light, the study of which historically derives from optics, has inevitably led me back to the "two cultures" of science and the humanities to which C. P. Snow's phrase gave notoriety, I mention it. I shall refer to these cultures again in my chapters, since Snow's phrase participates explicitly and implicitly in the cultural code of modernity.[9] Funereal poetry and the holdings of a medical library are hardly pure examples of different cultures, but their coincidence suggests that death is one place where the cultural fissure is bridged. Beyond the visage of death and the cessation of physical life, what is known is nothing, Spenser's "nought," which inevitably, humanly occasions imaginative construction. Ariès's *Hour of Our Death* is in a medical school for good reason.

Scholarly engagement with morbidity led, with similar inevitability, to contrasting and thereby defining terms—less stifling, more imaginative ones. The Spenserian stanza with which I began this introduction also returns in concluding Mutability's pageant to the imagined figure of human Life—"Full of delightfull health and liuely ioy"—and noticeably does so *after* treating Death, the last of Mutability's (and Nature's) witnesses. As Shakespeare's mad Lear remarks near Dover, "There's Life in't [yet] . . . You shall get it by running," a sequence, like Spenser's stanza, alluding to the commonplace that whatever lives moves and changes.[10] In the words of Spenser's narrator, Life is "lively," or *dynamically* joyous—moving and changing, more restless than inert.

Yet light, rather than life itself, has proved the more attractive focus for my project because of its extraordinary connection with figuration, from the Bible to philosophy and science. One of its most obvious and abiding connections in Western culture, of course, is also with life: *Fiat lux*. Another is with imagination, or *phantasia*, a term, as Aristotle noted, that was formed from *phaos*, "light."[11] Still a third is the bearing of light, sine qua non, on vision in its many, varied meanings. Fittingly, Hans Blumenberg has described metaphors of light as "incomparable" in "their expressive power and subtle capacity to change." Because of these features, he adds, metaphysics has "from its beginnings" used light "to give an appropriate reference to its ultimate subject matter," namely, Being, which is difficult to grasp in material terms. Hence, "The relation of unity to plurality, of the absolute to the conditional, of origin to descent—[have] all found a 'model'" in light.[12] Of course the moment we think of death as something beyond "the parting of the breath," it becomes figural as well, but light has infinitely more creative potential, whether cultural or simply biological. Sooner than later, it yields what Jacques Derrida has memorably referred to as the *heliotrope*, some form or other figuring a sun god.[13] This, too, is a tropic form and thus partakes of the archtrope, metaphor. Creativity in human culture is always *poetry* when this word is understood to mean "making, producing, modeling, metaphorizing, hypothesizing": again the Greek *poiesis*, from *poiein*, "to produce, invent, create, make." So understood, poetry bridges the cultures.

Having written a book on metaphor and another on allegory, or continued metaphor, I was further drawn to ancient, early modern, and modern controversies about analogy—a particular kind of metaphor and, still more broadly, of figuration—and to its embedding in the study of light and optics in early modern science. Analogy also occurs conspicuously in religion, philosophy, literature, and science, thus spanning the two cultures, and it has been a lightning rod for controversy in these fields, both in the past and the present. Analogy belongs both to mathematics and to language. Over centuries it has been, and still is, *the* connector of the known to the unknown, the sensible to the subsensible and infinite. In this way, analogy, particularly as it pertains to light, as opposed to death, became the next stage of my engagement with figuration, and light itself, including both its perceptual opposites, darkness and blackness, and its mediating forms, such as twilight and shadow, became focal as well. Such mediating forms separate and connect day and night, light and dark, life and death. Thus the origin story of my thematic concerns with death, light, and analogy. What follows details their further development, including brief summaries of chapters. It starts where this paragraph did—with the spanning of the two cultures, an offshoot of these concerns, and it relates such a bridge to creativity, whether in literature or science, poetry or physics.

Among the many excellent, recent discussions of relations between early modern literature and science, discussions that are pertinently engaged in my subsequent text and notes, one has particularly stood out for me: Fernand Hallyn's *Poetic Structure of the World*. Hallyn stresses creativity, or *poiesis*, as I do. His study differs from mine, however, in focusing not on the use of analogical metaphor but on narrative *mythos* in science—that is, on "heuristic fiction," as Hallyn broadly renders Paul Ricoeur's interpretation of Aristotelian *mythos* ("emplotment"). Hallyn also focuses on visual rather than verbal art.[14] Painting and drawing align with the relative visibility of geometric forms and their susceptibility of illustration while arguably finessing their abstraction as mathematical signs and as formulae that can both provide and structure empirical evidence. Johannes Kepler, a mathematician and an astronomer whose work occupies a chapter in my study, affords a suggestive example of the difference between such illustration and the evidence in his *Mysterium Cosmographicum* (1596/1621). In this treatise, Kepler, noting that so far in his argument "certain signs agree with the theorem proposed [and crucially pictured] and make it probable," turns next to his empirical data regarding "the distances between the astronomical spheres and the geometrical derivations" and then adds, "if they do not agree, the whole of the preceding work has undoubtedly been a delusion."[15] Kepler's mathematical data never caused him to abandon his cosmological theory of nesting geometrical solids after its pictorial realization in his treatise, but the flawed correlation of theorized depiction with the evidence left him still searching for a more fully satisfying version of intelligible design. A geometrical fiction thus drove his search, which returned again and again to constructive analogy.

Kepler's geometric forms are *iconic*, to evoke a term from C. S. Peirce's classification of signs. For Peirce, an icon resembles its object, having a community in some aspect with it. An icon also possesses its significance—its character as a sign—even if its object lacks existence: for an example, consider a lead pencil streak representing a geometrical line, that is, a continuous extent of length without other dimensions. Treating iconic representation in Kant, Paul de Man describes "the denominative noun 'triangle,' in geometry, . . . [as] a trope, a hypotyposis which allows for the *representation* of an abstraction by a substitutive figure" that is nonetheless "fully rational" and appropriate—indeed, correlatively proper.[16] His description applies equally well to the penciled shape of a triangle. Other, more mundane and, correspondingly, less purely mathematical, more purely metaphorical examples of an icon include a map (a representation of the earth or a portion of it on a flat surface), an onomatopoeic word (murmur, meow), or an iconic poem, such as George Herbert's "Easter Wings" or, more subtly, his "Church Monuments." Increasingly, such examples cross from mathematics to rhetoric and literature and do so on a continuum, a bridge. Of course Peirce and de Man are moderns, and the icons of the early modern past could make additional claims on reality, ranging from realms of physics and mechanics to those of mysticism and magic—a point that I make in passing before returning from light, analogical metaphor, and *poiesis* to death, the opening focus of this book.

In whole or in part, the first three chapters, which engage death in Spenser, Donne, and Milton, deal as well with evil and sin, which are inseparable from death in these religious writers. Not surprisingly, the figures of the Seven Deadly Sins and Spenser's Maleger, a reincarnation of the Pauline body of Death, recur in two chapters, and Milton's figures of Sin and Death recur in all three, each time within a different context and therefore from a different angle of perception. These deathly figures are intrinsic to discussion in these chapters. The subject of mortalism, the death of the individual soul, also pertains to two of these chapters and inevitably raises questions about individuation, selfhood, and knowledge, particularly self-knowledge. Both the soul and knowledge, variously involving the mind, inhere in such questions, and they return with the subject of death in the last two chapters as well. Epistemology in a broad sense intertwines with self-knowledge in these writers and does so in ways both on a continuum with modernity and different from it. A particular source of friction between the past and the present involves what is outside the mind and what is within it, as well as the bearing of each on the other. From another relevant angle, this friction occurs in the history of analogical theory, the subject of Chapter 4, where it involves the relation of imagination to science.

Reading, or interpreting, imaginative texts, another of my recurrent, open concerns in this and former volumes, responds to their invention and creativity. Such reading opposes the imposition of doctrine, content that is defined and fixed, from outside the workings of imaginative texts, as distinct from its incorporation, exploration, and revision within them. Sequential, contextual, and thereby situated reading, a particular interest of mine, is a distinct method of seeing, perceiving, and therefore interpreting. Textual *poiesis*, making, is also, ideally, an experiential process recreated and completed in reading, as Philip Sidney suggests when he describes the "imaginative ground-plot of a profitable invention" that a reader can find in, and take from, a poetic text and then act upon, whether intellectually, politically, aesthetically, or otherwise.[17] This process of reading includes what I'll simply call *textual respect*, at once encompassing verbal and rhetorical, intellectual and affective, historical and cultural dimensions. It is also attentive to the significance of form, which, certainly for each of the writers I treat in depth, inseparably and demonstrably participates in the meaning of the text—again, *each* of the writers I treat, Kepler included.

My fourth chapter is more abstract and theoretical than the preceding three on sin and death. It is a change of pace, a shift into another register and a refocusing of concerns from figures of death to the figure analogy. In terms of C. P. Snow's divide, to some readers it might even come as a culture shock, yet I hope an enabling one, since it once led and now leads to the chapter on Kepler. Treating analogy from ancient times to the present, Chapter 4 engages this tropic form as a broadly cultural concern and a method in language, rhetoric, religion, philosophy, and science. It also provides the pivot between my initial trio of chapters, in which the figuration of sin and death is a recurrent concern, and later ones that focus on light, analogy, or death, or else on some combination of these. This chapter argues for the fundamental relation of analogy to the figure metaphor and opposes the untheorized disparagement of *mere* metaphor in otherwise valuable studies of early modern science. Several of these studies treat the relation of science to literature in this period; others belong purely to the history of science.

My chapter on the history and theory of analogy maintains that the structure of this trope is definitively rooted in Aristotle's discussion of metaphor in the *Poetics*, of which analogy is a specific form with a basis in mathematical proportion (A:B :: C:D, or "as old age is to life, so is evening to day").[18] These Aristotelian roots later extend to Latin treatises on language, grammar, and rhetoric, and then, with the addition of Boethius in late antiquity, to Scholasticism, especially the philosophy of Thomas Aquinas in the Middle Ages and the influential, logical semantics of Thomas de Vio Cajetan in the Renaissance. Like Aquinas, Cajetan, writing in the Aristotelian-Thomist tradition, brings the otherwise dominant classical tradition into conversation with faith. The same structural roots are everywhere evident in early modern science, as well as in our own time. Connections with analogical thinking in modernity frame and occasionally punctuate this chapter, and its substantial middle treats the background and practice of early modern analogy. If expanded, the chapter title would summarily read, "Connecting the Cultural Dots: Classical, Medieval, Early Modern, and Modern Traditions of Analogy."

Even while poetic language does not differ in *kind* from ordinary language, it is distinguished by its pronounced tropicality—indeed, its metaphoricity, since *metaphor*, known also as *translation*, is the umbrella term for all the tropes.[19] All tropes involve an element of translation—transference, displacement, replacement, transfiguration (an exemplary, not an exhaustive list)—and therefore involve movement from one place or thing to another (Latin *translatio*: *trans*, "across," and *transferre/-latus*, "carry"). Studies that slight poetic language yet argue for *poiesis*, or creativity, in science are off to an unbalanced and unhistorical start. They reflect and often endorse a lack of connection between the two sides, or most fundamental forms, of culture, whether in the present or in the past. They replay a move that Francis Bacon made in the first quarter of the seventeenth century when, in terms strikingly similar to those of Sidney's *Defence of Poesy*, written late in the preceding century, he claimed poetic invention for the history of science and *opposed* it to literary *poiesis*.[20] Bacon's claim was itself a strategic instance of displacement—not to say, claim jumping—rather than bridging. Typically, like Bacon, such modern studies assume that further attention to poetic metaphor is superfluous while exhibiting an overly simple, or perhaps merely an unexamined or untheorized, understanding of its work. Assigning metaphor to the subordinate tasks of illustration, expansion, and education but excising it from serious thought, experimentation, and discovery, they overlook its defining work, which is creative and productive. Correlatively, they elide, ignore, or simply mistake the fundamental, structural presence of metaphor in analogy itself, wherever it is found.

To risk statement of the obvious, the signs of mathematics and of language have in common both the function of signifying and more precisely of doing so within a system—that is, as interrelating elements in a complex whole. Both are humanly constructed, and both have stakes in reference, imagination, and creativity. In a memorable phrase used by Donne to describe poetry and arithmetic, rhetoric and mathematics, both are "counterfait Creation[s]"—that is, counterfactual ones (French *contre*, "counter," and *faire*, "to make")—and the fact countered, or elided, is sensible, empirical, or simply actual.[21] Both linguistic and mathematical systems are abstractive—regularizing, conceptualizing, universalizing. The relation of any human system of signs to the physical world is precisely relation rather than identity, or sameness, and the historical dustbin of once-valid scientific theories that mathematics either informed or constituted testifies to its hypothetical character and tenuousness. Whether verbal or mathematical—Aristotelian logic, Pythagorean numbering, Platonic-Euclidean forms—signs have the capacity to shape worlds of their own that can float balloon-like into a space far removed from fact as well as (to paraphrase Bacon) fundamentally to alter what is considered fact, or, in Bacon's lexicon, "Nature." Like the broader term *truth*, *fact* is obviously another flexible, mutable word-concept whether in the early modern period or in those on either side of it.[22] *Fact* is also a term that straddles the cultures, both in its mutability and, seen historically, in its ever-developing forms. The relative weighting of counterfactual creations of either the verbal or mathematical sort shifts over time with changing conceptions of reality. Put otherwise, there are real differences between these two sorts of sign systems, but there are also connections, and at least as long as human beings have analogue, not digital, brains, both are indispensable to human vision.

Variously treating analogy, light, and death, the last three chapters of my study center on the writings of Kepler, Donne, and Milton. The first of these, "Proportional Thinking in Kepler's Science of Light," studies analogy in Kepler's treatises, especially on optics and light but with regard for their intersection with his astronomy and cosmology. Kepler's avowed motivation for studying optics, in fact, derives from his interest in astronomy. The second of these chapters, "Analogy, Proportion, and Death in Donne's *Anniversaries*," treats analogy in Donne's two most sustained achievements in poetry, which focally concern death. "Proportion," as a form of this word in the titles of my Chapters 5 and 6 signals, is another, familiar name for analogy from the classical through the early modern period and, in practice, still is.

Kepler discovered the inverted and reversed image on the retina that vision produces as well as three enduring principles of planetary motion: the law of the ellipse (planetary orbits are elliptical, with the sun at one of two foci); the distance law (the farther a planet from the sun, the more slowly it orbits), which is also known as the area law (the radius vectors of elliptically orbiting planets sweep out equal areas in equal times); and the harmonic law, which is fundamentally analogical (the cubes of the mean distances of any two planets from the sun are proportional to the squares of their periods of revolution). The harmonic law discovered by Kepler is also remarkably similar to Newton's theory of gravitation.[23] Kepler's optical discoveries affected the very notion of what it means "to see," one of the most familiar of figurative expressions and, again, one dependent on light, as is human vision. The investigation of the visible heavens, source of sunlight, moonlight, and starlight, coincides with that of optical phenomena in Kepler's writings, most conspicuously in his *Ad Vitellionem Paralipomena*, whose title merits translation in full: *Paralipomena* [*Supplement*] *to Witelo whereby the Optical Part of Astronomy is Treated; above all, on the Technically Sound Observation and Evaluation of the Diameters and Eclipses of the Sun and Moon. With Examples of Important Eclipses*.[24] According to William H. Donahue, a learned translator of Kepler's scientific writings, light is the "coherent theme" of the *Paralipomena* and optics its "proper subject" (xiv).

At this point, I want to make two observations about the focus and organization of scholarly argument and about academic pedagogy because these are relevant to my inclusion of a chapter on Kepler. Both observations also have some of their origin in my own experience, mentally processed over time. The first bears on the validity and value of the study of an individual author or phenomenon that goes into much depth and detail, compared to the broad survey of a subject, whether published or taught.

Ideally, of course, the latter at least touches on actual examples, and the former incorporates a broader context and implications, as does my chapter on Kepler. Both approaches are valuable, but neither has necessary priority over the other: that is, the particular can valuably precede the general as much as the reverse.[25]

Coincidentally, Kepler's method in the *Paralipomena* affords an extreme historical example first of an unfolding process, since Kepler even treats the reader to pages of false starts as well as to successful ones, and second of inclusiveness, since he discusses valuable, related material that is also tangential with respect to a narrow focus. Perhaps I should add "initially tangential" since it turns out to be both important and relevant. I shall have more to say about Kepler's method(s) later, but for now I am only proposing more attention to the *process* of reading and discovery—an interpretive process that is informed by cultural and historical breadth rather than replaced by it. By coincidence, numerous recent studies in the history of science have urged increased attention to method and process.

Although Kepler's influence on Donne is not my emphasis in the sixth chapter, right after the one on Kepler, Donne is likely to have had some knowledge of the *Paralipomena* or of another, related optical study by Kepler.[26] He was certainly aware of Kepler's work on astronomy, specifically referring to Kepler's *De stella nova* (1606) in *Ignatius His Conclave*, *Paradoxes and Problems*, and *Biathanatos*.[27] These references range from satire to matter-of-fact acceptance. Kepler corresponded with the English mathematician and astronomer Thomas Harriott, who, through Henry Percy, Earl of Northumberland and biographically associated with Donne, might have been known to Donne as well.[28] Kepler and Donne also met in Linz during the Doncaster embassy (1619), and, according to Kepler's correspondence, Donne agreed at this time to carry a letter to Kepler's London agent.[29] Kepler's correspondence refers to Donne as a Doctor of Theology, a profession to which the deeply religious Kepler had earlier aspired. Kepler had numerous other contacts with England: he was quick to make use of William Gilbert's discoveries concerning magnetism, and he was an admirer of King James, to whom he sent a copy of his *De stella nova* in 1607 and dedicated his major treatise *Harmonice Mundi* in 1619.[30] Henry Wotton, another Jacobean ambassador and one of Donne's correspondents, in 1620 visited Kepler in Linz and attempted to persuade him to come to England, proffering assurances of King James's eagerness to receive him, and he wrote to Francis Bacon about a camera obscura he had seen on his visit with Kepler.[31] Kepler also engaged and rejected in print some of the more eccentric theories of the English Rosicrucian Robert

Fludd, who objected in turn to Kepler's mathematical-mystical astronomy in Kepler's late opus *Harmonice Mundi*.[32]

Analogy, or proportion, is everywhere in Kepler's thought and practice, at once in his science, in his visionary speculation, and in the (formal) poems that sometimes occur in his writings. Hallyn relevantly observes of *Mysterium Cosmographicum*, Kepler's first publication, that "Above all" its subject "is *proportion*" (185). Initially published in 1596, this early treatise, which Kepler himself considered the theoretical seed bed of his career, was republished by him in 1621 with extensive notes that update, while the text itself preserves, his earlier views, chronicling developments in his thinking over the intervening quarter century. Kepler, like Spenser, Donne, and Milton, was self-consciously engaged with the process of thinking and discovering over time. He embedded this process in his creative, scientific works.

The word-concept "proportion" also occurs conspicuously over a dozen times in Donne's *Anniversaries*, mostly in the first of these, where it registers a loss occasioned by death but encompasses much more, including the loss of cosmic coherence, form, harmony, correspondence, and even comprehension. In *The First Anniversarie*, the lamentations of Donne's speaker for the young woman on whom, though a stranger to him, loss centers are so insistent as clearly to signal that something is disproportionate about them. This dramatized speaker stages a performance that shows him to be stuck in the past, the Old Testament, the body, and this ruined world. By the end of *The First Anniversarie*, however, he has finally overheard and tired of his own excess and has found a Mosaic sense of purpose, but like Moses, as is generally recognized, he has not reached the promised land. In *The Second Anniversarie*, Donne's speaker exhibits a new lease on life. He replaces his relentlessly repetitious lamenting with a dynamic renewal of vision that is fundamentally analogous in its extension and understanding and eventually becomes openly so. The word "proportion" appears only twice in this *Second Anniversarie*, but with its working presence, confident, constructive analogy has returned. In this *Anniversarie*, unlike its predecessor, the very physicality of death, captured in analogy, enables redemption. At the end of *The Second Anniversarie*, desire returns as the erotic, Christian-Neoplatonic connector between heaven and earth, between the soul's longing for God and God's for the soul. Desire becomes the affective realization of analogical construction.

My final chapter, "Milton's Twilight Zone: Analogy, Light, and Darkness in *Paradise Lost*," has a primary focus on light that brings with it reference to darkness, night, and death as inevitably as binarism has historically structured Western thought. The stages between light and darkness, such as twilight, dusk, and shadow—situations of between-ness—pertain suggestively to this focus as well, complicating and broadening it. Analogy, both biblical and not so and both implicit and explicit, is the major trope relating Milton's heaven and earth, and analogical parody recurrently extends this trope to hell. The structure of Milton's epic similes (more exactly, epic similitudes) is also distinctively analogical, and they effectually participate in the debate about the valid use of analogy that is evident from ancient times to the present.[33] Milton's epic is also known to be much concerned with perspective and more specifically with optics, notably regarding the telescope (and particularly in *Paradise Regained* the microscope), not to mention the pertinence of optics to Milton's own blindness. Milton belongs to the generation after Kepler's, but he is certainly aware of the great scientists and philosophers of his time, including Kepler, Galileo, Hobbes, and Descartes. Vision, in all the ramifications of this culturally generative concept—sight, perspective, imagination, interpretation, mysticism, speculation, contemplation, and prophecy—is another, multicolored thread that weaves through my chapters, and it is appropriately important in this last one. By the time of Donne and Milton, an increasing reliance on visualization, especially in early modern science, had made its reliability a basic crux, one that Milton's Satan exploits to the full in his temptation of Eve in *Paradise Lost*.[34]

This last chapter, the seventh, makes a particular point of beginning where Milton's epic begins, Books I and II, his books of darkness. That is, in a poem in which situation, speaker, and perspective, including those of its narrators, are fundamentally significant, the setting of hell and the viewpoint of its demons are fundamental to what follows and not merely the honoring of a conventional beginning in medias res. These opening books provide the principle of contrast so important to wisdom and choice and, more exactly for readers, the initial, experiential sight and locus of evil in the poem, with its origin later to come. In the initial section of this chapter, blackness and the sunlight that is figured in epic analogies are focal. The next section of the chapter concentrates on Light, especially in Book III but also in the creation narrative of Book VII, and it otherwise recurrently moves out from Book III to make connections with Night, twilight, and darkness in heaven and with the Night of Chaos in Book II. Increasingly in this section, the relation of Night to God becomes my concern. Ultimately, both the cave in the mount of Milton's God and the boundless deep of Chaos and Night participate in the most common pairing in Western myth, that of the womb and the tomb, a point about Chaos familiar to Miltonists attending to the decades-old debates about this place and its relation to deity. My argument goes further, showing that each, cave and deep, is both or, rather, potentially both womb and tomb; moreover, each is bisexually so, like the "two great Sexes [or principles that] *animate* the World," thus ensouling it, principles that find exemplary, analogical expression in the "Male and Female Light" of the sun and moon in Raphael's cosmological narrative in Book VIII (150–51; my emphasis). The final section of this chapter treats "Analogies of Being, Temptation, and Redemption," namely, God's analogy of redemption in Book III, the scale of Edenic being in Book V, Satan's parody of this in Book IX, and in Books XI to XII, the analogical reading of history. In the end, analogical figuration remains still with us, a creative, productive, nuanced *issue*, as it has been since ancient Greece.

CHAPTER 1

"The Body of This Death"

Donne's Sermons, Spenser's Maleger, Milton's Sin and Death

In Romans 7:24, Saint Paul, lamenting the conflict between his enlightened mind and his sinful flesh, utters a cry that has echoed down the centuries: "O wretched man that I am! Who shall deliver me from the body of this death?" Almost predictably, Paul's moving outcry attracted the sustained attention of the poets of early modern England, conspicuous among them Spenser, Donne, and Milton—predictably, not only because of Paul's anguish but also because of its unusual phrasing and figuration. Popular English translations of the Bible, such as Geneva and King James, in accord with the Latin Vulgate and the Greek, carefully specify "this death," which might suggest that the death at issue is spiritual and that it results from sin, not simply from the physical constitution of humankind.[1] This vital distinction becomes less distinct, however, the moment we consider that physical death itself is implicit in the Adamic sin that human beings inherit and is thus embedded in their fallen nature. Moreover, the sin to which Paul, in the voice of Everyman, refers is specifically that of his fleshly members, and it further indicates the inextricability of spirit and flesh. Not surprisingly, the sins of the flesh that Paul describes also infect, or perhaps simply override, his rational will: "For the good that I would I do not: but the evil which I would not, that I do" (7:19, 23).

The biblical origins of the translated terms *flesh* and *body* enforce these complications, as does the underlying fact that Paul was fluent both in Hebrew and Greek. The Old Testament Hebrew word *bāśār*, or "flesh," refers to the muscular portion of the body and therefore bears a relatively restricted, physical sense. Yet by extension this Hebrew term can (and did) also designate a living person since Semitic thought does not firmly distinguish between the physical and psychic aspects of existence.[2] In the Old Testament, the term *flesh*, whether understood in a physical sense or more holistically, is decidedly oriented to an earthly existence. Like Hebrew *bāśār*, or "flesh," the Greek word *sárx* signifies "flesh," the soft tissue of the body, but, according to Liddell and Scott, in the New Testament *sárx* also comes to mean "*the flesh*, as opposed to spirit," and it, too, can signify human nature generally.[3] Paul's *sárx*, or "flesh," in Romans thus suggests a basis for devaluing earthly existence even while paradoxically insisting on its irreducible reality, as in Hebrew: little wonder that Paul cries out for relief since it is his flesh, also identified as his very nature, that ensures the deadly vitality of sin.

In one of Lancelot Andrewes's sermons on "The Lord's Prayer," a sermon first published in the early Jacobean period, this is clearly his understanding of Paul's outcry. The learned Bishop Andrewes, himself a Bible translator, glosses the entreaty "Deliver us from evil" as a desire not simply to be freed "from those sins unto which our lust hath already drawn us away into sin" but more fully to be freed "from that *infirmity* of the flesh and *necessity* of sinning which doth accompany our nature, in regard whereof the Apostle saith, *Quis me liberabit de hoc corpore mortis*? 'Who shall deliver me from *this* body of death?'"[4] Notably, Andrewes has transferred the demonstrative adjective *this* from "death" to "body," a shift that results in phrasing less striking than Paul's. While further emphasizing the word "body," his shift also lessens the emotional immediacy of the phrasing "this death" and, to my ear, removes its whisper of mystery.

Here I want to introduce a further linguistic refinement. In New Testament usage the Greek word for *flesh*, *sárx*, is not the same as for *body*, Greek *sōma*. In Paul's culminating outcry, the word *body* distinctively carries a more holistic, less elemental force than would *flesh*. "The [Pauline] body [*tou sōmatos*] of this death [*tou thanatou toutou*]," which arises in and from the flesh, attributes to such death as this a substantive wholeness that threatens to become all-encompassing. Paul's phrase references not merely the body, but a body possessed by *this* death. This is death itself as and with body. In Paul's phrase, death, a nonentity and the absence of being, might thus be said virtually or, indeed, poetically to assume substance as the figure of Death. The Greek text of Romans actually encourages this figurative reading. The successive genitives in the Greek *tou sōmatos tou thanatou* enable three readings of these words: a genitive of ownership affords the meaning "death's body"; a genitive of description results in "deathly body"; and a genitive of hypostasis suggests that death has itself a body—"the body of Death" and, more exactly, "the body of *this* Death: *tou sōmatos tou thanatou toutou.*"[5] Evident here, in the suggestion that a personified Death has bodily substance, is the "creative poetic . . . force" (*dichterisch gestaltenden . . . Kräften*) that Erich Auerbach has more generally attributed to the figurative impulse in Paul's Epistles. In them, the lineaments of what Elaine Scarry more recently describes as a "fictionalized world" and carefully excludes from her account of the Old Testament are already beginning visibly to form.[6] Scarry's fictionalized world might further be described as metasomatic or, perhaps more precisely, as meta–*merely* somatic since it is at once fictive and real. As Auerbach insists, Paul's (re)configuration of history is, while purposive and creative, also actual, experiential, and, indeed, in Auerbach's own terms, concretely historical.[7]

Invoking Auerbach, I should clarify my use of the term *figure* and its cognates in this chapter. The meaning I intend is broadly figural. It includes Auerbach's conception of *figurae*, but it is not limited to this necessarily historicized biblical form. My conception is more *openly* fictive and imaginative than Auerbach's *figurae*. It certainly includes figures of speech, which, in Paul Ricoeur's words, have a "quasi-optic" or "quasi-bodily externalization"—as what we commonly call "images," for instance—and it extends to *personae* in narrative and in drama that is distinctively leavened with symbolism, whether as Spenserian epic or as Shakespearean tragedy.[8] Jean-François Lyotard's argument that dream-work inheres in substantive figuration is especially provocative in relation to the poetics of Paul's Epistles. Lyotard considers "an imaged text . . . [to be] a discourse which is very close to the figure." Its proximity resides in "the figurative power of a word . . . [, in] the rhythmic power of syntax, and at an even deeper level, [in] the matrix of narrative rhythm, what Propp called form. . . ." The figure, as form, "jam[s]" the communicative constraints inscribed in "any *language* . . . [b]y virtue of the fact that it sets up a closed circuit intercom system of the work with itself." Lyotard concludes that "language, at least in its poetic usage, is possessed, haunted by the figure."[9] For my purpose, the figure in question is the body of *this* Death, in which the interlinked figuration of dream-work, image, and form proves hauntingly memorable.

The overload of implication in Paul's outcry "Who shall deliver me from the body of this death?" predictably has had historical consequences, and the specifically figurative aspects of these are my subject. Fictive—in Sidney's sense, "poetic"—texts speak a language inseparably and significantly bonded to figure and form. Such figures and forms are not only cognitive but also imaginative, affective, mnemonic, and variously sensuous—temporal and rhythmic, spatial and imagistic—as variously recognized by Lyotard and Ricoeur. They are themselves distinctive ways of *perceiving* significance, and they cannot be replaced with more abstractive interpretations without alteration and loss. A poetic landscape, plot, dialogue, pun, image, or figure can simultaneously or sequentially keep alternative and even opposite possibilities in play without deciding between them whereas a purely logical argument cannot or, at least, cannot without invoking a different register of meaning, such as faith. The issue of poetic meaning—of *poiēsis*, "making"—will recur as I treat the figurative and historical legacies of Paul's outcry.

Like Andrewes, John Donne, whose sermons are often in the broad, Sidneian sense *poetic*, more than once replaces the Bible translators' "body of this death" with "this body of death," but Donne also goes beyond Andrewes to extend the implications of replacement and, in effect, to depict the physical body itself as death. The Pauline text is one of many that enforce the ancient correlation of moral with physical ills—conspicuous, for example, not only in Donne but also in the disease-ridden critiques of Shakespeare's Thersites or in Spenser's depiction of the Deadly Sins who counsel Lucifera, each of whom is marked by a fitting affliction: ulcers for Envy, venereal disease for Lechery, edematous dropsy for Gluttony, and so on. While Thersites's foul mouth in *Troilus and Cressida* is a very touchstone for the entwining of moral and physical disease, his utterances prove not merely symptomatic but also infectious among the Greeks, as instanced when Ulysses describes Achilles as being "so plaguy proud that the death-tokens of it / Cry 'No recovery.'"[10] Nothing, however, quite matches the concentrated physical corruption in Thersites's railings themselves, which punctuate the play and which, like Lucifera's councilors, imply a sordid political, as well as a moral, condition. In one particularly expansive display of venom, Thersites invokes as witnesses to perversion "the rotten diseases of the south, guts-griping, ruptures, catarrhs, loads o' gravel i'th' back, lethargies, cold palsies, raw eyes, dirt-rotten livers, wheezing lungs, bladders full of imposthume, sciaticas, limekilns i'th' palm, incurable bone-ache and the rivelled fee-simple of the tetter" (V.i.17–22). No "ounce of civet" to sweeten the imagination here.[11]

Physical disease weighs heavily on the psyche as well: for example, in Ben Jonson's poem "To Heaven," Paul's memorable outcry becomes a barely suppressed death wish growing out of both sickness and sin: "Yet dare I not complaine, or wish for death / With holy *Paul*," Jonson laments, as his poem nears its conclusion.[12] The Pauline text also takes especially powerful form in the figures of Spenser's Maleger and Milton's Death, who fully realize its impulse to figuration. In the rest of this chapter, I intend to explore what various receptions of Paul's text can tell us about sin, body, and death in the early modern period. Having discussed Jonson's poem elsewhere, I shall subsequently content myself with Donne, Spenser, and Milton, in this unchronological order.[13] Donne's engagement with the Pauline text affords a suggestive gloss on the power it assumes in the poetry of Spenser and Milton, and for this reason I start with it.

Donne, himself sick and near death, returns to Paul's text early in his last sermon, *Deaths Duell*, preached at White Hall in 1631. Donne employs the Pauline text to climax his argument that "The *wombe* whiche should be the *house of life*, becomes *death* it selfe, if God leave us there." After remarking Old Testament texts on the extremity of misery in "*a mis-carying wombe*," he concludes, "as soon as wee are . . . inanimated, quickned in the womb . . . our parents have reason to say in our behalf, *wretched man that he is, who shall deliver him from this body of death?* for even the *wombe* is a *body of death*, if there bee no deliverer."[14] While the womb without deliverer figures humankind without Savior, the connection between embodied life and the body of death—indeed the identification of human life with death from the very instant of its bodily inception—is as inescapable in Donne's argument as in the subtitle of his sermon, namely, *A Consolation to the Soule, against the dying Life, and living Death of the Body* (10:229).[15]

In *Deaths Duell*, Donne is a dying man; in Izaak Walton's memorable words, his own death is "his hourly object."[16] But in sermons that evoke Paul's text early in Donne's clerical career, a similar perception of the Pauline inextricability of sin, death, and the body is evident. This engulfing involvement is anatomical and physiological, structural and functional; in a word, it is quite literally *visceral*.

In a sermon on pureness of heart delivered at Paul's Cross in 1617, Donne describes a heart that is habitually sinful, and he warns, "when sin hath got a heart in us, it will quickly come to be that whole *Body of Death*, which Saint *Paul* complains of, who shall deliver me from the Body of this Death?" (1:192). This time, Donne assigns the demonstrative adjective to "Death," as Paul does, but he also subordinates Paul's "Body of this Death" to his own prior phrase "*that* whole *Body of Death*," thus again claiming a demonstrative adjective for "*Body*" and emphasizing the body itself as death and death as the body (1:192; my emphasis on *that*). Even minute changes in Donne's sermons, as in his poems, prove significant.[17] Simultaneously here, Donne also animates sin as both the getter and begetter, at once the possessor and the breeder, of a wicked heart: "when sin hath *got* a heart in us," he says (my emphasis). Such getting is at once infectious and prolific, for

> when it is a heart, it will get a *Braine*; . . . A Brain which shall send forth sinews and ligaments, to tye sins together; and pith and marrow to give a succulencie, and nourishment, even to the bones, to the strength and obduration of sin; . . . So also if sin get to be a heart, it will get a *liver* to carry blood and life through all the body of our sinful actions; . . . And whilst we dispute whether the throne and seat of the soul be in the Heart, or Brain, or Liver, this tyrant sin will praeoccupate all, and become all; so, as that we shall finde nothing in us without sin, nothing in us but sin. (1:192)

Once again, Donne's conception could hardly be imagined in terms more viscerally physical. The rapacious colonizer of his body operates with the relentless efficiency of a necrotic disease. Explanatory clauses that I have omitted from the quotation, such as "That's the office of the liver" or other physiological offices, which "the brain does to the natural body," both register Donne's interest in medicine and afford an outline of basic physiology: Dr. Donne, indeed.[18]

Donne invokes the Pauline text in another early sermon, which was preached at Lincoln's Inn in 1618 on one of his favorite verses in the Psalms: "*For thine arrowes stick fast in me, and thy hand presseth me sore*" (2:49; Ps. 38:2). Such arrows as tribulations and temptations are a kind of disease and sickness, Donne explains, and

> They stick to us so, as that we are not sure, that any old diseases mentioned in Physicians books are worn out, but that every year produces *new*, of which they have no mention, we are sure. We can scarce expresse the number, scarce sound the names of the diseases of mans body; 6000 year hath scarce taught us what they are, how they affect us, how they shall be cur'd in us, nothing, on this side the *Resurrection*, can teach us. They stick to us so, as that they passe by *inheritance*, and last more generations in families, then the inheritance it self does; and when no land, no Manor, when no title, no honour descends upon the heir, the stone, or the gout descends upon him. And as though our bodies had not *naturally diseases*, and infirmities enow, we contract

> more, inflict more . . . in *mortifications*, and *macerations*, and *Disciplines* of this rebellious flesh. I must have this body with me to heaven . . . And yet I cannot have this body thither, except as S. *Paul* did his, *I beat down this body*, attenuate this body by mortification: *Wretched man that I am, who shall deliver me from this body of death?* (2:62–63)

Now the demonstrative adjective *this* is back with the body, emphasizing its misery. Donne concludes that the human condition is "miserable, perplexed, [and] riddling" since "sin makes the body of man miserable, and the remedy of sin, *mortification* [that is, "deadening," from Latin *mors/-tis*, "death"], makes it miserable too." But Donne does not end with this conclusion. Instead, he amplifies it at length in order to show how "*plenty* is a misery, and *mortification* is a misery too; plenty . . . is a *disease*, a continuall sicknes, for it breeds diseases; And *mortification*. . . . is a disease too, a continuall hunger, and fasting; and if we consider it at best, . . . mortification is but a *continuall physick*, which is misery enough" (1:63). This is not the end of the sermon, of course, and Donne goes on to comfort and hope. In all the sermon passages I have cited, however, the effects of sin are specifically and vividly imagined in bodily terms to the extent that the body itself, rather than sin, momentarily becomes focal. Once again, Scarry's work on *The Body in Pain* comes to mind, in particular her argument that historically God's being is substantiated, his presence made visible, in the diseased, pained, or wounded body (e.g., 148, 183).

In the gloomy light of these passages from Donne's sermons, Spenser's depiction of the Deadly Sins in the House of Pride in *The Faerie Queene*, Book I, seems, well, *natural*. Consider the depiction of Avarice, for example: Avarice, like Spenser's Mammon, hoards and autoerotically fingers a "heap of coine" in his lap. His life is "nigh vnto deaths dore" since he scarcely ever tastes "good morsell," and his only care is "To fill his [money] bags," sparing "from backe and belly" clothing and sustenance alike.[19] Predictably and ironically, Avarice suffers from gout, a disease whose pain renders bodily touch—grasping—as well as movement or even standing painful to him. He is psychically fixated and physically fixed. The one precedes the other, yet they seem irrevocably joined simultaneously to imply a *necessary* linkage and to invite a reversed reading, in this instance from gout, the physical symptom, to greed, its moral cause, and thus from afflicted sufferer to guilty agent, outside to inside. Spenser's Envy, another Deadly Sin, inwardly chews "his owne maw" and harbors a bosom serpent (I.iv.30–32). He suffers from ulcers, as does his prototype Langland's Envy. The genealogy of Envy is extensive, however, and at least by Tudor times, this vice commonly implies treasonous political malice as well. Perched atop "a rauenous wolfe," implicitly Envy, too, like Shakespeare's Cassius, has "a lean and hungry look," and he is also "sad" or, in this instance, both actively resentful and passively depressive.[20] Again, his moral disease irrevocably implies his physical one, with the reversed reading from outer appearance to inner condition almost irresistible. As in Donne's sermons, the implication of interior and exterior, psychic and physical, morality and disease is thoroughly entangled.

Donne's sermons also shed light on Spenser's Maleger, whose name signifies both evil and disease—Latin *malus* and *aeger*, respectively: evil and sin are synonymous, and disease implies death.[21] Maleger is the primary foe of the House of Alma, or physical and psychic health, in Book II of *The Faerie Queene*, and he is a haunting, memorably spectral figure, for all the morality that surrounds his vicious attacks on Alma, who presides over the castle of the healthy, temperate body. He figures the sin and death whose inextricable entanglement causes the outcry of Saint Paul. His troops include temptations to the senses and the deadly sins that the senses sustain, and their hail of arrows symbolically "heap[s]" on Arthur's shield (II.xi.19).[22] Boney, "wan as ashes," withered, "cold and drery as a Snake," Maleger, too, is armed with arrows recalling those that "stick sore" in the Psalmist of Donne's sermon. Maleger's helmet is a death's head, "a dead mans skull, that seemd a ghastly sight," and his substance is subtle and unsound, "That like a ghost he seem'd, whose graue-clothes were vnbound" (xi.20, 22). Maleger is accompanied by Impotence and Impatience, the two hags who together embody the negation of action and passion, of assertion and suffering, of power and endurance, of the familiar topos *agere et pati*, and thus of the basic coordinates of human experience, here perversely undone.[23]

For a moment, before pursuing Maleger's significance, I want again to stress the contradictory, negating *inextricability* of sin, death, and body that issues in the outcry of Saint Paul. I also want to recall my earlier generalizations about the interpretation of form and figure in fictive, or poetic, texts and, as the case in point, to engage interpretations that identify Maleger exclusively or primarily as Death *or else* as Sin, whether actual, original, or both. These interpretations result from a failure to read either *The Faerie Queene* or the Bible in a truly, inclusively figural way, instead privileging Patristic theology or some other theological polemic. While valuing theological commentary, we, like good Reformation readers, can nonetheless prioritize the words of the Bible itself, which are often provocatively ambiguous, richly suggestive, and otherwise poetic: again, witness the Epistles of Saint Paul. Poetry is itself a mode of creative thought, not just embellishment, illustration, or passive reproduction.

Harold L. Weatherby's Patristic interpretation of Maleger rejects A. C. Hamilton's representatively Augustinian, or Western, interpretation of Maleger as inherited Sin and interprets this figure instead as inherited Death, in accordance with an alternative, pre-Augustinian Patristic tradition in theology, which also continued as the post-Augustinian Eastern tradition. In either Weatherby's or Hamilton's readings, however, the excluded alternative quickly returns as an entailment, a fact that itself suggests the silliness of a reading that is less faithfully figural—less inclusive, dynamic, and holistic—than Saint Paul's.[24] Be that as it may, in both their readings, the lake that swallows Maleger is "the *font* in which, by virtue of the Passion, Christ buries the curse of Adam," whatever it might be.[25] In Weatherby's version, however, the deathly Maleger is both captain of the passions and passion itself.[26] Indeed, he really *is* passion since subjection to passion and to death are "synonymous conditions" (347–48).[27] Then Alma, his victim, represents humanity untainted by Adamic sin but subject to its consequence, death, from the eternal curse of which Arthur frees her (350–51).[28] A failure to read figurally is evident here, and it will be an issue in the next chapter as well.

Alma is innocent of actual sin, yet, as I read her, she is woefully besieged, "dismayd," "affright[ed]," and implicitly vulnerable to Maleger and his agents—*factors* in Renaissance parlance. Maleger and Alma belong to each other when we find them in the poem, as do Maleger and his vicious troops. Neither figure comes alone. They share a single landscape, and they are part of a composite, interlinked episode, as indicated, for example, not merely by plot and narrative sequence but even by resemblances between images within Alma's House and outside it among Maleger's forces and by the presence within Alma's memory of wars, treasons, and other depressing records of human history.[29] Maleger, moreover, is no more simply passive than he is active and, indeed, aggressive—both *pati* and *agere*, like the perverse hags who accompany him—and although Alma, a virgin, "had not yet felt *Cupides* wanton rage," the word "yet" implies that in time she will, or at least could, feel passion (ix.18).[30]

Everything about Maleger is contradiction and negation. His ghastly embodiment fully actualizes the *figure* oxymoron: he flees and attacks at once and to his antagonist, Prince Arthur, presents

> Flesh without blood, a person without spright,
> Wounds without hurt, a body without might,

> That could doe harme, yet could not harmed bee,
> That could not die, yet seemd a mortall wight,
> That was most strong in most infirmitee.
>
> (II.xi.40)

Finally recognizing that Maleger is an extraordinary enemy, Arthur, who has variously responded to him "halfe in rage," in "astonishment," and in "great affright, / And trembling terror," throws away both his sword "And his bright shield, that nought him now auayled." Barehanded and thus symbolically naked as a warrior, Arthur fatally crushes the breath out of Maleger, only to discover that this adversary rises again, Antaeus-like, from "his mother earth" (xi.38–43). Now dismayed and desperate in his battle with "this [1590: his] lifelesse shadow," this "dead-liuing swayne" born of "th'earth," Arthur suddenly remembers Maleger's origin, and again squeezing him lifeless, carries him to a "standing lake" and throws his "carrion corse" into it (xi.44–46). On the basis of the word "his" in 1590, A. C. Hamilton observes in his second edition that Arthur's battle with Maleger is a *psychomachia* (268n44.3). But the correction of "his" to "this" in "Faults escaped in the print" (F.E.) of 1590 questions this point. Would the Elizabethan corrector have wanted to make the psychic dimension less apparent? Did he simply miss it? While I hardly want to dismiss the psychic dimension of Arthur's battle, I think it too limited by itself. "The body of this death" is *more* material, indeed more specifically corporeal, than *merely* psychic. It affects the mind, but it is emphatically, really, crucially body, not mind alone.[31] There is a wholeness here that is belied by psychic abstraction.

If victorious at the end, Arthur is also wounded, feeble, and unable even to mount his horse. For the second time in this episode, he needs the help of his squire, and when he returns to the Castle of Alma, she "comfort[s] him in his *infirmity*" (xi.49; my emphasis). Simultaneously the word *infirmity* signifies Arthur's frailty and is itself lexically weighted with theological resonance in this context; it is also an echo recalling the oxymoronic nature of Maleger, who has proved "most strong in most infirmitee" (xi.40).[32] In Andrewes's commentary on the Lord's Prayer that I cited earlier, an "infirmity of the flesh and necessity of sinning . . . doth accompany our [very] nature." Years ago James Carscallen observed that Maleger is "the natural man's sickness unto death." More exactly, Carscallen saw that Arthur meets in his "deeply baffling," malevolent adversary Maleger "the body of death itself, man's own substance as man's enemy."[33] In other words, Arthur encounters the embodiment of Paul's desperate outcry in

Romans for deliverance from the body of *this* death—not simply *the* body of death, as Carscallen's statement implies, but the inextricability of sin, death, and the body. Sin is the agent of *THIS* death.

The Pauline text suggests why Maleger is so profoundly and mysteriously persistent; why Arthur, who fights Maleger as a natural man, cannot unveil his magic shield or use his magic sword, and why he twice needs the help of Timias, whose first assist, also attributed to grace ("had not grace thee [Arthur] blest"), comes during the battle with Maleger and is directed against Maleger's supporting hags, and whose second assist comes only when Arthur's battle is finished (xi, 30, 48). Paul's words also explain why Arthur, although the best of knights, is so grievously wounded—so utterly weaponless and vulnerable (xi.48, vs. 3–7). Their encapsulation of human mortality yet further illuminates why Maleger's attacks are referred to the tactics of Tartars and New World Indians, thus glancing at an inclusive expanse of lands, times, and peoples (xi.21, 26). Here, in contrast to Spenser's first Book, Arthur, not Redcrosse, is Everyman, and Arthur's battle more closely enacts the anguish of Paul's Epistles than Christ's triumphant revivification of humankind. At the same time, however, Arthur, as a redeemed Adam, is Christlike, sharing in Christ's death, burial, and mortification of the flesh and thereby dying to sin, a process, in Calvin's words, that "we pursue day by day and which will . . . be accomplished [only] when we pass from this life."[34] Although cleansed of guilt and its resultant condemnation, those baptized "are still besieged by sin and still carry sin" with them. The body of this death is theirs to battle, suppress, and, for any who are fully Christlike, perhaps even to eliminate. Perhaps: such fullness would be tantamount to the restoration of Edenic purity.

Aside from Arthur's need to separate Maleger from earth, symbolically from the source, the very material, of physical life, it is particularly significant that Maleger's body should be carried beyond the length of a dragon's tail—that is, three furlongs, as in Book I, Canto xi.11—and then finally buried from sight in stagnant ("standing") water.[35] These two eleventh-canto battles, that of Redcrosse with the dragon in Book I and that of Arthur with Maleger in Book II, are related. The baptismal well of I and the lake of II differ as well as reflect each other, as do a font and a natural body of water: in poems, the vehicles of metaphor bear on their tenors, *pace* the biblical exegetes of this world.[36] The waters of I and II inversely mirror each other: redemptive, revivifying, life giving in Book I and standing, lifeless, dead and deadening, and, indeed, in Donne's apt word, *mortifying*, in Book II.

Mortify, a word Donne employs conspicuously in his sermon on the Psalmist's lament that "*thine arrowes stick fast in me*," is exactly right to describe what finally happens to Maleger: *mortify*, "to subdue" and "to deaden," derives, via Latin *mortificare*, "to kill" or "to mortify, subject," ultimately from Latin *mors/mortis*, "death," as earlier noted less expansively. Mortification—again, literally "deadening"—the disciplining of the flesh and the remedy of sin, is itself a misery, as Donne describes it: "sin makes the body . . . miserable, and the remedy of sin, *mortification*, makes it miserable too."[37] Understood as mortification and, further, as a response to Paul's outcry, "O wretched man that I am! Who shall deliver me from the body of this death," Arthur's victory over Maleger is qualified at once by his own resulting and remaining infirmity when Timias assists his return to Alma's castle and by a shadow of uncertainty about the efficacy of Maleger's drowning. In actuality, Maleger just sinks from sight.

Insofar as the figure of Maleger alludes to the mythological Antaeus, there could be further reason for unease about his elimination. Antaeus is a Giant descended from Earth and Poseidon, god of the sea and of waters in general. If earth is Maleger's element, always sustaining him during his battle with Arthur, his mythological heritage suggests that in time water might turn out to be his element as well. One thinks of Guyon's passionate lapse at the nymphs' flowing fountain in the Bower of Bliss, soon to follow.

Yet Maleger's apparent elimination necessarily precedes the adventure of this Bower in Canto xii, as the canto's opening stanza clearly signals:

> Now *ginnes* this goodly frame of Temperaunce
> *Fayrely* to rise, and her adorned *hed*
> To pricke of highest prayse forth to aduaunce,
> Formerly grounded, and fast setteled
> On firme foundation of true *bountyhed*;
> And this braue knight, that for that vertue fightes,
> Now comes to point of that same perilous sted,
> Where Pleasure dwelles in sensuall delights,
> Mongst thousand dangers, and ten thousand Magick mights.
>
> (II.xii.1; my emphasis)

"Now" but only "Now ginnes this goodly frame of Temperaunce / Fayrely to rise . . . Formerly grounded, and fast setteled / On firme foundation of true bountyhed." Presumably the line "On firme foundation of true bountyhed," or "bountyhood," looks back to Arthur's defeat of Maleger. In it, the word *bountyhed*, which is the true foundation here, simply means "goodness," as Hamilton notes in his edition, but *bounty* alone means this.

The suffix *-hed*, which is punningly the noun *head*, may suggest mind over matter ("good head," *bonne tête*), but in this context it more strongly indicates a source and, to play on Spenser's verb "ginnes"—"Now ginnes" this frame to rise—a fundamental grounding and firm [*be*]*ginning*. The verb "ginnes," combined here with "rise," introduces an action that is logically incomplete, just as this verb does in Middle English. The rhyme word "hed" earlier in the stanza, where it is clearly the noun "head" ("her adorned hed"), further enforces the likelihood of a pun on this noun when "hed" recurs as a rhyming suffix ("bountyhed").[38] As head or source, *bountyhed* therefore recalls such analogous, familiar composite nouns as *wellhead* and *fountainhead*, sources of the fresh, running, living water that starkly contrasts with the "standing" water into which the disabled, mortified Maleger drops from sight. Maleger's defeat and all that it means thus enable Guyon's foray into the Bower.

Although the beginning of Canto xii is overwhelmingly positive and promising, buried—perhaps the better word is *submerged*—within the word "Fayrely" that modifies the rising of the "goodly frame of Temperaunce" is the hint of caution that theologically befits a sequel to Maleger's evident disappearance: "Now *ginnes* this goodly frame of Temperaunce / *Fayrely* to rise." While *fairly* typically conveys senses like "fully, actually, and really" and does so here, its lexical history and suggestive potential nonetheless include ambiguity, ranging, as the *OED* puts it, from a "very different," elusively qualified sense to the specific meaning "speciously" (from Latin *species*, "appearance").[39] A goal achieved *fairly well* still has room for improvement. The later lines of the stanza, which open into the Bower of Bliss, are no longer preoccupied with Maleger's apparent mortification. Anticipating instead the "peril," "sensuall delights," and "thousand dangers" still awaiting in Acrasia's Bower, they signal caution and incompletion more openly.

Leaving Maleger submerged and Guyon afloat with his Palmer in *The Faerie Queene*, I want now to turn to Milton's *Paradise Lost*, where the Pauline text resounds recurrently and brings with it memories of Maleger. The fundamental recollection of Paul's outcry comes with the figures Sin and Death in Book II. Yet the first sight, or rather the first sound, of Milton's Satan early in Book I is resonant with death and specifically with the memory of Hector's death. Satan's first words, addressed to Beelzebub, recall those of Vergil's Aeneas to the ghost of Hector on the night Troy falls: "If thou beest he; But O how fall'n! How chang'd / From him, who in the happy Realms of Light / Cloth'd with transcendent brightness didst out-shine / Myriads though bright."[40] If death is a denial of being, its shadow thus accompanies Satan from the very start, yet when Satan actually encounters a *relatively* more substantial realization of death and more exactly of *this* Death that arises from Sin, his sin, he fails to recognize its consanguinity with him—if such a word can be wrenched to describe this bloodless blood relation. Satan's humiliating, not to say mortifying, failure speaks volumes about the distance he has yet to fall into self-recognition and thereby into self-knowledge—for him, a grasp of nothingness.

Sin, both the brainchild and the incestuous sexual partner of Satan, is also the mother of Death, Satan's equally incestuous son/grandson, who subsequently fathers the howling pack (or "cry") of hellhounds that torment his mother/half-sister ceaselessly, both within and without (654). The enclosed circularity of all these relationships embraces solipsism, autoeroticism, and narcissism as well as incest; all too graphically it substantiates the intertwined relation of mind and will with flesh and of body with sin and death that is found in Saint Paul.[41] Sin's hellhounds are figures especially apt to this inextricable intertwining. I want to stress again that this relation is not just psychic or just fleshly, just inside or just outside, just subjective or just objective. Inside and outside, physical and psychic, "hourly conceiv'd / And hourly born," they embody the unceasing, temporal circularity of Sin's birthing, which is spatially duplicated in their encircling her middle and later reduplicated in the compulsive, despairing circularity of Satan's affecting soliloquy at the outset of Book IV: "Which way I flie is Hell; my self am Hell" (II.796–97, IV.75).

"Then when lust hath conceived, it bringeth forth sin: and sin, when it is finished, bringeth forth death": James 1:15 is often cited to gloss the relation of Milton's Sin and Death. While surely relevant, this grim, formulaic rendering of cause and effect lacks the emotional depth, inextricable entanglement, and explicit bodily figuration of sin and death in Paul's affective outcry, and it lacks as well its evident fascination for imaginative writers of the early modern period. The verse from James also refers specifically to lust, concupiscence, or the desire for carnal things, which is the first bad fruit rather than the root cause of Satan's rebellion in Milton's epic, as my third chapter will argue. The linearity of the verse from James, rather than its circularity, is further striking.

Shape without form and shadow with only virtual substance, Milton's figure of Death recalls both the Pauline Epistle and the oxymoronic contradictions and negations of Maleger and his hags.[42] He experiences and expresses wrath, lust, emptiness, and insatiable hunger. He is founded on the utter emptiness of desire per se, that is, desire in itself. Satan, though not the most reliable perceiver, calls him a detestable "Fantasm," an apparition or ghost, a name that links him to fantasy and thus suggests that he is at this point another offspring of the evil brain (II.743).[43] When first encountered in Milton's Hell, he therefore appears essentially to be the Pauline "body of *this* death," the specter born specifically of sin rather than of the fallen human body. His appearance actually *precedes* mortality, or death of the body, as *we* know it. Marked by conceptual indefinition and corporeal insubstantiality in Book II, his bodily reality as Digestive Apparatus ("Maw") and as a "vast unhide-bound *Corps*"—Latin *corpus*, "body"—is yet to come (X.587, 601; my emphasis). Only the Fall will actualize it.

These distinctions are neither fully, nor long, to be distinguished, however. As intimated early in *Paradise Lost*, Death's body becomes ours, and our mortal body death's, in the course of Milton's epic. Even before we reach the survey of biblical time in the final books, this physical embodiment of the Satanic "Fantasm" has been realized in the material history of Earth. The birth of Sin from Satan's head analogically parodies Miltonic "brooding" on the abyss in creative harmony with the Spirit at the outset of the poem since it perverts both mental and physical conception, and the birth of Death from Satanic Sin continues this analogous, parodic perversion. Death lacks a proper body, however, until "Earth trembl[es] . . . from her entrails, as again / In pangs, and Nature g[ives] . . . a second groan, / . . . at compleating of the mortal Sin / Original" (IX.1000–4). This description in Book IX eerily, memorably, significantly recalls Sin's account in Book II of Death's portentous birth from her "entrails": she feels "Prodigious motion . . . and rueful throes" as "this odious offspring . . . breaking violent way / Tore through . . . [her] entrails" to distort and transform her "nether shape" (780–85). Initially the realization of Satan's sinful, phantasmic conception, Death thus achieves a newly substantial existence with the Fall of humankind. Earth's birth pangs and groans witness and seal the *embodying* of sin in and as death. At this moment, the earthly body of *this* Death is born.

In explicitly giving to Earth the echo of Sin's distorting labors in the birthing of Death, however, the poem displaces this from Eve and declines to offer the immediate analogy otherwise suggested, namely that Eve gives birth to death as Sin gives birth to Death. It thus places the physical embodiment of Death at one remove from Eve's role in human procreation, as is appropriate to Milton's valuation of "wedded Love" even after the Fall (IV.750). Incidentally, Earth's monstrous birth also recalls Maleger's mother Earth in the epic of the Faerie poet whom Milton described to Dryden as his "Original."[44] It additionally recalls my earlier renderings of the Greek text of Romans 7:24 that would fully embody death—readings ranging from ownership by death, to description of death's effect, to the hypostasis of Death and finally its actual fleshing out as the human body.

Predictably, when Milton's Adam at last sees death corporeally realized—specifically embodied—in the full panoply of historical forms taken by human death, this is no longer simply the phantasmic or monstrous body of death but overwhelmingly death in all the physical agony arising from bodily maladies and diseases. Identified as multiple manifestations of a single vicious cause, intemperance, these maladies and diseases include "gastly Spasm[s]," "qualmes / Of heart-sick Agonie," a variety of fevers,

> Convulsions, Epilepsies, fierce Catarrhs,
> Intestin Stone and Ulcer, Colic pangs,
> Daemoniac Phrenzie, moaping Melancholie
> And Moon-struck madness, pining Atrophie,
> Marasmus, and wide-wasting Pestilence,
> Dropsies, and Asthma's and Joint-racking Rheums.
>
> (XI.481–88)

By the end of this list, which Milton expanded between his 1667 and 1674 editions, the pernicious alignment of vice with physical affliction is as evident as it was in the curses of Thersites or the procession of Deadly Sins in Spenser's first Book of Faerie.[45] When Adam questions the reason, and thus the justice, of this deforming and disfiguring of the human *imago dei*, the "Divine similitude" in humankind, the Archangel Michael tells him that "Thir Makers Image . . . Forsook them, when themselves they villifi'd / To serve ungovern'd appetite, and took / His Image whom they serv'd, a brutish vice" (XI.512, 515–18). "Villifi'd"—etymologically turned into "villeins/-ains" like the brutish hordes Spenser's Maleger commands—these Miltonic sinners likewise turn allusively into Circean beasts ("brutes").[46] They disfigure "Not Gods likeness, but their own, / Or if his likeness, by themselves defac't," since they have "pervert[ed] pure Natures healthful rules / To loathsom sickness" (521–24). In the mouth of the Archangel, these words convey an authority to which the fallen Adam, albeit at moments a straw man in these final visions, quickly and significantly enough assents. This exchange between Michael and Adam appears to be a thinly masked reflection of Milton's own views, as expressed elsewhere in his writings and as evident, for immediate examples, in the Archangel's typically focal, high valuation of the virtue of temperance in the last two books of the epic.[47]

Yet we still might wonder about the relation of a blind poet to views that blame humanity for all its maladies and might ponder this in further relation to other potential self-indictments by Milton, whether in relation to the vehemence with which his Son of God rejects the learning of Greece in *Paradise Regained* or even to Milton's belief that he lost his sight by straining his weak eyes in what turned out to be an evidently futile political effort.[48] Of course, there are more religious and more idealistic understandings of such Miltonic moments and passages readily available, and, by and large, Milton himself appears to have subscribed to these more positive understandings sincerely. Moreover, affliction and disappointment offered opportunities for the virtuous exercise of patience and fortitude; they could be welcomed (and rationalized) as the trial and reassuring evidence of faith: witness Prince Arthur's painful battle with Maleger. Yet one might still wonder: even St. Paul cried out in pain, as does Milton's blind Samson.

Probably nothing troubles the sublation of human misery, including its special and literal sublation within faith, more than pain, illness, and death. Guilt, implying responsibility and therefore agency, rationalizes these. It makes sense of them, so to speak, and does so evidentially, palpably, and immediately when the manifestations of guilt and punishment are seen to be written on the sick, deformed, and dying body. The evil cause is moral and spiritual, but its effect, like its sign, is physical. Historically, the reversed reading, from effect to cause, seems inevitable, as does the transfer of this death from sin to the death of the body in receptions of Paul's outcry that span the longue durée of the early modern period in England. This transfer is metaphorical, a simple analogy underlying it: sin:death :: cause:effect.[49] Metaphors and analogies are the means by which we make sense of life and death.

CHAPTER 2

Mutability and Mortality in *The Faerie Queene*

Spenser's Mutabilitie Cantos engage the subjects of sin and death from start to finish—from the introduction of Mutability herself through the end of her pageant. Descended from Mother Earth and the rebellious Titans, Mutability desires "Rule and dominion" over Earth, her maternal inheritance, "That as a Goddesse, men might her admire, / And heauenly honours yield" (VII.vi.4). To this end, she has perverted the order of Nature, quite altered "the worlds faire frame," and made all those "accurst / That God had blest" (vi.5):

> Ne shee the lawes of Nature onely brake,
> But eke of Iustice, and of Policie;
> And wrong of right, and bad of good did make,
> And death for life exchanged foolishlie:
> *Since which*, all liuing wights haue learn'd to die,
> And all this world is woxen daily worse.
> O pittious worke of *MVTABILITIE*!
> By which, we all are subiect to that curse,
> And death in stead of life haue sucked from our Nurse.
>
> (VII.vi.6; my emphasis, l. 5)[1]

Such havoc accomplished, Mutability decides "T'attempt th'empire of the heauens hight, / And *Ioue* himselfe to shoulder from his right" (vi.7). With her decision thus to reclaim the inheritance on her father's side as well, the action of the Mutabilitie Cantos begins.

Mutability's pride and ambition, her deleterious effects on nature, and especially her responsibility for the curse of mortality associate her with the Fall.[2] But if she has broken the "lawes of Nature," she does not rebel against Nature as such but rather seeks the authority of Nature herself in Canto vi and in Canto vii calls upon natural elements, cycles, and patterns to attest to her dominance in the world. Her name suggests a phenomenon, a principle, an archetype, perhaps a force that is larger, more inclusive, and variously other than the Fall, though also related to it. Should we want a single word or phrase to gloss her significance, we might try the one Spenser gives us in the opening lines of the Mutabilitie Cantos, namely, "*Change*," or "the euer-whirling wheele / Of *Change*," imagery that associates her with fortune, in all the challenge of this concept for a universe of law and order.[3] But the use of a personifying pronoun to refer to Mutability in my preceding sentence is somewhat misleading. The name *Mutability* does not occur until the fourth line of the opening stanza and then only as a transferential synonym for *Change*. She does not exist as a character before lines 5 through 9, at which point her form, emerging dramatically and sequentially, is conspicuous as a staged poetic construction rather than as the cipher for a prepackaged entity. Process, development, movement in the whole and the part, the poem and the stanza—these are the first point.

Neither a simple mirror of the Fall nor just a metonymic encoding of it, Spenser's Mutability is at once highly original and genealogically complex. Her figure and story—even the entirety of her Cantos—draw on numerous and diverse renderings of change, those of Ovid, Lucretius, and Boethius salient among them, and they do so without merely repeating, or figuring, any single one. In fact, the Fall seems more an expression of Mutability in her Cantos than is she of the Fall; she represents a concept of which the Fall is a historical expression. She is the larger term within the poem, and the Fall is an important manifestation of her in the Cantos bearing her name yet not an exclusive or complete manifestation. This is the way Spenserian allegory works: consider Duessa, double being, *duo esse*, whose fundamental principle of duplicity finds and interrelates many expressions in nature and history, the psyche, philosophy, theology, the Bible, the polity, and the church. To read Duessa as simply one of these would be to make her a one-to-one, metonymic substitution for it rather than a figure, a complex poetic construction, in tensive relation with it.

Indeed, as the evil antagonist of Una, or oneness, truth, and unity in Spenser's foundational first Book, Duessa, too, might even be seen as both cause and result of the Fall. Additional candidates could be discovered, suggesting a need to discriminate poetic details and whole stories in *The Faerie Queene* rather than simply equating them with doctrinal or other genera.

In some ways, Mutability also appears oddly out of sync with the Fall: for example, she is beautiful, as numerous readers have noted, and the damage she does to the world *precedes* her rebellion against the gods. Moreover, she damages a world in which Bellona, the goddess of war, and the infernal Hecate are *already* empowered and, indeed, afford role models for her to emulate (vi.3). Consider also the foolish satyr-god Faunus's successful temptation of the nymph Molanna with flattering words, "red Cherries," and theologically (and politically) redolent "Queene-apples"; this comic Ovidian vignette, which concludes the first Canto of Mutability, is variously interlinked with Mutability's disruption at its outset, and its folly teasingly suggests the Fall as much and as partially as does her figure (vi.43, 46).[4] The line I cited earlier about Mutability's foolish exchange of death for life certainly expresses her strongest tie to the Fall; nonetheless, Death, far from being the same as Mutability, is a comparatively minor figure in the poem who is paired with, and balanced by, Life near the end of Nature's (and Mutability's) pageant in the second of the Mutabilitie Cantos. There, in the single stanza with which my introduction began, Life and Death first share one line and then each gets an additional four, with Life getting the final set and thus, in effect, reversing Death's negation (vii.46). Mutability actually encompasses both life and death, if "all that moueth, doth mutation loue" (vii.55). Everything that lives, moves, according to this commonplace of natural philosophy in the Renaissance.

The reversion of Death to Life at pageant's end also accords with the continuous cycling of generation and degeneration and of spring and harvest in the eternal mutability of the Garden of Adonis in Spenser's third book. Not surprisingly in this connection, the figures Life and Death immediately succeed various representations of time in Nature's pageant (Day, Night, the Hours) and with these similarly recall Time's scythe and Venus's sorrow amid the "laughing blossoms" of Book III's fertile Garden (vi.42). In Mutability's closing argument, as in Nature's, Change, now along with Time, again becomes an exchangeable name of Mutability: the Titaness observes the reign of "*CHANGE*" over all creatures,

> For, who sees not, that *Time* on all doth pray?
> But *Times* do change and moue continually.
> So nothing here long standeth in one stay:
> Wherefore, this lower world who can deny
> But to be subiect still to *Mutabilitie*?
>
> (VII.vii.47)

The shift in Mutability's rhetorical questions from the personified abstraction *Time* to the depersonified *Times*, from singular concept to plural, phenomenal occurrences, suggests another kind of change that italicization only accentuates here. The measure of time, its reckoning into the minuscule and large periods that are its reality, is humanly perceived and extensively constructed. It is no coincidence that another personified construction, "an hory / Old aged Sire, with hower-glass in hand, / Hight *Tyme*," sits by the gates of the Moon Goddess Cynthia in Canto vi (8). The popularized Father Time, no less than Chronos and Saturn or Jove and Mutability herself, is humanly conceived and figured as a personification.[5] Nature's warning to Mutability that she seeks her own decay by desiring supremacy refers to the upstart's disappearance as a figure, as a concept, and, presumptively, as a distinguishable phenomenon: if she is all things in Nature, she is no thing. Of course, the judicial figure of Nature herself then vanishes, as if dissolving her own hypostasis into the words of her verdict, the conspicuously judicious, all-containing truth she has spoken.[6] Her departure is the final irony that accompanies and qualifies the affirmation in her judgment.

Maintaining that Spenser's Legend of Mutabilitie is his "only sustained allegory of the Fall," Harold Weatherby notes that in characterizing Mutability herself, whom he identifies as Death, Spenser's narrator never uses the word *sin*, as distinct from both the words *bad* and *wrong* and from reference to her deleterious effects on the natural and instituted orders (321). He acknowledges that Mutability is "a rebel in her own right" as well as through her Titanic bloodline, that she is thus "associated with the primaeval rebellion" of Satan and Adam, and that "rebellion is tantamount to sin," as the words *bad* and *wrong* might indicate. Yet, he adds,

> we normally think of sin as more nearly a condition than an act (rebellion) and more nearly the cause of "bad" and "wrong" than their equivalent. Sin is an entity or hypostasis, something which, as Saint Paul says, "*entered* into the world" (Romans 5:12) as a power or person might—something easily personified, as Milton's Sin, sprung from Satan's head, indicates. (321–22; emphasis mine)

For Weatherby, Milton's personification of Sin signals allegiance to an Augustinian view of the Fall, in contrast to Spenser's, and once again he concludes that Spenser has reversed the relation of sin to death by showing that "death (Mutabilitie) entered into the world [first] and [then] sin (at least 'bad' and 'wrong') by death." In short, "we sin because we die" (322, cf. 327, 330–32). If for Saint Paul, "the wages of sin are death," for the Eastern Fathers, as Weatherby interprets them, the wages of death are sin (Romans 6:23).[7] If we think otherwise, we are under the Augustinian spell of a "strenuously ethical understanding of the Fall," along with Aquinas, Luther, Calvin, and Milton (322–24).

Yet *bad*, *wrong*, and rebellion have become virtually indistinguishable from *sin* in Weatherby's own argument, and the characterization of sin as condition rather than act, if Milton is the exemplar, simply backfires. In *Christian Doctrine*, Milton defines evil—that is, sin for Milton—as an oblique or perverse *action*, which can include words, thoughts, or even the omission of a good action. Privation is the resulting punishment of evil, or sin, in Milton's view: writ large, this privation is death, indeed, the figure Death in Milton's epic.[8] Raphael's account in *Paradise Lost* of Satan's rebellion dramatizes the rebellious angel's deviant, progressive act of sinning, and Sin herself is the full manifestation, the realization and figurative embodiment, of Satan's evil *action*. Death is the ultimate and seemingly inevitable end of his evil—*end* in both its temporal and teleological senses. The point is that Satan's evil act, its figuration as his brainchild, and Death, its final, spectral, shapeless expression, are on a continuum. Satan's Sin holds the key to hell, suggesting that he is self-confined, albeit with God's cooperation. When Sin opens the gates of hell in response to Satan's evil desire, her action means that he sins again, and his doing so thrusts him into the worse hell to which he gives voice at the outset of Book IV. The continuity of Satan, his Sin, and Death is that of allegorical narrative, and it is belied by reduction to isolated figures and arrested moments.

Weatherby's reading of Milton's narrative, like his reading of Spenser's, affords a strong example of the more general problem of an interpretation that displaces narrative process with translative assertion, depends heavily on concordancing alone, and by such means supplants poetry with doctrine, religious or otherwise. Valuable as are verbal concordances, they are no substitute in themselves for figural and sequential reading. Glancing back once again to the opening stanza of the initial Canto of Mutabilitie, for example, we there observe first the invention, or poetic construction, of Mutabilility as a figure and therefore as the entity or hypostasis that Weatherby requires for an Augustinian and Miltonic reading but rejects because she is not named *Sin*. Only after this opening construction do we learn about Mutability's genealogy and rampages to date, follow her developing adventures, and fully recognize that neither *Sin* nor *Death* is a sufficient name for her.

I will return to Mutability's connection with mortality, but I want first to consider related passages on sin and death in Spenser's poetry, especially in the first two books of *The Faerie Queene*: Book II has particularly notable ties to the subjects of time and mortality in the Mutabilitie Cantos. Briefly, I will follow Weatherby's lead in considering passages he cites for support before departing from it to pursue additional episodes and broader implications. Again, his larger claim, whose influence persists in critical essays and editorial annotations, is that Spenser rejects an Augustinian, or Western, conception of original sin as inherited guilt and subscribes instead to an earlier Patristic tradition in which the Adamic inheritance is death alone; crucially for his claim, Mutability figures Death.

Weatherby considers two passages in *The Faerie Queene* open to an Augustinian conception of the Fall, however. One passage occurs when Redcrosse sees the New Jerusalem built for God's

> chosen people purg'd from sinful guilt,
> With pretious blood, which cruelly was spilt
> On cursed tree, of that vnspotted lam
> That for the sinnes of al the world was kilt.
>
> (I.x.57)

The other passage occurs in the chronicle of *Briton moniments*, which Arthur discovers in Alma's brain-turret (II.ix.59). Dating events in the chronicle, the narrator remarks "What time th'eternall Lord in fleshly slime / Enwombed was, from wretched *Adams* line / To purge away the guilt of sinfull crime" (II.x.50). In both passages, Weatherby holds that there is no compelling reason to understand the sinful crime that is washed away as Adamic guilt and instead understands it as actual sins, as distinct from original, or inherited, sin. (One could ask for a compelling reason not to understand it the Augustinian way instead, seeing that the "sinful crime" to be purged comes from "*Adams* line," a rather pointed way of referring to humankind.) He makes the same choice for other passages where the words *sin*, *guilt*, or *crime* occur relevantly. The "only unequivocal statement of inherited guilt" that he finds in the Spenser canon occurs in "An Hymne of Heavenly Love," where "Christ's blood flows to 'clense the guilt of that infected cryme, / Which was enrooted in all fleshly slyme'" (167–68, cf. 141–42).[9] While he does not note the similarity in wording between this passage and the one in *Briton moniments*, he stresses a distance in dating and a variance in theology and philosophy between Spenser's *Hymns* and his romance epic (324–26).

Going beyond the terms of examination that Weatherby sets for Spenser's Book I, I want to shift from lexical concordance alone to figural elements that initiate and define the quest of Redcrosse. These involve his battle with Error, his temptation by Archimago, and his encounter with Fradubio. In these, the question of nature and particularly of the nature of Everyman who dons the armor of Ephesians (6:11–17) is fundamentally at issue. The figural elements representing these natures can hardly be separated, still less omitted, from the evidence of lexical patterns. The red cross on the knight's breastplate marks him as a Christian, signaling his adoption into the faith and thus his baptism.[10] What does it mean, then, that Error confronts Redcrosse in a form that is not only serpentine and sexed, and not only religious and specifically biblical, but also present in the rising of the Nile—in the origin of life, in generation and birth—and even in the noisome gnats whose murmurings simultaneously accompany the gentle shepherd and distract the knight from battle? This Error, deviant and perverse like the evil Milton describes in *Christian Doctrine*, is expressed in physical, social, and cultural terms, including "books and papers," and looks a good deal more like sin than death, even if succumbing to her power would surely bring destruction (I.i.20). The figure of Sin in *Paradise Lost* has long been known to allude to Spenser's Error, of course, and with good reason.[11]

The relation of the tempter Archimago to Redcrosse further reveals the nature of this knight and his dark night of dreaming in Archimago's hermitage, where the imagery of his battle with and temporary victory over Error haunts the dreamworld of Morpheus's cave. Archimago's name connects him with magic, with images, and with the human imagination. Recurrent, often clustered epithets and descriptors further identify him, as most definitively do his actions: Archimago is old, insistently old. He is the "aged sire" and, most tellingly, "the old man" when Redcrosse accompanies him to "his guilty sight" of the false Una locked in lust with the false squire. This is a duplicitous sight if ever there was one since it is at once the sight seen and the sight seeing, the sight at once inside and outside Redcrosse (I.ii.5–6). Openly alluding to the old man of Ephesians 4:22–24, also known as the old Adam, the figure of Archimago, who has access to the knight's imagination, is not only an external, diabolical tempter but also on a continuum with the post-Edenic, Adamic, sinful nature that Redcrosse inherits. He, too, is inside as well as outside Redcrosse, and his wickedness in conjuring the sprights and lustful dream, then fashioning the false couple, precedes Redcrosse's conscious volition. If Redcrosse finally assents to "his guilty sight" and thus actually and consciously sins, the sight that affects him initially derives from his fallen nature, the lustful flesh that he inherits.[12]

Relying exclusively on lexical concordance, a reader can miss the poetic forest for the ideological trees, a particular temptation for a criticism that does not prioritize narrative process, figuration, and poetic detail. But trees can escape notice as well, such as Fralissa and Fradubio, arborealized figures of fallen humanity, who need to be "bathed in a liuing well." Fradubio's lust and loss of faith leave him trapped in the defining form of a tree, which figures nature, or the Pauline flesh, in the early cantos of Book I. The bleeding bough that Redcrosse plucks from this tree man seals their *brotherhood* ("fra") with blood. Fradubio's story, which is becoming Redcrosse's, although Redcrosse doesn't know it, implies an encompassing, paralyzing guilt because of sin. Fradubio may speak, remember, and feel loss and pain, but he is immobilized, unmoving, and therefore symbolically dead as well, since "all that moueth [and thus liveth], doth mutation loue." According to the spell controlling his fate, however, he is recoverable, not lost forever. While Fradubio (like Adam) has committed actual sin, he ends up a figure of fallen nature, a kind of reversion to archetypal form, from which only living water, the renewal of spiritual life, can now free him. Taken whole, his figure, like Saint Paul's Epistles and the leading Reformation theologies, does not offer a clean distinction between sin and death, whether actual or imputed, symbolic or physical. Although Fralissa, too, suffers the same fate as her mate Fradubio and awaits the same release, aside from her entrapment in fleshly nature she appears actually, or otherwise, innocent. By her name, signifying Frailty, and by the ties that bind her to Fradubio, she is guilty by heredity and association. Mutatis mutandis, if you'll pardon an ironic allusion to mutability, she, too, is condemned, or spiritually dead.

Like Error, whose cave lies deep within the Wandering Wood, Fradubio and Fralissa share an imprisoning nature that is better understood as revealed or fulfilled than as simply imposed by Duessa. Yet Duessa herself is a further complication: who or what is she, aside from Una's opposite in this instance—again, *this* instance? How did she first *happen to* (or happen onto) Fradubio and Fralissa? Do we see only a doubling of Redcrosse's experience in this arborealized couple or something that underlies and precedes it? Fradubio's fate is a warning (unheeded) to Redcrosse, but it is simultaneously an image of entrapment that has been sealed in *blood*—the blood of his brother in doubt, namely, *Fradubio*. Blood, clearly relevant to a question of inherited sin and death, will figure again in later discussion.

In the story of Fradubio and Fralissa, we further glimpse the difference between poetry and systematic theology, as, indeed, between poetry and abstraction. As earlier noted, the wages of sin are death, and of death sin, depending on whether one looks to the Western or Eastern Fathers. But Spenser, instead of hitching his poem to the finer points of doctrinal dispute, which he actually mocks with Error's vomit of books and papers, appears more interested in the fact that these wages come out either way in the wash, or the living well in this instance.[13] Theology strives for logical clarity; Spenser's poetry, indeed his allegory, engages the complexities and perplexities of lived experience. Within Paul's Epistles the distinction between inherited sin and death, as between the corrupt flesh and the baptized Christian, has proved notoriously and historically elusive as well as richly, wonderfully, and poetically suggestive. The broad significance of Erich Auerbach's attributing "creative poetic . . . force (*dichterisch gestaltenden . . . Kräften*)" to the figurative impulse within these Epistles should never go unnoticed.[14]

In Spenser's Book II, issues of innocence, blood-guilt, sin, and mortality again loom large, and, as earlier indicated, they bear suggestively both on figuration and on the Mutabilitie Cantos. The indelible stain on Ruddymane's hands provides their initial focus: the name *Ruddymane* signifies "red hand" (*ruddy* and Latin *manus*), to which I would add, "remain red" (*ruddy* and Latin [*re*]*manere*). Once again, in a reading whose influence persists, Weatherby rejects an interpretation of this stain as original sin and its penalty and argues for it as death, an inherited curse without the imputation of sin. He stresses the *alternative* that Ruddymane is innocent of what was done ("As [if] carelesse of his woe, or innocent / Of that was doen": II.ii.1) by his parents, but he does so without noticing that both alternatives are *counterfactual*, that is, "as if," as my bracket has indicated. The implication of Spenser's conditional phrasing is that Ruddymane is neither without woe nor without nocence—harm, stain, evil. Next, Weatherby rightly identifies Mordant as death giving and finally notes that Ruddymane is "bred" in his "dead parents balefull [evil, harmful, ominously *nocent*] ashes." Italicizing the word "dead" while ignoring the word "baleful," he explains that "bred" means "nurtured"; this is a lexically likely sense, yet "bred" could also mean "born," as in "breed," or "reproduce," a meaning that would more strongly imply the reproduction of parental nocence, the deadly evil of lost innocence, in the infant's flesh. Although the word "borne" occurs in the line preceding "bred" ("Ah lucklesse babe, borne vnder cruell starre, / And in dead parents balefull ashes bred": ii.2), context challenges a variation of meaning here: the sense of "bred" as "reproduced" is also justified both by an Elizabethan and a specifically Spenserian penchant for doublets and by the practical need of a rhyme word and, more tellingly, by the *emphasis* on inherited, *inbred* stain that proximate recurrence affords. The fact that the ashes of Ruddymane's parents are conspicuously figurative is further relevant. Traditional signs both of penitence and mourning for sin, not merely of death, come to mind (for example, sackcloth and ashes: I.x.26). Thus contextualized and glossed, in the first two stanzas one piece of evidence, "bred," could, but need not, support Weatherby's position (338), and two other pieces, "as [if]" and "baleful," oppose it. The issue of the larger context then comes into play.

Weatherby considers the death of Mordant another allegory of the Fall, and therefore Ruddymane's counterfactual innocence of it (or carelessness about it) a decisive argument for the Eastern version, in which the bloody-handed babe would be unstained by Adam's transgression. Yet Mordant's story, like Fradubio's, is one of lust and faithlessness, and Amavia's one of loss and despair. Although Mordant leaves Acrasia for Amavia, he is nonetheless doomed. Ruddymane is innocent of his father's *actual* sins, lust and faithlessness, and of his mother's despairing suicide, but he nonetheless remains stained by their blood. Bloody hands generally symbolize blood-*guilt*, not simply death; witness the torments of Lady Macbeth. *Red-hand* (modern *red-handed*), moreover, does not necessarily imply murder or death; according to the *OED*, it can refer to robbery, an archetypally diabolical, hence sinful, act that recurs in depictions of the Fall: Langland's *Piers Plowman*, for one, provides readily accessible analogues.[15] Ruddymane's indelible stain, which surely indicates blood-*guilt*, would also appear to represent what underlies, precedes, and glosses the sinful actions that ultimately lead to his parents' deaths. The stain looks a good deal more like original sin than like death to me, although, of course, it has ultimately led to death in the instance of the actually sinful Mordant and Amavia. Mordant's vulnerability to Acrasia's charm suggests that he has neither escaped her nor the fallen flesh to which she has appealed so fatally.[16]

Of course the washing of hands is an ancient and abiding cultural symbol, one found in pagan ritual, in the Bible, and in Christian ceremony, for example. Spenser's Pilate in Book II's Cave of Mammon merits a relevant glance, since his obsessive effort to wash his "filthy feculent," or bloody, hands looks back allusively to Guyon's futile attempt to cleanse Ruddymane's (vii.61). *Pace* A. C. Hamilton's note, "feculent" (with *OED* support) means "bloody" here.[17] Pilate is coupled with Tantalus in the depths of the Cave, or rhetorical House, of this Mammon of unrighteousness because Pilate shares his presumption, selfishness, and, above all, his blood-guilt—Pilate for the unjust execution of Jesus and Tantalus for dismembering Pelops, his own son.[18] Pilate and Tantalus are defiled by actual sin, but their significance is far more archetypal. The narrative sequence of Tantalus's plea for help, Guyon's self-righteous rebuff—"Nay, nay, thou greedy *Tantalus* . . . Abide the fortune of thy present fate"—and then, immediately after Guyon thus washes his hands of Tantalus, the irony of his seeing Pilate's hands "aboue the water . . . on high extent," faining "to wash themselues incessantly" of guilt, tells quite a different story. Narrative sequence dramatically signals the paradoxical connection between Guyon's actual innocence and Pilate's actual sin that invites our recollection of that between Ruddymane's actual innocence of his parents' actual crimes and an inherited stain on his hands that is deep and original and from which actual human innocence in the fallen world is never entirely free.[19] Guyon's connection with Pilate's guilt is primal, archetypal, original, and it recalls Ruddymane's stain. It is also analogical, as are the relation of Pilate to Tantalus and numerous other relations of figures and concepts in *The Faerie Queene*.

Weatherby reads the nymph's fountain that declines to wash the stain from Ruddymane's hands as an Eastern allegory of inherited death alone, without inherited guilt. The passage that narrates and focalizes Guyon's bewilderment when water will not wash the blood from Ruddymane's "guiltie handes" is so important to this argument that I want to cite it in full:

He wist not whether blott of fowle offence
  Might not be purgd with water nor with bath;
  Or that high God, in lieu of innocence,
  Imprinted had that token of his wrath,
  To shew how sore bloudguiltiness he hat'th;
  Or that the charme and veneme, which they dronck,
  Their blood with secret filth infected hath,
  Being diffused through the sencelesse tronck,
That through the great contagion direful deadly stonck.

(II.ii.3–4)

I agree with Weatherby that the nymph's inefficacious fountain cannot be considered baptismal.[20] But where I think of a Chaucerian analogy suggesting overdetermination or undecidability, he chooses one of the options, of which the stanza affords at least two and likely three. My Chaucerian analogy involves the selection of the Knight, who just happens to be of highest rank, as lead tale teller from among the Canterbury pilgrims, a selection that is comically determined by "aventure, or sort, or cas," that is, by luck, or fate, or chance, and good luck in distinguishing these.[21] In my reading, the first two lines of Spenser's stanza could themselves be an option (water is unable to clean this kind of offence, a reading that would acknowledge Reformation refusal of efficacy to the baptismal water and accompanying words per se), or they could be merely an introduction of the alternatives that follow. The construction *or/or* (lines 3 and 6) could be taken at face value to mean *two additional* possibilities, or it could be taken to imply a Latinate construction—glossed for modern readers as *either/or*—implying a total of *only two* possibilities, both in apposition to the first two lines. Ockham's razor would suggest the first of these readings and therefore a total of three explanatory options.

But I must also confess that I do not follow the two options that Weatherby offers for lines 3 to 9 on the basis of what the stanza says. In the stanza cited, the first option affords a dominantly natural explanation (lines 1–2); the second surely looks theological and specifically like inherited guilt (the Augustinian tradition), as Weatherby indicates (lines 3–5); but the third looks to me like physical poison, witchcraft, or actual sin rather than like the Eastern tradition, namely the curse of mortality alone, which Weatherby finds here (lines 6–9).[22] The charm and venom of these lines, now applicable to Amavia as well as to Mordant ("they dronck"), which have destroyed Ruddymane's parents, tracks back to the witch Acrasia, and its effect, "secret filth," "infect[ion]," and "contagion," looks more immediately like the result of sin than of the death to which it leads. Whereas infection conceivably could be limited to Mordant, the "secret filth" that affects him and spreads to Amavia is *contagious*—transmissible by contact or association (from Latin *contingere*, "to touch"): shades of Fralissa. Death is not contagious, except through metalepsis, elision of the disease that causes it. The disease here is sin. By despairing, Amavia is more culpable than Fralissa, but she, too, is primarily the victim of collateral damage.

The Palmer's immediate response to Guyon's indecision with respect to the interpretive options the narrator offers also signals that we are not to choose among them. In fact, what the Palmer does clearly indicates that our focus should be on interpretation per se and on the futility, or perhaps just the unimportance, of choosing among those offered. The Palmer deflects our choice to an entirely different, if also reasonable, way of understanding the inefficacy of the fountain's water and by doing so renders his own highly rhetorical act of explanation, which substitutes for the narrator's two or three options, very conspicuous indeed.[23]

The Palmer's lengthy explanation also has points of contact with the story of Faunus, Molanna, and Diana in the Mutabilitie Cantos. The Palmer recounts a history of Faunus, ever passionate, and a nymph chaste and fearful of shame, whose waters, once she is transformed by Diana to a wellspring, remain "Yet colde through feare, and old conceiued dreads" (ii.8–9). His story offers extreme opposites, inflamed lust and frigid purity, as an explanation of why this water offers Ruddymane no cleansing.[24] But the Palmer's manner of presentation is so insistently careful, formal, and analogical that it calls attention to itself. "But know," the Palmer commences; then he continues, "Such is this well"; later he adds, "Lo now she is that stone"; and he concludes, "From thence it comes, that this babes bloudy hand / May not be clensd with water of this well" (ii.5–9). His procedure is complete with the Aesopic transfer from myth to landscape ("Lo now she is that stone from whose two heads, / As from two weeping eyes, freshe streames do flow"), which is intended to *realize* his story as a phenomenal fact. When the following canto gives us Braggadochio's mythmaking to rationalize his cowardice before Belphoebe—"For from my mothers wombe this grace I haue / Me giuen by eternall destinie, / That earthly thing may not my corage braue / Dismay"—a brighter spotlight illumines and potentially compromises the Palmer's interpretive myth. Like Braggadochio's craven excuse, the Palmer's constructive myth attempts to rationalize appearances, tightly structuring them (iii.45). The same retrospective illumination also casts a shadow of irony over the jumble of interpretive choices the narrator has attributed to Guyon regarding the babe's bloodstained hands, all of which I consider pertinent and plausible. In the episode of Ruddymane and the fountain, narrative sequence has proved crucially defining.

In the eighth canto of Book II, untreated by Weatherby, Arthur providentially arrives to rescue the fallen but *actually* innocent Guyon and then has a relevant exchange with Pyrochles and Cymochles about guilt and vengeance. The pagan brothers intend to avenge themselves on Guyon's body, supposed by them to be dead, because "The trespas still doth liue, albee the person dye": shades of original sin, it would again seem, as well as of unredeemed justice (viii.28). Arthur answers,

Indeed, . . . the euill donne
Dyes not, when breath the body first doth leaue,
But from the grandsyre to the Nephewes sonne,
And all his seede the curse doth often cleaue,
Till vengeaunce vtterly the guilt bereaue:
So streightly God doth iudge.

(II.viii.29)

Soon after this exchange, Arthur battles and defeats the pagan brothers. His restoration of Guyon to "Life"—a rebirth—refers to redemption by Christ, but instead of aligning with the presence of Christ in biblical time, as do Arthur's defeat of Orgoglio and rescue of Redcrosse in Book I, it realizes Christian redemption in the present and person of Guyon.

The inset lines most recently cited obviously bear on the question of an inherited Adamic "trespas," and like it, "the euill donne" suggests the guilt of the inherited sin of the actually innocent but nonetheless physically fallen Guyon. Yet the passage is otherwise problematical for either an Augustinian or an Eastern reading. The words *often* and *utterly* ask for apology from anyone alert to theological resonance. *Often* suggests the possibility of exceptions—unlikely, barring the instance of Jesus's immaculate conception—but it also invites a closer look at the verbal homonym *cleave*, which means either "split" (or "double, render duplicitous") and relatedly "penetrate," or else alternatively, it means "stick fast" (to). While the alternative *stick fast* might occur more readily to a modern reader, the absence of a preposition like *to* from the clause at issue makes this intransitive form less likely. The senses "split" and "penetrate" suggest an effect less absolute, more receptive to degree, and thus better fitting the problematical *often*. The word *utterly* is easier to deal with, insofar as it is an absolute that leaves open, or arguably implies, the possibility of individual variations or degrees and stages over time. After all, Guyon, here the generic "Nephewes sonne" immediately concerned, has proved both heroic and finally helpless, Christlike but in the end fallen in adventuring Mammon's Cave, and his rescue by Arthur ends the story of neither of these figures in Book II. More adventuring comes relevantly for both, albeit beyond the scope of present discussion.

Yet another possibility is that the problematical words in stanza 29 are deliberately so, an attempt, like the alternative interpretations of Ruddymane's stain, to frustrate any single, facile allegoresis and to insist on *the primacy and process of poetic narrative*. The poet of *The Shepheardes Calender*, with its games-playing glosses, could be capable of such an attempt. This is also the poet known to use pronouns with deliberate, significant illogic and to have offered his readers a whole Book (IV) whose nominal heroes function almost exclusively in digressions and exit with two-thirds of the Book still to come. Moreover, this is the poet who ironized his use of the most obvious symbol of Christianity, the cross, by dressing Archimago in the armor of Ephesians early in Book I and wryly, pointedly applying to him the same words and syntax used of Redcrosse in the opening stanza of the poem, again suggesting the consanguinity of Redcrosse's nature with this old man's, along with the unreliability of appearances (I.i.1, ii.11).

The other major image of mortality that Weatherby treats is to be found in Arthur's battle with Maleger, indeed, in the figure of Maleger himself. In the preceding chapter, I offered an interpretation of Maleger's demise as mortification (numbing) and rejected the reduction of his figure simply to death, as I have Mutability's in her Cantos. Maleger and Mutability are otherwise significantly related. The Earth is Maleger's mother, as she is Mutability's, and insofar as his figure alludes to Antaeus in his battle with Arthur, Maleger shares more mythological genes with the upstart "*Giantesse*" of Book VII (vi.13). Mutability's father is a Titan, descended from Earth and the god Uranus, and the Giant Antaeus descends from Earth and the god Poseidon. This is the note on which I would return to the figure and progressive figuration of Mutability, Maleger's half-sister on his mother's side.

I have argued that Mutability fails to afford a sustained allegory of the Fall, although her figure surely glances at it, and that she more broadly suggests the disruption of order and law—*logos* and *nomos*. Before concluding, I intend a further look at Mutability's connection with mortality. Besides Mutability's manifestations at the outset of Canto vi and the pairing of Life with Death in Canto vii, as earlier discussed, Mutability's depiction of "Earth (great mother of vs all)" in the opening argument of her trial recurrently mentions mortality, death, and decay, the last in concert with "mortall crime" (VII.vii.17–18).[25]

Presenting her claim to Nature (ambiguity intended), Mutability offers Earth, planet and element alike, as her initial evidence since Earth "only seems vnmov'd and permanent" but is everywhere subject to change (vii.17). She then devotes a stanza to the cycling from growth to decay and death, which fleetingly glances again at human history in the Garden of Eden. In her rendition, dead things revert to "earthly slime," a translation of the Vulgate's *limus terrae*, which is used in Genesis to describe God's material for the fashioning of humankind and which variously and relevantly recurs in *The Faerie Queene*: for example, "fertile slime," "*AEgyptian* slime," "fleshly slime."[26] Out of this ooze and the aforementioned "mortall crime" included in Mutability's account, new creatures arise abiogenically that are "Vnlike in forme, and chang'd by strange disguise: / So turne they still about, and change in restlesse wise" (18). Thus described, they sound less like Edenic creation than like the excrescences that arise in the ebb of flooding Nilus during Redcrosse's battle with Error (I.i.21). Mutability's presentation, which *is* her perception of Earth, recurrently exposes her ugly side, even as it reveals her fertility, growth, patterns, and beauty. As ever in the Cantos, the pun evident in the concept *still turning* in the alexandrine of stanza 18 fittingly participates in an undercurrent of paradox: "so turne they still about."

Mutability's next stanza is the one that most interests me since it suggests Spenser's engagement with mortalism, the death of the soul—alternatively, of the individual, or personal, soul, according to the Averroist tradition—along with the body, a concern that makes sense as an offshoot of Spenser's pronounced and recurrent engagement with materialism in his epic.[27] (This is another preview of Milton as well.) Mutability now ups the ante: she moves from generalized references to Earth/earth to more specific reference to animals and human beings. She refers to animals as the massacred servants and subordinates of human beings and then moves swiftly, associatively to us, traditionally specified as rational animals:

> And men themselues doe change continually,
> From youth to eld, from wealth to pouerty,
> From good to bad, from bad to worst of all.
> Ne doe their bodies only flit and fly:
> But eeke their minds (which they immortall call)
> Still change and vary thoughts, as new occasions fall.
>
> (VII.vii.19)

Crossing in a series of analogies from the physical into the economic and moral, Mutability begins to reassume her identity with Fortune, which was emphatic at the beginning of the Cantos before she took further shape and another nomination: Fortune's wheel is implicit in the second and third lines cited above. But Mutability delivers what may be her greatest challenge to any sort of idealism or immaterial formalism, be it Platonic, Aristotelian, Neoplatonic, or simply Christian, in the last three lines of the stanza. From the beginning, Mutability has been a representative of a chaos force from below, one notably similar to the inescapable Mammon of materialism, to the "huge eternal *Chaos*, which supplyes / The substaunces of natures fruitfull progenyes" in the Garden of Adonis, to the ambivalent, egalitarian Giant of materialism in Book V.ii, and even to her own half-brother Maleger, resurgent, resilient, monstrous son of the Earth (III.vi.36). If indeed Earth is "great mother of vs all," Mutability's consanguinity extends even further. She would sweep into her inheritance any kind of pattern, form, or order that attempts to transcend a chaotic, material world. *Mens*, the intellective mind—not just *ratio*, abstracting reason, though including this—is the object of her challenge, which is all the more alarming for its almost casual, matter-of-fact delivery at the end of the stanza: after all, who knows not "Earth (great mother of vs all)"? In Mutability's assertion, all intellectual thought becomes relative to occasion, to temporal, material causation: "But eeke their minds (which they immortall call) / Still change and vary thoughts, as new occasions fall." Occasion is time, fortune, happenstance, expediency, and now Mutability, or so she intimates in the stanza at issue. Occasion becomes the latest of Mutability's shifting identities, her shifting sameness, her still turning, moonlike constancy. Thus seen, Mutability indeed makes her claim to Nature.

Mutability's claim recalls Book II, where the hag Occasion is fettered by Guyon in the fourth canto. The arrest of Occasion, simultaneously an attempt to stop time and to isolate a personified abstraction from the movement of narrative, leads to impasse: characters standing around talking (and talking and talking . . .) without forward movement until Atin, the emissary of irrational, disorderly, appetitive impulse and himself, like Ate, an ambassador from the material depths, erupts onto the scene. Occasion, hag though she is in Book II, might also be called Opportunity or Fortune, as Paul Alpers demonstrated long ago.[28] She represents the material, the many, and the spontaneous, which come with positive as well as negative aspects. In time, her binding, or arresting, ironically becomes itself an occasion of strife, as it should and must. *Temporarily*, it suspends Guyon's quest to avenge "guiltie blood" and to establish the "goodly frame of *Temperaunce*" (II.xii.1; my emphasis). *Temperance* has its very root in *tempus*, "time," as implicitly does Guyon's virtue.

In closing, I want to align Mutability's earthborn challenges, which reflect forces, factors, and material phenomena throughout *The Faerie Queene*, with Lodowyck Bryskett's depiction in *A Discourse of Civill Life* of his friend Spenser's investment in the relation of the mind to matter, especially as it bears on the immortality or mortality of *mens*, the intellective power, and thus on mortalism.[29] Although Bryskett's *Discourse* is an imaginary dialogue and the part he assigns each of his friends is discretionary,

Spenser's role is nonetheless suggestive relative to *The Faerie Queene* and particularly to Mutability's directly challenging the soul's immortality: "But eeke their minds (which they immortall call) / Still change and vary thoughts, as new occasions fall."[30] Bryskett's *Discourse* will also prove further relevant to my later chapters on Donne and Milton—to Donne's treatment of the relation of the soul to the body in his *Anniversaries* and to Milton's explanation of the continuity of body and spirit in *Paradise Lost*.

No less than five times in the *Discourse*, the speaker designated "Maister *Spenser*" urges questions that exhibit his concern about the soul's immortality, and at the end of the dialogue, right after Bryskett's extensive attention to this subject, we are reminded that Spenser is the instigator of the whole disquisition. Spenser's first contribution to the subject of immortality challenges Bryskett's conclusion that the possible understanding, or possible power of the intellective soul, is immortal and impassible rather than mortal and corruptible like the passible understanding or passible soul, which is also identified as fantasy or as the imaginative or cogitative power. Citing Aristotle, Spenser objects that the possible understanding of the intellective soul "suffreth in the act of understanding" and must therefore be mortal since "to suffer importeth corruption" (200). It is tempting to read "importeth" here as a pun, both as "implies, signifies" and as "brings in," corruption. Spenser's point is that the possible understanding is affected, or acted upon, by fantasy and by sensation, which variously interact with the material world and consequently are corruptible and mortal. Bryskett rejects Spenser's argument by elaborating and refining his own, now distinguishing between the natures of suffering by the possible and passible understandings and thus between their operations; since his possible understanding is "the place of the Intelligible formes," his distinction looks like an effort to combine Plato with Aristotle (199–200). He also compares the relation of matter to form to the relation of the possible understanding to the agent (active) understanding, both within the intellective soul. This time the authority he cites is Averroës, Islamic Spain's commentator on Aristotle, whose view of individual immortality was otherwise thought to conflict with that of the Christian tradition. In Averroës's influential view, as understood and much noted in the Renaissance, single human intellects perish; they are mortal.

Spenser next questions whether human understanding is immortal because Aristotle has said that we have no memory after death. In response, Bryskett takes Aristotle's view to indicate instead that the intellective soul is precisely *not* mortal since memory is part of the sensitive, not the intellective, soul, and he further explains how the understanding, or intellective soul, acts in the imagination, which is another sensitive faculty, to "rememorate," or recall (201). For a moment, perhaps, the wars and treasons in Alma's memory or the links between Maleger's forces and the imagery of her fantasy (Phantastes) might again come back to mind.

Following another exchange between Bryskett and a second questioner, Spenser presses again at the relation of the intellective soul to the body and more generally to matter. How, he asks, can the intellect be both immortal and as multiple (numerous) as human beings are? He then invokes Aristotle's view that multiplication proceeds from matter and clarifies what still weighs on his mind by remarking, "things materiall are alwayes corruptible" (201–2). Now he gets the answer that his objections are inapplicable to what is "Intelligible and spirituall, such as is the understanding"; this "might remaine after the matter" is gone and thus survive the death of the body. Still unsatisfied, "Maister *Spenser*" now questions the immortality and impassibility of the soul on the basis of felt experience, for we see "by experience . . . dayly that she is troubled [later, "overcome"] by Lethargies, Phrensies, Melancholie, drunkennesse, and such other passions"—all the tribe of Maleger. Bryskett explains to the probing Spenser that such passions belong not "to the Intellective soule" but to "the vertue cogitative, fantastike, or imaginative, called by Aristotle . . . the passible understanding." The passible soul he describes as an inward sense that is tied to the body and feels its passions, whereas "the understanding feeleth not" (202).

Spenser's final question, which Bryskett declares a philosophical and Christian heresy, is, therefore, whether there are two separate souls in human beings, "the one sensitive and mortall, and the other Intellective and Divine" (203).[31] Such a view would endorse a pronounced dualism—*duo esse*, shades of Duessa, free-floating idealism without basis in matter, or, arguably, an extreme Eastern spirituality, or *theosis*. The answer he receives at some length is a resounding negative. "The Vegetative and Sensible soule, or their powers," are in the Intellective soul, "as the triangle is in the square": here is the tie to the celebrated geometrical stanza describing the Castle of Alma (II.ix.22), for which Jerry Leath Mills has ably argued (203).[32] The Intellective soul is the "more perfect" and therefore contains "the natures [I would prefer to say the *capacities*] of both" the lesser souls (203). Properly and immediately, this greater soul also possesses its own higher and lower powers, namely, *mens*, the contemplative faculty, and the lesser *ratio*, which functions as the intellective soul's more mundane power of reason.[33] In a living person, the intellect maintains a functioning relation to the sensible soul subordinate to it, whose nature, capacity, or power but not whose passibility or feeling and connection to sensation it includes. The intellective soul, "the particular and proper forme" of every human being, thus remains "everlasting, impassible, not mingled with the bodie, but severed from the same, simple and divine, not drawne from any power of matter . . . [or] ingendred by [sexual] seede" (204). Crucially, Mills has observed, Bryskett's conception "implies the Understanding (Intellective soul) must answer [before God] for the entire soul, passible and vegetative soul included."[34] The arresting statement in Spenser's Book II that the virginal Alma "had not *yet* felt *Cupides* wanton rage" comes back to mind, its implications enhanced (ix.18; my emphasis). Whether Alma informs the body or stays aloof from its frailties, she, together with Fralissa and Amavia, is still implicated in them.

Whatever the relation of the Spenserian poet to the role Bryskett assigns "Maister *Spenser*," the *Discourse* is a pertinent contemporary document treating the intellect, and it suggests dramatically what is at stake in Mutability's claim that the mind is subject to whatever "new occasions fall"—occasions not restricted by Mutability to those that "befall it" in its own conscience, household, or hamlet. Near the end of the sixteenth century, when such occasions included changes of authorized belief with every alliance or death of a sovereign and changes of official policy with every Lord Deputy in Ireland, not to mention the additional occasions evoked by the looming, material Giant of Book V, Mutability's claim would be alarming enough even without a belief in the immortality of the soul.[35] As before with the question of Eastern and Western readings of the legacy of the Fall, Spenser's poem raises and shapes the many issues, examines and complicates them, and then leaves us puzzling over Nature's verdict and his own concluding stanzas in Canto viii, which are as full of perplexities and possibilities as ever. This is the way of his metaphorizing, allegorizing, analogizing poetics, which comprehends a good deal more openness, inclusion, and uncertainty than doctrinaire readings would allow. More to the immediate historical point, however, his poetics shares a good deal with Milton's, as subsequent chapters will indicate.

CHAPTER 3

Satanic Ethos

Evil, Death, and Individuality in *Paradise Lost*

The organization of this chapter on *Paradise Lost* might be considered a diptych. The first of its panels—Part I—shows the origin of evil, hence sin, in Satan's envy when the Son is exalted, an envy that underlies Satan's self-authoring pride.[1] This panel focuses primarily on negation and death. Part II, the second panel, which examines God's terms of exaltation in greater detail, focuses on the inseparable questions of individuality, allness, and pride, each of which, even pride, has a more positive potential.[2] My argument in both panels grows out of two crucial speeches in Book V, examining them in reverse order. In the first speech I treat, Raphael recounts Satan's response to the Son's exaltation; in the second, the proclamation of Sonship itself, God reveals the ultimate purpose of this manifestation, which has already been foretold in the unity, or allness, envisioned in Book III. Recurrently, the following chapter also engages a circularity that exists in the poem and in the intellectual culture of which it was and remains a dynamic part. Satan's ethos—that is, his character (Greek *ēthos*) as an ethical, or, as it happens, an unethical, being—is central to this circularity.

In traditional Christian theology, pride and to a lesser extent envy are Satan's characteristic sins.[3] Both are quickly evident in Book V of *Paradise Lost* when Raphael introduces Satan with a sketch of his status and response to God's manifestation of the Son as "King anointed." Although Satan is

of the first,
If not the first Arch-Angel, great in Power,
In favour and præeminence, yet fraught
With envie against the Son of God, that day
Honourd by his great Father, and proclaimed
Messiah King anointed, [he] could not beare
Through pride that sight, & thought himself impaird.
Deep malice thence conceiving and disdain,
Soon as midnight brought on the duskie houre
Friendliest to sleep and silence, he resolv'd
With all his Legions to dislodge, and leave
Unworshipt, unobey'd the Throne supream
Contemptuous.

(V.659–71)[4]

Citing the end of this passage, Stella Revard identifies intellective pride as "the first cause of the war in Heaven" and notes that pride is "repeatedly named from Book I on as Satan's main characteristic and leading motive" (28). Pride is what "cast him out from Heav'n," in the narrator's judgment, and in Satan's own, "Pride and worse Ambition" wrought his fall (I.36–37, IV.40). Revard later describes pride and envy as alternative sins of Satan's soul and adds that while both are spiritual offenses, Milton differentiates sharply between them: "Pride arises when a creature values his own self-will above the will of his Creator; envy arises when one creature willfully resents the person or the accomplishments of another" (67, 79). These statements generally represent established views of the causes of Satan's rebellion in Milton's epic and in theological tradition, yet they fail adequately to address the role of envy in the lines from Book V that I have cited and to examine and account for the origin of evil in them.

Prioritizing envy in this way, I do not mean to deny the importance of pride in Satan's sinning or in his character more generally: the continuum between envy and pride is recurrently evident in Satan. Tracing Satan's sin from its inception to its fully blown expression as self-authoring pride, I will nonetheless argue that the essential negation belonging to envy is his ethical core. Recognition of envy as origin is fundamental to Milton's fully realized conception of evil itself as the necessary consequence of a sinful act. This consequence is deathly negation, nought, not-vital, not-good, in *Paradise Lost*. Appositely, as Milton observes in *Christian Doctrine*, "all evil, and whatever is seen to lead to ruin, is encompassed summarily in Scripture under the sign of *death*."[5]

For traditional theology, the categorization of sins is reasonable, useful, and necessary, yet the dynamism of actual life tends to elude its order and clarity. Although epic narrative is hardly a photocopy of life, to a *significant extent* the movement of its heightened art is similarly elusive. Brian Cummings's view—in a book aptly subtitled *Grammar and Grace*—that poetry is not "a passive recipient of doctrine" pertains as forcefully to the present chapter as to the last one. If, in Cummings's words, "The perplexity of theology itself is profoundly related to the perplexity of language," the categories of theological science (as it used to be known) understandably seek to abstract, fix, and otherwise simplify such perplexities.[6] Accordingly, the following section of this chapter is a series of snapshots of envy that is intended to suggest its roots and depictions beyond theological conceptualizations and moral taxonomies, although hardly insensitive to these or unaffected by them. Provided with these snapshots, I will return to the crucial lines from Book V already cited in order to examine the Satanic inception of evil with the benefit of the perspectives they afford. My primary concern in assembling these snapshots is not to establish sources and influence; rather, it is to suggest methods, frames, and lenses for reading.

Part I: The Circle of Envy, or the Hole at the Center of Selfhood

Snapshots of Envy and Death

My snapshots begin with Aristotle and Cicero, but they are arranged topically rather than chronologically insofar as one from Francis Bacon precedes those from Langland and Spenser. Aristotle's fullest treatment of envy in particular is not in the *Nicomachean Ethics* but in the more pragmatic *Rhetoric to Alexander*, which advises how in a peroration to "win a favourable hearing for ourselves and an unfavourable one for our opponents."[7] Pursuit of the latter specifically includes the arousal of envy toward opponents through appeals to fear, resentment, hatred, and anger, "for envy is very near hatred."[8] Centuries later, Cicero's Antonius in *De Oratore* similarly treats the stirring of emotions in legal and political argument but also attempts finer distinctions. He first treats the incitement of hate or, where "really bitter hate" is improbable, then "at least [the incitement of] a negative emotion that is not unlike envy or hate." Next he treats fear, whether communal or personal, noting that the latter "penetrates more deeply." Then he gets to envy itself, "the most vehement emotion by far" as well as one both "highly common and persistent." Having tried to distinguish hatred and fear from envy, Antonius now concentrates on forms of envy aimed at the advantages of superiors, equals, or even inferiors, the last signaling the great extent to which assumptions pertaining to social rank are operative for him.[9]

Bacon's *Historie of the Raigne of King Henry the Seventh* affords a quite different political example in the late Jacobean period. Toward the end of *The Historie*, Bacon observes that John Morton, Archbishop, Chancellor, and Cardinal, willingly deflected envy from King Henry to himself. Morton

> was willing . . . to take envy from the King more than the King was willing to put upon him. For the King cared not for subterfuges, but would stand envy, and appear in any thing that was to his mind; which made envy still grow upon him; more universal, but less daring.[10]

Although Morton willingly shielded the King against envy, taking it upon himself, Henry characteristically bypassed his mediating shield to act openly and boldly, "which made envy still grow upon him." Bacon's phrasing is distinctive. In it, envy, without specific attribution to persons, has forceful, objective existence and even a kind of life; envy grows. This near-figuration enforces envy's significance—at once its impact and its meaning—in the passage, which hovers between "ill will, malice" and "odium, opprobrium." Notably, however, this envy is neither provoked by nor aimed at the "superior advantages possessed by another," which is nowadays the meaning we generally assume for this word, as anciently did Cicero's Antonius.[11] Envy in Bacon's account is not, in a manner of speaking, so reasonable, so clearly tied to fantasied betterment and in this sense so purposeful. In contrast to the Baconian example, our current, popular sense of *envy* verges on covetousness.

Two allegorical poems, Langland's late medieval *Piers Plowman* and Spenser's Elizabethan *Faerie Queene*, offer moral examples that specifically involve characterizations of the Seven Deadly Sins, of which one is envy. These examples are at once conceptual and embodied, and they are involved in figuration and movement rather than being simply abstract and static. What both poems demonstrate is the extent to which, given any ethical or characterological embodiment within narrative, including mere personification, one sin develops multiple aspects or simply slides into another. Distinctions among spiritual, intellectual, passionate, and appetitive expressions of sin also blur, an effect that monism, even an arguably equivocal monism like Milton's, would only encourage.[12] As earlier indicated, taxonomies of sin have proved indispensable for theological and casuistical analysis but seldom a neat fit for living.

Langland's Envy, miserable, pale, and thin, is a figure of withered health, "as pale as a pelet [sheep's pelt]" and shriveled "As a leek that hadde yleye [lain] longe in the sonne."[13] His fist clenched, he just waits to avenge himself on those around him, all the while biting his lips in frustration, his body swelling with wrath. His words are the tongue of an adder, chiding and challenging, backbiting, reproaching, spreading calumny. He tells damaging tales, bears false witness, and sows discord, causing loss of life and limb. In church, instead of praying for blessings on others, he prays for their misfortunes. He covets others' possessions—"turne I myne eiȝen, / And biholde how Heyne hath a newe cote; / Thanne I wisshe it were myn"—laughs at their losses, and weeps for their gains (V.110–12). He is loveless and suffers from ulcers, a form of self-consumption, for "enuye and yuel wil is yuel to defie [digest]" (V.122). This is Envy in action, and his loveless, self-destructive nature extends to wrath and covetousness but specializes in meanness and malice. With the momentary exception of his coveting Heyne's coat (2.5 lines out of a portrait of almost sixty), he quintessentially consists of denial, refusal, and negation. He would be at home in Satan's hell, particularly when Beelzebub, fronting for Satan, proposes the malicious plan that participants in the Great Consult embrace, thereby defining themselves.

Spenser's Envy in Book I, like Langland's, ulcerously chews "his owne mawe": "For death it was, when any good he saw . . . But when he heard of harme, he wexed wondrous glad."[14] Outwardly he chews a toad whose venom runs down his jaw, symbolizing hateful calumny (from Latin *calvi*, "to deceive"), and inwardly he harbors a bosom serpent with mortal sting; his figure is deathly. He gnashes his teeth covetously at the gold of Avarice, the Deadly Sin who rides aptly "next" to him, and "grudge[s] . . . at the great felicitee / Of proud *Lucifera*, and his owne companee," the latter an ambiguous phrase that suggests his self-loathing along with his social treachery (iv.31). Backbiting and spewing poison from his leprous mouth, "So euery good to bad he doth abuse" (32). His complementary female manifestation in Book V similarly "feedes on her owne maw vnnaturall" and externally chews on a poisonous snake, which, though "halfe gnawen," she uses literally to backbite her knightly victim (V.xii.31, 39). She, too, "inly fret[s], and grieue[s]" at any good news and ever takes pleasure in another's harm or loss (32). Again, as with Langland's, everything about

Spenser's Envy is supremely negative—malicious, malevolent, malignant, deadly.

In Book I, Spenser's embodiment of Envy appears in the pageant of Sins in the House of Pride, over which Queen Lucifera presides. Envy sits astride a wolf, one of six Deadly Sins atop iconic beasts that draw the coach of Luciferan pride. Satan perches on the wagon beam, whipping the sins to propel this ungodly coach and six into action or further to accelerate its movement. Visually and operationally, the six sins are thus connected to Satanic Pride, Lucifera, as she is to them, their implication inextricably mutual. Precisely who is primarily or finally in charge—the councilors and their bestial drawers, the Satanic impeller, or the Queen in the coach—is uncertain and meant to be so.

All six sins counsel Lucifera, but Envy's connection to her is special. For all Lucifera's glitter, riches, and sunlike radiance, "her bright blazing beautie did assay / To dim the brightnesse of her glorious throne, / As *enuying* her selfe, that too exceeding shone" (I.iv.8; my emphasis). Competing for preeminent brightness with her own throne, whose glitter envies her even as she envies it and whose excess is mirrored in hers even as it mirrors hers, Lucifera is a restless image of discontent and vicious circularity. The ambiguous, irrational syntax describing her in the lines I just cited befits the instability of a condition in which she competes with her own seat of power, with her own situation, and indeed with herself. Underlying her proud show, envy gnaws at her "felicitee," as already noted in the portrait of this treacherous counselor. Her pride is the surface of disease. Hardly by accident Redcrosse, the hero of Spenser's Book I, will find Sans Joy (Joylessness) in the House of Pride, and this place, indeed this topos, will turn out to be merely the surface of the inner hell eating away at him. In time, of course, his submerged hell will issue in fully conscious despair, as will Satan's in Milton's epic.

A religious encyclopedia from fifteenth-century England and a secular monograph from the twenty-first century can serve as bookends for this section on envy. Both emphasize the downward movement, the destruction, the deathly and deadly negation that characterize envy, very much in contrast to the overreaching, ambition, and even aspiration of pride: from this vantage point, envy is the denial to pride's assertion, the negative to its excess. Whereas pride is epic, envy is mean—more exactly, mean spirited, degeneratively spiteful, hateful, and cruel. In *Jacob's Well*, a medieval encyclopedia, "Pride is the root of all sins. [But] Envy is called the worst of all sins because it is contrary to all virtues and all goodness."[15] Joseph Epstein's contemporary monograph particularly stresses the malice of full-blown envy: "Malice that cannot speak its name, cold-blooded but secret hostility, impotent desire, hidden rancor, and spite all cluster at the center of envy."[16] In Raphael's account of Satan's rebellion in the fifth book of *Paradise Lost*, envy is incipient rather than full-blown, but its presence at the first instant of the inception of evil is utterly *vital*, a word I use here with ironic awareness of its etymology. In the inceptive context of Satan's rebellion, death already resides within life, waiting for birth.

When this birth actually occurs in hell, Satan's hell, the resultant figure of Death is a shape without form and a shadow without substance. Like envy, Death is wrathful, lustful, empty, and insatiably hungry. As yet his bottomless desire lacks even an unsatisfying object of consumption. To Satan, he is ironically an abhorrent "Fantasm," the product of fantasy originating in an evil brain (II.743).[17] Death, Satan's incestuous son/grandson, has fathered the hellhounds that endlessly torment his mother/half-sister, both from the inside and the outside (654). As I remarked in Chapter 1, the enclosed circularity of all these relationships embraces solipsism, autoeroticism, and narcissism. By these, it viciously perverts both self-reflexivity and creativity. Sin's hellhounds, implicitly both physical and psychic, figure these perversions in the ceaseless, temporal circularity of Sin's birthing, which is later reiterated in the circularity of Satan's soliloquy near the beginning of Book IV: "Which way I flie is Hell; my self am Hell" (II.796–97, IV. 75). Satan always circles back to himself.

"Fraught / With envie"

Returning to the lines from Book V with which this chapter began, I want first to emphasize that they are spoken by Raphael and second to reiterate that the initial response he attributes to Satan is envy rather than pride: Satan is "fraught / With envie against the Son of God, that day / Honourd by his great Father." In contrast, at the outset of the epic, while envy is also mentioned before pride, the latter alone is identified as the cause that earlier "Had cast him out from Heav'n, with all his Host / Of Rebel Angels." Thus, although the motivating envy the narrator first mentions has its roots in Satan's rebellion, here, coupled with revenge ("Stird up with Envy and Revenge"), it pertains contextually to the deception of Eve (I.35–38). As Revard observes, such an earthly-oriented envy moves Satan to action "only after he has lost the favors of heaven" (78). In sum, the primary reason the narrator assigns to Satan's loss of heaven is pride. With the narrator—or one narrator—thus giving primacy to rebellious pride and another narrator, Raphael, giving priority to envy, Milton is accommodating two explanatory traditions, much as he will by doubling narrators to accommodate the two accounts of creation in Genesis, giving one to Raphael in Book VII and the other to Adam in Book VIII. Numerous forms of double voicing, or equivocation, recur in *Paradise Lost* and greatly enhance its openness to interpretation. They might be said to model a process of dialogic possibilities.[18]

In Book V, however, the poem is also making a point about the origin of evil, whose significance calls for attention. Starting with the motivation of envy, Raphael is starting with negation, not assertion, the latter of which, whatever its basis, is typically more positive. Deficiency is envy's trademark, not the excesses of pride. In the lines cited from Raphael's opening account, Satan's envy—if only for a sufficing moment—seems to come out of nowhere. Its occurrence is unreasoned, if not quite irrational or uncontextualized, and at once both affective and conscious; or rather, its occurrence has already been so since "fraught," a past participle, signals a fait accompli, whether accomplished because the event itself *is past*, at once present and absent within the narrative, or because for Raphael as narrator, not subject, this interior event must ever have been so. "Fraught"—"filled, laden," literally "freighted," like a vessel weighted with cargo—also signals that the envy Satan has experienced is no passing twinge.[19] It has weight and substance, qualities that invite, without closing, interpretation. They imply an unspoken background—an earlier beginning or root, perhaps a potential. All these impressions are important: if the incipience of evil were really explicable or fully comprehensible, it would appear to have reason. Instead, envy is occasioned; it happens. It is triggered but not really caused: "fraught / With envie against the Son of God, that day / Honourd . . . and proclaimed" is not quite the same as saying "envious *because* the Son was honored," and the noticeable inclusion and placing of the phrase "that day" could also suggest the actualization of something earlier *in potentia*—something (the exaltation) in the works, so to speak—and even implicit and desirable in the ardent mystical vision of a preeminent archangel or two.[20] Such vision, understood as leading to greater intimacy with God, would not have exceeded the prescience of a Florentine Neoplatonist, a subject to which I'll return in Part II of this chapter.

Less speculatively, the envy that occurs that portentous day is "against the Son of God." When the *OED* authorizes as its third definition a use of *envy* that could easily lead to covetousness—"the feeling of mortification and ill will" at "superior advantages possessed by another"—the prepositions it recognizes (*at*, *of*, *to*, *upon*) do not include "against." The definition that "against" best matches (though not exemplified in the *OED*) is the first: "Malignant or hostile feeling; ill will, malice, enmity"—period. Envy *against*, like *toward*, is in Latin *adversus* and in its nominal English cognate becomes the "adversary." The instant of Satan's advent as the Adversary lies in his envy. His response is adversarial, hostile, negative—*against*. For Milton, in *Christian Doctrine*, sin starts with ill will ("evil desire, or the will to do evil": *malevolence*), and it is expressed in oblique or perverse action, which can include words, thoughts, or even the omission of good action, such as gratitude, obedience, generosity, or even enlightened interpretation (right reading), since "Reason also is choice" (III.108).[21] For Satan, sin thus starts with envy, or malevolent negation.[22]

Within the same passage of two sentences in which Raphael first recounts Satan's envy, he identifies pride as the motive that kicks Satan into decisive action: "With all his Legions to dislodge, and leave / Unworshipt, unobey'd the throne supream / Contemptuous." As others have noted, the verb *dislodge* is at once potentially intransitive ("remove, depart") and transitive ("assault his throne").[23] Thus the act of omission, or departure, on which Satan resolves already carries within it the threat, indeed virtually the promise, of aggression. Yet the motive of pride that resides within this decision by Satan to act openly, or externally, is *continuous* with his originary envy in the sentence just before: "yet fraught / With envie against the Son of God . . . [he] could not beare / Through pride that sight, & thought himself impaird." As distinct from the originary incidence of Satan's envy, the initial stirring of his pride has reason here, if only fancied, namely, his impairment or reduction; this reason is actively "thought" rather than more passively "fraught," as was Satan first with envy.[24] Yet the rhyming of the active "thought" with the passive "fraught" simultaneously marks and offers to override their difference. Unlike the typical excesses and assertions of pride, its self-assertive overreaching, Satan's reason entails belittling and lessening—impairment. His reason, too, is negative because of the company it keeps with envy, and if the sequence from "could not beare / Through pride" to "impaird" is reexamined, it suggests that pride accounts for Satan's lack of forbearance while his sense of impairment reverts to his envy. The line that follows this account of Satan's impairment now reaffirms the seemingly inextricable co-operation of his envy with his pride—"Deep malice thence conceiving and disdain"—a combination of envy's virulence with the haughtiness of pride and thus, more simply, of malice with disdain. Moreover, Satan's awakened pride leads at this point preeminently to denial and refusal—"Unworshipt, unobey'd"—and contempt, of course, or, scorn and spite, even loathing. We are not so far from

Langland's Envy. Like the verb *dislodge*, the adjective "Contemptuous" that concludes the passage cited is potentially double in reference; it modifies either Satan or "the Throne supream," namely God, and attributes spiteful malice and haughty disdain to both. Briefly, Raphael's syntax seems to mirror Satan's confused perspective.

But Raphael, who, we need remember, cannot know another's inmost thoughts, since this power is God's alone, and who has already warned us of the need of accommodation in his account of rebellion, is hardly finished with Satan's character.[25] His narration proceeds next to Satan's infusion of "Bad influence into th'unwarie brest / Of his Associate," by inference Beelzebub, who does the dirty work, telling "the suggested cause, and . . . [casting] between / Ambiguous words and jealousies, to sound / Or taint integritie" (V.694–96, 702–4). Satan's target is the breast, which is traditionally both moral center and seat of affections, rather than the mind of his Associate, who then arouses jealousies to test or taint wholeness, innocence, unimpairment—or, to borrow a word from Milton's God, "individuality" (undividedness, unity). Satan's targeting another's breast establishes the paradigm of incipient corruption in the poem, which effectually shares his own experience, as we will subsequently have reason to observe again.

Having led his legions into the North, Satan then addresses them and for the first time expresses his rebellion openly. Even within this first speech, his argument moves from objection to assertion. At first he speaks of the Son's engrossing power and eclipsing "us" (V.776). Further heightening his rhetoric, he objects to "Knee-tribute" and "prostration vile," whether offered to the Son or to the Father (782). Before he finishes, he is arguing both for "Orders and Degrees" yet against Monarchy, which introduces "Law and Edict on us, who without law / Erre not." He ends with the ringing assertion that he and his followers are "ordain'd to govern, not to serve" (772–802). Clearly, Satan has progressed in rebellion, moving from festering refusal to brave assertion. His assertion, however, is suspect, indeed, based in the heightened rhetoric of rhetorical questioning, not surety, and Abdiel rises to challenge it.

Abdiel's celebrated challenge has two bases. The first is God's justice and omnipotence, and its rhetorical form parries Satan's questions even as its argument counters them:

> Shalt thou give Law to God, shalt thou dispute
> With him the points of libertie, who made
> Thee what thou art, and formd the Pow'rs of Heav'n

Such as he pleasd, and circumscrib'd thir being?

(V.822–25)

The second basis is God's generosity and beneficence, his providence in making the angels not less but more "exalt[ed] / Our happie state under one Head more neer / United" and "more illustrious made, since he the Head / One of our number thus reduc't becomes"; in the latter explanation, "reduc't" indicates either that "Head / One" is lessened in order to elevate others, or if "reduc't" is taken instead to modify "number," that "our number" is reduced, literally led back (Latin *reducere*), to their higher origin, their "Original" in Miltonic usage; here, as cited, they become more *exalted* and *illustrious*, more radiant because nearer the source of Light (829–31, 842–43). This is the dynamic of redemption recurrently explained in *Paradise Lost*: the Son stoops that others might be raised in him.[26] Abdiel's phrasing "more neer / United" affords further play insofar as "more neer" can be taken adverbially as "more nearly," "more intimately," simply "closer," or else adjectivally in the same senses if "neer" is taken to modify either noun in the line. Especially when read as "more nearly," the angel's assertion accommodates individuation, here understood now to include some degree of distinction.

Abdiel evidently recollects God's earlier revelation that all shall abide under the Son's reign "United as one individual Soule / For ever happie," and Abdiel's words also bear on God's chronologically later (narratively earlier) promise that "God shall be All in All" (V.610–11, III.341).[27] Abdiel's trust in God counters Satan's distrust, his confidence Satan's disbelief—his *providere* (providence) Satan's *invidere* (envy), so to speak. A spiritual catalyst, Abdiel provokes the self-assertion of Satan's pride not only to deny God's creation of him but also to claim his own total autonomy, "self-begot, self-rais'd" by his "own quick'ning power, when fatal course / Had circl'd his full Orbe" (V.860–62). Satan embraces an origin and resulting disorder in which things just happen; in effect, he formulates a consequence of what we earlier saw enacted in his figure when his envy first occurred, notwithstanding nebulous hints of its deeper origin. By the end of his response to Abdiel, Satan has directly challenged the Son to a contest. With others, I would reemphasize the progressive nature of Satan's rebelling.[28] His initial envy, or ill will, resentment, feeling of impairment, and the consequent assertion of his pride give way to grandiose claims of self-empowerment. Beneath this surface, however, negation gnaws enviously, ulcerously. In short, the malignancy of Satan's envy both underlies and issues in his self-authoring pride.

Two occurrences of the word *envy* (or its cognates) in Book VI register the further development of its co-operation with Satan's growing pride. Early in this book, Raphael describes the rebels' intention "To win the Mount of God, and on his Throne / To set the envier of his State, the proud / Aspirer" (VI.88–90). Here, "envier" and "Aspirer" alike express the desire, albeit evil, for something higher, quite in contrast to the envy and impairment of pride with which the Adversary begins. In a later passage in which Raphael recounts the response of the rebels to the Son's approach in God's Chariot, they "at the sight / Took envie, and aspiring to his highth, / Stood reimbattell'd fierce" (VI.792–94). Now hardened in sin and desperate, the rebels again combine their enmity with emulation, as they will in their relatively more positive moments in hell.

When *pride* and its cognates occur by themselves in the war of Book VI and pertain to Satan, they usually refer to his martial exploits and operations, that is, to his active aggression. Once he is defeated and has fallen from heaven, however, Milton's God characterizes him merely as "our envious Foe" (VII.139). Satan's pride deflated (at least for now), his enmity appears to have been peeled back to envy, its original, negative core. In a shameless (but not wholly un-Miltonic) pun, I'll add that this core leads in time to the fatal apple, a development I want briefly to pursue through hell into Eden and then back to hell. My pursuit will further characterize the debilitating progression of Satanic envy, which is finally just (and also justly) its deathly circularity.

Satan's Envy in Hell and Eden

Satan struts his "Monarchal pride" aplenty in hell, along with his vaunted power, but as earlier suggested, the truly (re)defining moment for him and his rebels comes at the end of the Great Consult when they all choose more than "Common revenge": either the destruction or the usurpation and corruption of the created world and its inhabitants, their goal being merely to "interrupt" God's imagined joy in their ruin and to raise their own joy "In his disturbance" (II.371–73, 428). In this significant moment, the rebels' choice defines them and their evil as truly diabolical. Such utter malice, to no end but waste and pain, is envy writ large and rendered absolute.

Satan, early in his celebrated soliloquy at the outset of Book IV, blames "Pride and worse Ambition" for throwing him down, but note the absence of the possessive *my* and the personified agency of Pride and Ambition, at whose paired hands he becomes a victim (40). Through line 69, Satan's examination of his conscience makes him sound like a skillful casuist until his logical conclusion that "Heav'ns free Love [has been] dealt equally to all" suddenly leads to a startling result: "Be then his Love accurst, since love or hate, / To me alike, it deals eternal woe" (68–70).[29] This moment is a striking, dramatic recurrence of his initial rebellion. From somewhere within him, hateful malevolence issues explosively. Now his willful, irrational resolve leads to powerfully affective despair and to his penultimate consideration of repentance, the embrace of which a prideful word, *Disdain*, forbids him, as does his prideful dread of shame among his followers. Yet again he reconsiders repentance, only to affirm its futility: "how soon / Would higth [1667: highth] recal high thoughts, how soon unsay / What feign'd submission swore" (94–96). The word "height" refers to Satan's restoration to his place in heaven, and the phrase "high thoughts" suggests his aspiring pride, yet both also recall the Son's superiority and Satan's envy of it. Satan's sense of futility passes seamlessly into his conviction of his own insincerity and then into his assertion that reconcilement can never "grow / Where wounds of deadly hate" so deeply fester (98–99). His moving, self-destructive soliloquy begins with pride and apparent honesty, only to end in envy and despair, as the narrator emphasizes immediately on its conclusion: "Thus while he spake, each passion dimm'd his face / Thrice chang'd with pale, ire, envie, and despair," a noxious mix recalling those in the snapshots of envy with which Part I of this chapter began: the intersection of anger and hatred with envy in Aristotle, Cicero, and Bacon, the paleness and self-consumption of Langland's envy, the envious discontent *underlying* Spenser's Pride (114–15).[30] While Satan is willing to acknowledge his pride and ambition, both relatively heroic excesses, as the causes of his fall, what most basically makes him tick (hint of mechanical compulsion intended) again turns out to be envy—negation and ultimately death. Notably, his envy at this point is also both dominantly and openly a passion. He *suffers* its necrotic recurrence. Both poignant and parodic, this is his passion play.

With the partial exception of Satan's confrontation near the end of Book IV with the angels guarding Eden, an encounter that engages his envy but more conspicuously his haughty pride, envy is generally what motivates him in the Garden. Now, however, it is preeminently a response to Adam and Eve, and it is especially envy of their wedded bliss. By contrast, their bliss sharpens Satan's ever-present awareness of hell, "Where neither joy nor love, but fierce desire . . . Still unfulfill'd with pain of longing pines": desire per se, emptiness, death (IV. 509, 511). Yet Adam and

Eve finally have no interest for Satan except as his means of revenge; they are merely surrogates for the Godhead:

> Since higher I fall short, on him who next
> Provokes my envie, this new Favorite
> Of Heav'n, this Man of Clay, Son of despite,
> Whom us the more to spite his Maker rais'd
> From dust; spite then with spite is best repaid.
>
> (IX.174–78)

This is how the world now looks to Satan: envy and spite, motivation and mirror, a self-reflexive circle of hate. Outside his circle, there is no reality for him.

Traditionally, Envy is associated with vision, necessarily including perceptual point of view. This association is found both in its etymology, Latin *in*, "upon," and *videre*, "to look," and in the "evil eye" of which scripture speaks (for example, Matt. 6:22–23, 20:15). In Book IV, envy's "evil eye" focuses with a vengeance on the Edenic couple's lovemaking when "the Devil" turns away "For envie, yet with jealous leer maligne . . . [Eyes] them askance" (502–4).[31] In Eden Satan's envy becomes increasingly vicious, degenerate, and self-consuming. Without the envy that defines him, he can only be "Stupidly good." Briefly stupefied, he is "abstracted . . . From his own evil, . . . of enmitie disarm'd, / Of guile, of hate, of envie, of revenge"—abstracted at once from envy itself and from all the attributes that traditionally accompany it (note the association of Spenser's Envy with deception, or guile). With blatant irony, he can be "good" for a "space"—"spaced out" in strikingly apt modern parlance—if also disoriented, unsituated, and unreal. Abstracted from his own essence as "the Evil one," here a single-minded, demonic abstraction like Sin and Death, he has become risibly oxymoronic (IX.463–66). The language of the poem emphasizes the play of irony here: momentarily, in a parody of self-recovery and reunion with the original source, what is now the real Satan vaporizes into self-canceling goodness.

Satan's own envy may be why he uses envy so insistently and successfully in his temptations of Eve. When Satan first overhears Adam explaining God's prohibition of the apple as the sign of the couple's obedience, he either hears the explanation selectively, or else for his own entertainment he performs his hearing of it simply as a prohibition of knowledge and therefore as an expression of divine envy: "Why should thir Lord / Envie them that? Can it be sin to know, / Can it be death?" (IV.516–17). Of course he also uses the verb *know* intransitively, leaving the object of knowledge wide open, and he focuses instead on the penalty of knowing, death. By this thinking, the penalty becomes envious, and death itself expresses envy rather than justice. Ironically, this is exactly what death expresses, but the envy belongs to Satan, who is mindful here, if unwittingly, of his incestuous son/grandson. He has been so from the start of the poem, again unwittingly: I would recall both that his first words in hell resonate with the memory of death, since they echo those of Aeneas to Hector's ghost on the night Troy falls, and that death awaits birth in the envious inception of Satan's rebellion.[32] If death is the scriptural sign of "all evil, and whatever . . . leads to ruin," as Milton observes in *Christian Doctrine*, it is also the signature and defining feature of Satan. This feature is evident at once in the deformed offspring of his mind and in his ever circling back to himself and thus to a dead ending.

While the extent to which Satan controls his own interpretation of God's prohibition of the apple right after he overhears Adam remains questionable, I am inclined to see him performing—staging—his (mis) interpretation of it for himself and beginning to fashion an illusion into which he can enter and therefore can credibly enact. By performance, he actualizes—realizes—his illusion to and for himself. Of course this illusion also reflects his own envious character: here as elsewhere in *Paradise Lost*, the work is an image of the worker. Actualizing illusion, his figuration looks forward to the reptilian metamorphosis of Book X. When Satan is discovered at the sleeping Eve's ear, he is "Squat like a Toad" (plausibly another recollection of Spenser's memorable Envy, down whose jaw drips a toad's venom, sign of calumny, or slanderous deception), and in the dream he inspires in Eve, his phantom wonders rhetorically whether knowledge is "so despis'd" or whether "envie, or . . . reserve forbids" her to taste the apple (IV.800, V.60–61). When later in Book IX, in order to tempt the waking Eve, he assumes the body of a serpent, a more familiar emblem of envy, he first asks, once again rhetorically, whether God's prohibition is envious, and then, with a self-reflexive irony that he either dubiously or ludically controls, he questions how envy could ever dwell "In Heav'nly brests" (IX.729–30).

Satan's culminating metamorphosis into a snake completes and comments on his path from the first inception of evil to its realized form, the reptilian "icon of his envy."[33] Simultaneously, this icon is his essence fully manifest and the precipitate of his envy. Although "some say," this humbling metamorphosis is meant to dash his "pride, and joy for Man seduc't," as it surely must, the tormenting compulsion he suffers with his fellow snakes looks more like the fulfillment of their envy (X.575–77). Like Milton's figure of Death, they are driven relentlessly by their lusts and appetites to seek an illusory satisfaction that only increases their frustration and pain. Worse, they apparently retain some awareness of the futility of their own actions and responses: their imaginings multiply the trees "to work them furder woe or shame; / Yet parcht with scalding thurst and hunger fierce, / Though to delude them sent, [they] could not abstain" (X.555–57). In the conspicuously braided syntax of these lines, what the snakes do and feel themselves cooperates with its retributive purpose ("to work . . . shame," "to delude them") to double their misery. Their condition at once recalls Satan's earlier characterization of hell, where "fierce desire . . . Still unfulfill'd with pain of longing pines," and earlier poets' figures of self-torturing Envy: ulcerous, unnourished, rapaciously empty, and, in short, malignant—deathly (IV.509, 511). As I have recurrently indicated, death, Satan's final and finally real production in human history, is already implicit, or present as envy, in the inception of his rebellion, which is the original sin.[34]

Part II: The Complementary Circularity of Self-Authoring Pride

Individuality and Allness

The other pertinent thread of my argument, Satan's pride, also begins with God's crucial proclamation, consequent on the Son's exaltation as "King anointed," that all shall be "United as one individual Soule / For ever happie" (V.610–11). This promise complements God's earlier one in the narrative sequence of the poem that "God shall be All in All" (III.341). Satanic pride, which near seamlessly succeeds envy when God elevates the Son and promises greater unity, has not hitherto been the major focus of my discussion. This more positive and heroic aspect of his evil character is what initially attracts many of us, and it is most evident in the first book of hell, in the account of war in heaven (Books V and VI), and, to an extent, in Book IV, Satan's opening soliloquy and subsequent confrontation with the angels guarding Eden. The self-authoring aspect of Satan's pride, which is its apex, is central to the rest of my argument regarding his character. This aspect is inseparable from the questions of individuality and allness, in which, like Satan's envy, it is rooted. All these issues are profoundly linked in the narrative of God's exaltation and promise and Satan's response to it. Turning now to the further consideration of Satan's pride, however, I do not mean to lose sight of envy as the original, ulcerous cause of his rebellion, his fundamental negation, and his parodic circularity.

Whereas Part I of this paper began with snapshots of envy, I start here with an admission; like the snapshots, which highlight the difference between reading poetic narrative and theological doctrine, my admission arises from and bears on the experience of reading. Excepting Raphael and Abdiel, I have always had difficulty remembering which good angel is which at what point in *Paradise Lost*, despite my having read and taught the poem for decades. I have little trouble remembering which bad angel is which. I think this curious mnemonic failure, which is hardly unknown to others, has something to tell us. It sets two different conceptions of character in sharp relief. Historically, these interrelate with conceptions of "self" and "individual."

The *OED* first recognizes *self*, signifying "one's own intrinsic being," in Traherne in 1674, although this modern sense is arguably found earlier in Spenser, Sidney, and Shakespeare.[35] *Individual*, as already indicated, means "unified," "undivided," or "indivisible" when Milton's God uses it in proclaiming the Son's manifestation. According to the *OED*, however, from 1613, the same word also signifies "existing as a separate indivisible entity," that is, as "the individual" as distinct from the general, or as "this or that particular man." This is the sense of Milton's Latin in *Christian Doctrine*, where "man" (a living soul, *anima*, or the biblical *psuchē*) is defined as "a living being, intrinsically and properly one and individual (*individuum*), not compound (*duplex*: 'double') or separable, . . . a substance individual (*individuam*), animated (*animatam*), sensitive, and rational."[36] In a late Jacobean sermon (1620?), John Donne anticipates Milton's Latin when he asserts that "*Ego*, I, I the same body and the same soul, shall be recompact again [at the Resurrection], and be identically, numerically, individually the same man."[37] Donne means both indivisibly and intrinsically, uniquely. By the 1630s and 1640s, the *OED* recognizes—belatedly, given Donne's sermon—the use of *individual* to signify "self-identity," "self-sameness," or "distinguished from others by attributes of its own," and at roughly the same time, the logical and metaphysical uses of this word, again overlapping with Milton's Latin, include the signification of a singularity that cannot be subdivided into others of the same species.[38] Such usage recalls earlier, richly various conceptions of individuation in medieval Scholasticism and Renaissance humanist philosophies, whose roots go even deeper into the past. Ernst Cassirer's classic study *The Individual and the Cosmos in Renaissance Philosophy* has shown how profoundly the status of the individual psyche (individual thinking consciousness, individual thinking lifeform) is implicated in the whole "subject-object problem," the problem of self-consciousness, self-awareness, from the quattrocento to the seventeenth century—from Nicholas Cusanus to René Descartes.[39] In what follows, I have used the term *character* in an effort to contain this problem, to keep Milton's poetics in the forefront, and, at least for now, to simplify discussion.

One conception of character is evident in Abdiel's trust in God's plan for union and wholeness, for all-ness, in-dividuality, un-dividedness, and the other is evident in Satan's distrust of this plan, which culminates in his assertion that he is "self-begot, self-rais'd" by his "own quick'ning power" (V.860).[40] Abdiel's heroic trust and loyalty are gloriously undercut and applauded, their importance simultaneously diminished and enhanced, when he returns to God. As Abdiel's name, "Servant of God," signals, his action fulfills—realizes—his being. In comparison, Satan's claim, in historically loaded words, is bold and brave and liberating, whether or not it is true. Simultaneously idea and act, it is in both senses a *realizing* of autonomy. It not only expresses Satan's ontological claim but also more crucially actualizes his ethical being as a character—his *ēthos* in the richly suggestive terminology of Aristotle's *Poetics*.

Diana Treviño Benet's fascinating essay "'All in All': The Threat of Bliss" pertinently reflects on how "the prospect of union [promised by Milton's God] seems a prospect of diminished selfhood or an omen of personal obliteration." Benet observes that "the poem never overcomes lingering images of union as a frightening reabsorption of creature into an originary being"; I would add to her insight, "be this reabsorption angel into God, Eve into Adam, or perhaps even Death into Satan."[41] To juxtapose a familiar biblical passage to Benet's reflection, however, is to highlight both the meaning of the terms *self, person*, and *creature* that she opposes to *union* and *originary being* and their centrality to the issue at hand.[42] Matthew 16:25–26 reads,

> For whosoever will save his life shall lose it: and whosoever will lose his life for my sake shall find it. For what is a man profited if he shall gain the whole world, and lose his own soul? Or what shall a man give in exchange for his soul?

Variations of this passage occur throughout the gospels, signaling its importance: Matthew 10:39, Mark 8:35, Luke 17:33, John 12:25. In the passage cited itself, *life* and *soul* translate the same Greek word *psuchē*, what animates one's whole being in the usage of the synoptic and fourth gospels.[43] A peculiarly modern version of the Evangelists' point that I have often read, heard, and spoken is that of losing the self to find the self, with the crucial modifier "true" silently inserted before the second occurrence of "self."

I want to invoke a conflict in Renaissance Neoplatonism for the provocative light it throws on the all-embracing unity of Milton's God and on the tendency of Milton's good angels to blend together, in radical distinction from the whole demonic lot of bad apples. My purpose is *not* to claim influence or to invoke an entire Neoplatonic system but only to explore certain possibilities and implications, the most salient of which is the connection between individuation and kinds of knowing, whether through knowledge as such, through love, or through both.[44] The conflict in question pertains to the disparate beliefs of Marsilio Ficino and his sometime student Giovanni Pico della Mirandola regarding immortal felicity. Insofar as Pico's belief reflects the thinking of Averroës, Islamic Spain's great medieval commentator on Aristotle and a huge influence on the scientific thinking of Renaissance Padua, where Pico also studied, it has ties with mortalism, to a version of which Milton subscribed.[45] Averroists held that the individual, or personal, soul dies with the body but that an eternal and immaterial Intellect that is single, individual (un-divided), and impersonal is immortal. In early modern England, various radical movements and individuals, whom Norman Burns terms "experimental Christians," believed that "the personal soul dies with the body, while only the divine essence of the soul survives as an undifferentiated part of God"—swallowed in deity as a drop in the ocean. By a mystical route, they thus reached a position approximating that of the Averroists.[46]

Although the Averroist Intellect is not the Christian God and Satan is an angel rather than a human being, the Averroist combination of personal obliteration with universal Intellection has more than a little resemblance to Satan's worst fear. As Cassirer puts it,

> Averroism . . . explains thinking consciousness in such a way that it makes it disappear *qua* consciousness. At best, its unitary "active intellect" can be considered a cosmic being and a cosmic force; but this force lacks precisely that moment which could transform it into self-consciousness, transform it, that is, from a mere "being-in-itself" to a "being-for-itself." (136)

A "being-for-itself" in a God-centered universe could take us back to a definition of satanic pride as the sin that "arises when a creature values his own self-will above the will of his Creator," Revard's definition cited earlier, which is consonant with the view of Augustine and Aquinas. Implicitly, Cassirer associates such pride with the birth of the modern self-consciousness that he admires—a modern self with a heightened awareness of "one's own intrinsic being." This is the awareness that Satan powerfully exhibits in his opening soliloquy in Book IV, his first speech without an audience within the narrative. Not surprisingly, Descartes, with whom Stephen Fallon, for one, has associated Milton's Satan, makes several cameo appearances in the final chapter of Cassirer's study (21–29, 204). With respect to the circularity of Cartesian logic, that of Sin's hellhounds particularly comes to mind. To revert, not to say cycle back, to the opening of Part I of my argument, "Snapshots of Envy and Death," Satan always circles back to his sole self: "Which way I flie is Hell; my self am Hell" (IV.75).

In contrast to Ficino, Pico believed that the *summa felicitas*, the highest form of felicity, entailed "mystical self-annihilation," in the words of Edgar Wind. It was to be a "personal 'extinction in God,'" an ecstatic blindness that was identical to Orphic night.[47] In the *Oration on the Dignity of Man*, Pico can still sound relatively Ficinian in speaking of an ultimate "peace, [an] . . . indivisible bond, the friendship which is one soul, the friendship whereby all minds do not merely accord in one [Averroistic] intellect that is above every intellect but in some inexpressible fashion become absolutely one." Sentences later, however, he compares the soul to a bride who forgets not only her birth family but also herself, longing "to die in herself that she may live in her bridegroom."[48] In the later *Heptaplus*, Pico sounds further radical. He defines felicity "as the return of each thing to its beginning. . . . [for] the end of all things is the same as the beginning of all: one God." He continues, "Created things can acquire this good in two ways, either in themselves or in itself," and he soon distinguishes these ways by their value, discounting the first way (148). He explains that nothing can rise above itself through "its own strength (otherwise it would be stronger than itself)," and therefore "nothing relying on itself can attain a felicity any greater or more perfect than its own nature" (149–50). Although Pico carefully allows that "we reach God either in the creatures in which he participates or in himself," what he stresses is "that through their own powers, created things cannot achieve this ultimate felicity, but only the former," namely God in themselves, and that this is "rather the shadow of felicity than true felicity" (150). For a moment, a reader of Milton might think of Abdiel's understanding of the plan Milton's God announces when he proclaims his Son "King anointed." Yet what Pico actually has in mind is a loss of self in God, and this is unlikely to be the full extent of Abdiel's meaning.

Ficino's view of the highest form of felicity affords another possibility. For him, too, it is blinding but with an exhilarating excess of joy (*gaudium*). As for Pico in the *Oration*, it is friendship but it is also mutual friendship, a dynamic reciprocity that Pico came to deny on the basis of the immeasurable inequality of mortal and god.[49] In the *Platonic Theology*, Ficino asserts that "God loves what is reciprocally loving Him" and that "the life of God can be communicated to the intellect for the very reason that it is loved by the intellect, since the goal of friendship is sharing life [*vitae communio*]."[50] The life that is desired is perpetual. Cassirer makes the obvious point that Ficino's conception of friendship is clearly tied to subjectivity, or conscious self-reflection, to which I would add the radical dynamism of eros in Neoplatonic thought (132–33).

Ficino's God is at once transcendent and immanent, and, to borrow a word from Milton's lexicon, the created world is full of his traces. Ficino held that the two ways of knowing God, in oneself or in himself, which Pico distinguishes as shadow and substance, are actually one, because "only by reverting to God [compare Abdiel's reduction as a leading back: *reducere*] do men achieve 'not falling off from themselves'. Since transcendence . . . [is] the restorer of immanent virtue, man should find, not merely lose, himself in God."[51] Whereas Pico compares the inferior felicity creatures find in their own natures to "The linear motion by which the elements are carried to their proper places"—their own places, themselves—he considers "Circular motion, through which a body is carried around to the point from which it started, . . . the most express image of the true felicity" (*Heptaplus*, 151). In sharp contrast, Ficino, sounding much like Augustine, here as elsewhere an influence on his thought, describes how the perfected intellect itself is "reflected into itself" and how, given this characteristic of self-reflection, it "exists and remains within itself" while not, like Satan's intellect, becoming self-enclosed there. He adds that the intellect is "entirely incorporeal and simple" and "goes forth from itself in a circular motion," ensuring its perpetual, dynamic movement and thus that "it can always act and be alive."[52] In another relevant contrast, Wind finds Pico seeing in "Ficino's optimism . . . a Narcissus-like self-love through God," whereas Ficino finds God both in himself and in perfected friendship among men.[53] Perhaps it is not too far-fetched to recall a passage from Milton's *Reason of Church Government* about self-esteem and self-reverence (self-regard or self-respect):

> And if the love of God, as a fire sent from heaven to be ever kept alive upon the altar of our hearts, be the first principle of all godly and virtuous actions in men, this pious and just honoring of ourselves is the second, and may be thought as the radical moisture and fountainhead whence every laudable and worthy enterprise issues forth.[54]

Sufficient self-esteem and self-reverence are what Raphael and the Son accuse Adam of lacking or, once fallen, of having lacked.

With respect to eternal felicity, Ficino, unlike Pico, believed in a perfected self-preservation, one "conformable" not only with the aspirations of the intellect but also with "the free nature of the will" and even with the inclinations of the perfected body: "The condition of the everlasting [human] soul which seems to be in the highest degree natural is that it should continue to live in its own body made everlasting"—that is, in every respect as its self. Ficino continues, "the immortality and brightness of the soul can and must at some time shine forth into its body," for "in this condition alone, the highest blessedness of man is indeed perfected."[55] Milton, equivocal monist visionary and proponent of a materialized vitalism, is not just a Neoplatonist, yet Neoplatonism is a significant part of the mix in his culture, and it enters his poetry.

In *Paradise Lost*, God's statement that He "shall be All in All" invites more than one reading, as does his statement that all will be "United as one individual Soule," which I earlier considered in its bearing on the immediate context of Satan's rebellion in Book V. "All in All," unlike a declaration that all shall simply be one, implies a collective, or a unity of members like the resurrection body of Christ in Paul's epistles. Pertinently, "as one individual Soule" could be read either as "unified into absolute nondifferentiation" or as "comparable to a unified lifeform," which consists of various harmonious parts, not unlike music.[56] The word "as" itself invites further play, prepositionally yielding "in a manner similar to" or "the same as" or conjunctively (with silent "is" assumed) yielding "in the manner that." Yet another conjunctive possibility is that "as" bears the counterfactual sense "as if." That "God shall be All in All" obviously differs as well from the statement that "All shall simply be folded into God"; indeed, God's statement could conceivably suggest a further extension, not only a further intension, of radiance.[57] The ambiguity of God's statements simultaneously contains, or includes, and refuses to limit possibility, indeed, divine potentiality.[58] By this same reasoning, the univocation, or union, God proclaims is also equivocal. His statements are suggestively consonant with either Pico's or Ficino's understanding of supreme felicity, with either Abdiel's or Satan's response, and with different views of the self, the individual, the knowing and willing self-reflexive subject. To look ahead to my next chapter for a moment, a condition that lies between univocation and equivocation, sameness and difference, also engages the central philosophical issue concerning analogy. This issue is simultaneously ontological and epistemological, at once about origin and about knowing.

"Know ye not mee?"

In *Paradise Lost*, the only time Satan actually interacts with unfallen angels after his expulsion from heaven and his brief meeting with Uriel is at the end of Book IV: it is this interaction that initially led me to suppose that innocence and fallen experience are connected conceptually with angelic characterization, or angelic ēthos, since the names of the unfallen angels involved are (at least for me) especially elusive. Conspicuously, this interaction also involves knowledge and specifically self-knowledge and awareness, focally raising issues of innocence, experience, and ēthos that will later pertain tensely and urgently to Adam and Eve. It concludes, as does Book IV, with a prodigious battle not joined and the much-discussed epic simile of "the careful Plowman" and "his hopeful sheaves," neither of which makes much sense until the pointed undercutting of the faithful Abdiel's news about Satan's coming attack in the books that follow.[59] At the end of Book IV, antiheroic anticlimax thus foreshadows Abdiel's reasoned and voluntary self-subjection to God in the narrative sequence, which is simultaneously his refusal of Satan's self-authoring pride and his own triumphant fulfillment: in sum, losing the self to find it. Abdiel's return to God also reflects and reflects on the *relative* facelessness, the *relative* absence of autonomous individualism, among the unfallen angels. Two conceptions of individual character, or individuated selfhood, are in active and sometimes explosive tension in the culture and society of the seventeenth century and not least in Milton's epic.

When two of the angels guarding paradise discover the toadlike tempter close at Eve's ear in Book IV and cause him to reassume his own likeness, they do not recognize him except as one of the rebel angels. In prideful scorn and startled disbelief, Satan asks twice in succession, "Know ye not then . . . Know ye not mee?" as if he cannot even utter "mee," the absent object of recognition, the first time. He adds with wonderfully epic arrogance, "Not to know mee argues your selves unknown" (827–28, 830). Of course he means, "You two young angels are nobodies," but as his subsequent arguments make clear, he also means, "You don't know your own selves." His assertion pointedly raises the question of selfhood and does so explicitly in the context of self-knowledge, self-awareness, self-consciousness. With respect to the unfallen angels he confronts, his assertion is easily dismissed, or perhaps not so easily, once Raphael recounts the inception of Satan's rebellion—the origin of evil as envy—in the next book. Satan's epistemological assertion is based on his experience of himself both in that episode and in its more recent sequel (chronologically, but narratively its prequel) in his soliloquy at the opening of Book IV. It is based on his self-knowledge of the self he has chosen to be and has now become. The very change in his outer shape shows, in the words of Neil Graves, that "formal essence [form, simultaneously material and intrinsically connected to *essentia* in Milton's thought] is determined by ethics"—by *ethos/ēthos*, in Greek the cognate terms designating both ethics and character (14).[60] In Milton's epic, ethos determines what a thing properly is; it is as fully real as the very substance that it informs and renders singular, individual, numerically one. Evil is therefore really, not just metaphorically, a necrotic disease, and here it *transfigures* Satan's whole being. Not just a rhetorical figure, an appearance, or a figment of fantasy, his transfiguration is vital. This point is not lost on him, reluctant as he is to embrace it in the fourth book. He grieves enviously at the loss of his former shape and chafes at others' noticing it, but he is not now unaware of his diminished radiance in others' eyes: "abasht the Devil stood, . . . and pin'd / His loss; but chiefly to find here observd / His lustre visibly impar'd" (846–50).

Satan's assertion regarding self-knowledge has another dimension as well. Where I have used the word "recognize" (and its cognates) to describe the response of the unfallen angels to Satan, he emphatically repeats "know" (and its cognates). Confronting Satan tempting Eve, the unfallen pair of angels do recognize his wickedness, but in the ensuing dialogue first with them and then with Gabriel, Satan insistently tries to refocus the issue on what real knowledge is and in what sense it includes his own undeniable presence in the Garden and thus, in actuality, himself—his deathly negation, his depletion, his evil.[61] Again and again, he argues from personal experience and implicitly urges it—experience that is direct not vicarious, actual not hypothetical, personal not general, intrinsic not extrinsic. Such experience, as Cassirer would have been quick to tell us, is properly subjective and self-conscious.[62] Satan can only validate his own self-knowledge, and indeed himself, by drawing others to embrace and know these in the fullest participatory sense. As before in heaven and hell, for all his proud independence, he becomes once again an analogically parodic mirror of the Godhead who offers those who abide in him allness and individuality. Alternatively, of course, we could side with Descartes and Cassirer—more exactly Cassirer's Descartes—and *see* that it is really God who enviously and contemptuously mirrors Satan. Milton has set such provoking choices ever in the eyes of his audience. From the "sight" of "*Messiah* King anointed" that Satan first envies in Book V, Satan's envy has pointedly been associated both with vision and with death, with the way what is real is seen, perceived, and understood.

Looking back at my first two chapters and the present one, then ahead to the next, which shifts to the topic of analogy and treats it at length, I want briefly to clarify the relation of allegory to analogy, both terms that have recurred in discussion of Spenser and Milton to this point. Allegory and analogy are specific forms of metaphor, related in genealogy and in rhetorical theory. In practice and over time, figures or tropes such as these overlap, blend, and otherwise develop, a subject that will return in later chapters. Some analogies might also be considered allegorical or even allegories, and many allegories considered analogies: for obvious examples, Spenser's Redcrosse is to Una as holiness is to truth, Guyon to his Palmer as temperance to reason, Milton's Satan to Death as his God to Life (A:B :: C:D). More definitively, an elaborate allegory like Spenser's is a continued form of metaphor, one that is greatly extended and modified in narrative as it develops. Allegory occurs more briefly but explicitly in Milton's initial representation of Sin and Death in *Paradise Lost*, although, like a kind of residue, it afterwards trails Satan through Chaos to the created world—an argument I have made elsewhere and occasionally recalled in this chapter.[63] Analogy can specifically be defined as a form of metaphor that consists of a structurally proportional relationship. It is especially present in Satan's parodic imitation of God and his Son and more broadly evident throughout *Paradise Lost*: the relations of Milton's hell to earth and both to heaven afford other, familiar examples. The final chapter of this volume, returning to *Paradise Lost*, treats the typically analogical structure of epic similes and their participation in the tradition of analogy in multiple fields of knowledge in the early modern period. This tradition itself is the subject of my next chapter.

CHAPTER 4

Connecting the Cultural Dots

Classical to Modern Traditions of Analogy

Changing focus, this chapter engages the history and structure of analogy. The change is pronounced—from sin and death to rhetoric, from poetry and belief to science and methodology. In manner, the chapter is historical, analytical, and abstract, in effect a further shift. These changes threaten to confirm the very chasm between science and the humanities against which I argue. Yet my first three chapters have treated matters that relevantly recur in this one, such as physical and intellectual vision, body and mind, imagination, knowledge, figurative illumination, and Neoplatonism. More importantly, the argument of this chapter enables a theorized broadening of concerns from negation to construction, sin and death to life and light, and it offers to get poetry and religion under the same canopy of creativity as science without masking the difference between them. This is possible only with a considerable degree of abstraction, starting with verbal and mathematical languages—sign systems—which fundamentally underlie the connections I trace. Analogy, whose other name is traditionally proportion, is as familiar to mathematicians and scientists as to poets. As the Introduction to this volume explained, my route to analogy and science began with the preceding three studies of death, which led in contrast to the principle of light and thence to constructive analogy, pointedly including the science of light. With the first three chapters and the present chapter as backdrop, the final three chapters—first on Johannes Kepler, then Donne, then Milton—will focus partly on analogical practice in science and poetry and partly on my other related, thematic concerns with light, death, or both.

The convoluted roots and widely divergent applications of analogy in the early modern period have drawn the attention of historians of science, rhetoric, literature, philosophy, and religion, yet the study of these has often occurred in isolation from, or narrow relation to, one another's concerns. Instead, my argument emphasizes the bond among disciplines in their use of analogy for exploration, experiment, and discovery. It addresses questions of continuity and change between early modern theories and uses of analogy and those of its ancient predecessors and, as a framing device, between both of these and modern ones. In short, it seeks to connect the cultural dots: classical, medieval, early modern, and modern. Science and rhetoric are two of its major foci.

The theorized roots of analogy in the early modern period are Aristotelian, as, indeed, is the psychological basis of analogy in the imagination.[1] According to Aristotle, moreover, analogy is a subcategory of metaphor that has a technical relation to, and a basis in, mathematics. The relation of analogy at once to metaphor and to mathematics is vitally important with respect to its roots and their development from classical times to our own. Further important is the fact that metaphor, Latin *translatio*, or "translation," is, in enormously influential Latin rhetorics, the archtrope, the generic umbrella term for all the tropes, as well as itself a specific trope. Metaphor not only relates to other tropes (cousin-like) but also inheres in them (gene-like). Generically, the tropes are translative; they carry meaning from one place to another (*trans*, "across," plus *ferre/-latus*, "to carry"). Specifically, however, metaphor, the single trope (apart from the generic umbrella term), is also, definitively, a source of creativity and a means of construction that differs in these respects from (typically metonymic) substitution. Analogy is an especially close relative of metaphor and is the trope that Aristotle specifically defines in the *Poetics* as a version of metaphor, a point to which my next section returns expansively.

An application will suggest the pertinence of family relationships among the rhetorical tropes. Recounting Copernicus's discovery of a heliostatic universe, Fernand Hallyn invokes mathematical analogy and distinguishes it from rhetorical likeness, effectually separating mathematics from rhetoric and analogy from metaphorical translation: accordingly, Copernicus "reduces" analogy, or proportion, "to its pure mathematical form," and rhetorical similitude (*similitudo*) is "thrust aside" in the process. In a cultural transfer that Hallyn later terms metonymic, cause is thus "substituted" for effect, and mathematics for appearance, or mere sense impression.[2] Yet Hallyn's argument bears extension. Reduction of analogy to the pure abstraction of mathematics does not simply cancel its metaphoricity, specifically the fact that the analogy remains a translation into, or more complexly within, a language of signs. Moreover, even if Hallyn, in his role as cultural historian, understands this Copernican transfer to be metonymic—merely "substitutive"—it is also generically metaphorical, insofar as metonymy itself is a form of translation—indeed, a distant relative of metaphor, the archtrope. Mathematical signs are purer and more abstract, stable, and exact than verbal ones, but they share the general problem of representing something other than themselves that is more real. The heliostatic universe of Copernicus, after all, was not the heliocentric one.

Departing from Aristotle's view of analogy as metaphor, the modern psychologists Dedre Gentner and Michael Jeziorski, in an essay tellingly titled "The Shift from Metaphor to Analogy in Western Science," make a move similar to Hallyn's. They describe analogy in current terms "as a kind of highly selective similarity" that is based on "relational commonalities," "relational structure," or "a system of relations."[3] Such modern analogy differs from metaphor, in their view, on the basis of its structural content, which is consistent, relational, and systemic, requirements that do not exactly sound un-Aristotelian or even un-Scholastic. In our own time, deconstructive and neocognitive discussions of metaphor alike have surely established that conceptual metaphors in philosophy are systemically significant, not random vestiges of etymology.[4] The elements of a verbal language are relational and systemic, and, when disciplined by standard definitions and other procedures of logic, reasonably consistent. In other words, shared properties, relational structure, and systemization do not in themselves neatly divide mathematics from verbal language. The structural *content* that Gentner and Jeziorski consider definitive of analogy, moreover, is not the same thing as the *structure* of analogy that was Aristotle's concern and now mine: namely, A is to B as C is to D.

While content is the positive determinant of scientific analogy for Gentner and Jeziorski, their exclusions further define its boundaries. Excluded are causal implications, extraneous associations (especially thematic ones), and also mixtures of separate domains or, in the instance of comparisons between two domains, none allowable that fails to "convey" a coherent system—in short, further kinds of content (450). Negations, dependent on the negated other, thus sharpen the edges—limits—but do so in terms repeatedly requiring judgment calls, and these open the door to something more psychological or subjective. For Gentner and Jeziorski, early practitioners of science like Robert Boyle (1627–91) and Sadi Carnot (1796–1832) afford positive examples of the shift to modernity, in contrast to the negative examples of alchemists as a group, with Theophrastus Paracelsus (1493–1541) as the prime example. Gentner and Jeziorski also invoke Kepler (1571–1630) and J. B. Van Helmont (1580–1644) as transitional figures who approached analogy more critically or with greater discrimination (472–75). Curiously, if not contradictorily, Gentner and Jeziorski belatedly acknowledge that early modern scientific analogy is indeed a "special case" of metaphor but quickly wall it off by reasserting the structural content they have earlier defined and distinguished from metaphor (452). The relation between the structure of analogy and the structural relationship(s) it represents are at best unclear in their essay.[5] Reference to their essay will recur since it affords a strong version of the claim both for the novelty of early modern analogy and for a corresponding disparagement of *mere* metaphor in the early modern period that I consider oversimplified and misleading with respect to the historical roots and more modern uses of analogy.[6]

Before turning to other modern and early modern assessments of analogy, however, I want to consider a modern conception of analogy that, unlike Gentner and Jeziorski's, makes no claim for progressive enlightenment or historical difference. It offers a competing and clarifying point of theorized reference based on modeling and pointedly includes metaphor. David Douglass, a specialist in rhetoric and media studies, derives this conception from the influential analytical philosopher Max Black and uses it in his analysis of "I. A. Richards' Tenor-Vehicle Model of Metaphor."[7] For my purposes, Douglass represents a current connection to Richards and more broadly to rhetorical studies that Black alone would not. In further contrast to Gentner and Jeziorski, Douglass additionally affords an alternative that not only admits but also admires the cultural and psychic creativity of metaphor.

Douglass follows Black in distinguishing four kinds of models, three of which afford a physical representation of a concept or intelligible phenomenon.[8] These four kinds are the scale model, the analogue model, the mathematical model, and the theoretical model, the last specifically including metaphor. The scale model is iconic, and it attends to "geometrical magnitudes and ratios." The analogue model is more isomorphic in the mathematical sense than is the scale model; it therefore represents "relationships or functions present" in its subject. If it is also iconic, it is so "more abstractly" and shares "the same structure or pattern of relationships" instead of the "identical proportionality" of the scale model (Black, 222–23). Black mentions the use of electrical circuits in computers and hydraulic models of economic systems as examples of the analogue model. These are operative analogies par excellence.

The mathematical model, which Black terms "something of a digression" from his argument and not a true model, maps an "'object system'" onto one or more "'mathematical systems or models'" (223); his example comes from the practice of social scientists, instanced by studies of population growth. This sort of model, which lends itself to mathematical calculations, seems to be "*simpler and more abstract*" than both its source domain and the analogue model. Its rhetoric (for example, source domain, mapping, target, and base) and its penchant for graphs somewhat disconcertingly resemble those of Gentner and Jeziorski. But for Black, in the specific practices he references, it is likely to be "an illusion"—"a kind of ethereal analogue model, as if the mathematical equations referred to an invisible mechanism whose operation illustrates or even partially explains the operation of the original social system under investigation" (223–24). The offending illusion confuses "accuracy of the mathematics with strength of empirical verification" (225). Black's critique has far-reaching implications for early modern efforts to connect the systematism of mathematics with the social, cultural, and physical world.

As distinct from Black's scale and analogue models, his theoretical model, which includes metaphor, is "not literally constructed" (229). Douglass, apparently sensing a problem in Black's use of the word "literally" (that is, "by letters") to approximate "physically" or, more likely, "really," wisely has recourse to Black's own words at this point: "'the heart of the method consists in *talking* a certain way,'" in introducing "'a new language or dialect,'" with the result that "'The theoretical model need not be built; it is enough that it be *described*.'" In traditional terms, Black's model here belongs to the order of saying rather than to the existential order, a useful distinction I stress in anticipation of later discussion. Not surprisingly, the reality to which this sort of model refers, like that of the illusory mathematical model, can become "mysterious" in Black's view, but used properly, it can also yield path-breaking insight and "novel hypotheses and speculations" (Douglass, 418; Black, 230, 233). It becomes misleading if its inventor starts to assume its existential reality and then makes it a substitute for what is truly real. From a rhetorical perspective, I would add, its assumed reality then becomes metonymy, the trope of substitution, an assertion to which I will return again, only noting here the possibility in any given case of a shifting relation over time between a trope of metaphor and one of metonymy—that is, between construction and substitution, the latter, in effect, mere coding.

Metaphor resists exclusion from all Black's models, as both the beginning and second half of his argument suggest (231–42, cf. 25–47). He acknowledges that "To speak of 'models' in connection with scientific theory already smacks of the metaphorical" (219). As Black surely knew, metaphor can be based on function as well as on sight, and on mathematical proportion as well as on tactile properties, as the history of analogy indicates. Perhaps another point to notice here is that Black's synchronic models are themselves imaginative, hypothetical acts of verbally constructive speech.

Gentner and Jeziorski's requirements and exclusions, what they label their "framework" and identify as the "modern principles of analogical reasoning" already manifest in the new science of the seventeenth century, do not align nicely with Black's more metaphorical models (448–50). Although his analogue model at first might seem to fit their framing principles, theirs are narrower and more restrictive and fit only selective early modern examples. His scale model is too iconic for Gentner and Jeziorski's abstraction—his being more suggestive of an architectural miniature or a perspective box than of the graphs they use to illustrate relational structures. His theoretical model they would banish with the alchemists. Applied to early modern practice, moreover, their principles would soon encounter the "illusions" of Black's mathematical model, a misleading amalgam of mathematical certainty and empirical observation, even in as exemplary a figure as Kepler. To consider Kepler transitional, as they do, rather than representative or integrative is effectually to neutralize him, along with a host of likeminded scientific practitioners: a transitional figure, subject simply to what comes before and what after, lacks integrity; our stock phrase "merely transitional" is hard to shake. To deny the tie of analogy to metaphor, moreover, is precisely to deny its counterfactualness, the part of it that is not identity (in metaphor, my love *is not* a rose, even if it *also is* one). To downplay the fundamental metaphoricity of analogy is to obstruct its larger, broader cultural significance, including its simultaneous tie to (verbal) language and to mathematics. Yet the basic issue at this point is not whether Black is right or Gentner and Jeziorski are; instead, it is that an abstract, modern frame like theirs or even a more imaginative, heuristic model like his prepackages early modern practice. The norm of our modern culture is to separate the concerns and methods of science and the imaginative arts and, more fundamentally, of things and words. Often unwittingly, as in the instance of Gentner and Jeziorski, this norm has shaped our investigations of history.

In contrast to such modern framing—to the selective and systematic content Gentner and Jeziorski require of analogy and to the specific distinctions offered by Black and endorsed by Douglass—a brief introduction to three grouped articles on early modern "Imagination and Analogy" in *Isis*, a journal devoted to the history of science, medicine, and technology, identifies analogy broadly and inventively, if also ahistorically, as "*the* trope of resemblance," of which metaphor (Latin *translatio*) is merely one kind, others being *exemplum*, *emblema*, and *similitudo*, "among other things." The same introduction, preparing for the articles it precedes, defines analogy "as that which connects the unfamiliar to the familiar, the insensible or abstract to the sensible."[9] This definition shares perceptive and provocative ground with Black's models. By it, even the concerns of logic and metaphysics, extending to what was known well after Aristotle's time as the science of theology, would be admissible. (The old meaning of *science*, deriving from Latin *scire*, "to know," was simply "knowledge," and natural philosophy was considered a branch of metaphysics.)

Acknowledging modern claims for progress like Gentner and Jeziorski's, the pursuit of precision by Douglass and Black, and the ahistorical inclusiveness of the *Isis* introduction—ahistorical in subordinating metaphor to analogy—my position is that the application of an ancient rhetorical technique is not to be uprooted without a trace or, indeed, without leaving a hole in its grounding. This is especially true if its root is as deep and as rhizomatous as that of analogy and if it is, contrary to Gentner and Jeziorski's suggestion, also psychological, as the historical study of language indicates (447, 477–78). Their suspicion of a common denominator, such as psychology, imagination, and indeed metaphor, for the various historical applications of analogy underlies this denial, whose basis is their own formulation of the "modern principles of analogy." Targeting insufficiently qualified claims for historical novelty, the philosopher and historian of science Thomas S. Kuhn was right to caution us about the incommensurability (not the incomparability) of divergent systems of thought having their own internal and terminological coherence, even while reminding us that "the ontology of [modern] relativistic physics is, in significant respects, more like that of Aristotelian than that of Newtonian physics."[10] Kuhn's remarks might serve as a motto for the rest of what I have to say about the history of analogy.

Early Modern Analogical Science

The three learned articles that follow the introduction in *Isis* all center on imaginative analogy in early modern science: specifically on Francis Bacon (1561–1626), Galileo Galilei (1564–1642), and René Descartes (1596–1630), respectively by Katharine Park, Lorraine J. Daston, and Peter L. Galison. "Bacon's Enchanted Glass," the title of Park's article, derives from Bacon's metaphor for the human mind, which, although also influenced by Galen, is basically Aristotelian (290–302). Bacon thus conceives a mind that depends on the human imagination, which is constitutionally liable to deception. It is Aristotle, after all, who acknowledges in *De Anima* that, whereas "sensations are always true, imaginations are for the most part false."[11] Park's admirably balanced argument is that Bacon, for all his suspicion of the imagination, whose offspring is analogy, ascribed to it a "positive and permanent place" in the area of scientific discourse (301–2). I quite agree, although I think Park downplays Bacon's suspicion of the imagination, which is more pronounced and open than his trust. Evidence of Bacon's suspicion is especially clear in his view of the historian's task but not surprisingly so given the older view that history, like the new science, dealt primarily and relatively directly with objective reality—or, in the memorably naïve words of that vintage American crime show *Dragnet*, "just the facts, ma'am." Science, in Bacon's hands, was in danger of heading in this direction.[12] His suspicion of verbal signs extended to mathematical ones, and his distrust of theory, deduction, and partial knowledge underlay his failure to appreciate the early work of William Harvey (on the circulation of the blood) and William Gilbert (on magnetism). In short, Bacon, for all his acumen, too severely restricts the imagination. In contrast, Kepler, a distinguished, contemporary exemplar on the Continent, both admired and employed Gilbert's work.

Revealingly, Bacon opposes the historian to the poet on grounds of the former's truth—his fidelity or allegiance—to the existent world, rather than to one that is imagined. In open contrast to Sidney's *Defence of Poesy* (1595), he distinguishes between "true history" and untrue history, which he denominates poetry, or "nothing else but *Feigned History*." Bacon's negative description of poetry is a textbook description of the human imagination; just like this faculty, poetry "being not tied to the laws of matter, may at pleasure join that which nature hath severed, and sever that which nature hath joined, and so make *unlawful matches and divorces* of things."[13] Here Bacon describes the innate capacity of the Aristotelian imagination not only to receive but also to combine and recombine images, a capacity that, in metaphorical language evoking the early modern ideology of marriage, he would bind to the natural world and to the laws of matter as these really are—in his view, of course. Here, with nature, marriage, law, and matter all on the Baconian table, we have already plunged into cultural complication, not to mention systemic profundities. And there is yet another complication: in Aristotle, the active power that mediates the relation of the sensory images for the intellect as a whole by disposing or disciplining them is called *nous poiētikos*, productive, indeed poetic intellect, or, in L. A. Kosman's resonant translation, "the maker mind"; this poetic mind "turns the receptive intellect into something that actually thinks."[14] Memorably, Sidney calls the poet "a maker" and relates this term to the Greek name for a poet, namely *poiētēs*, from the verb "*poiein*, which is 'to make.'"[15]

What Bacon has to say about a novel branch of natural history in his classification further illuminates his understanding of the relation of mind to matter and of human art to nature, as well as our awareness of the discipline to which he would subject the human imagination. He designates this branch the history of "nature altered or wrought" and, alternatively, the history of mechanical and experimental arts—in short, the history of science.[16] In *De Dignitate et Augmentis Scientiarum*, however, Bacon speaks of the mechanical and experimental arts much as Sidney speaks of poetry. The *transfer* (my own metaphorically weighted word derived from Latin *transferre*) from the verbal to such scientific arts that Bacon thus attempts is at once illuminating and deceptive. His contemporaries would have recognized it as a translation/*translatio* (likewise from Latin *transferre*), that is, as a metaphor, the other name for which in Bacon's time is a *translation*. Like any other metaphor, Bacon's simultaneously is and is not (my love is a rose; my love is not a rose: A=B; A≠B), and it both reveals and conceals (my love is sweet and beautiful; my love is thorny and short lived). For Bacon, the role of the mechanical and experimental arts, that is, of scientific arts, is to constrain and to mold nature, making her "as it were new by art and the hand of man." The reader might forgive a fleeting memory of Pygmalion here; the personification of nature as a woman, while culturally unexceptional, invites it. Expatiating further, Bacon first rejects a distinction between nature and art in the context of the experimental arts and then attacks the more pernicious error "of considering art as merely an assistant to nature, having the power indeed to finish what nature has begun, to correct her when lapsing into error, or to set her free when in bondage, but by no means to change, transmute, or fundamentally alter nature."[17] More memories inhere in this language, not least the freeing of the fair damsel from draconian bondage: for historians of science, these memories are tactical or pedagogical, but they are also indistinguishable from Bacon's Elizabethan-Jacobean habits of mind. The visionary impulse in Bacon's thinking sits uneasily with his desire to fetter the imagination.

Sidney likewise describes the poet's transmuting and perfecting power to make "things either better than Nature bringeth forth, or, quite anew" (100). Where Sidney imagines the poetic maker rising above physical nature to utilize the higher nature that God has given him alone of all earthly creatures, Bacon imagines an art that not only corrects natural imperfections and exceeds natural limits but also escapes the very fundaments of nature (or feminized Nature) herself; at the same time, he insists on the binding of such art by natural laws and to natural things, and he really means "things"—material objects. He wants a metaphysic tied—Bacon elsewhere suggests the use of weights, hence "tied down"—to the physical world, as, somewhat ironically, the Aristotelian intellect is fundamentally moored to the sensory images of the imagination. The tensions, equivocations, and outright contradictions within this desire are evident at once in Bacon's own thought and more generally in the intellectual culture of his period. The product of the maker mind, *nous poiētikos*, is not so easily severed from poetry, that is, from *poiēsis*, particularly in the cultural imaginary. Moreover, the differential tensions evident at once in Bacon's own thought and more sharply between his thought and Sidney's look ahead to C. P. Snow's notorious phrase "the two cultures" of scientific and humanistic pursuits.[18] In addition to versions of them that are early modern, they also look back to ancient, late antique, and medieval versions, as will be evident in subsequent argument.

The title of Daston's "Galilean Analogies: Imagination at the Bounds of Sense" is ready made for my transition to her essay, the second in the *Isis* trio (302–10). Daston stresses Galileo's distrust of the imagination, which Galileo, too, understood within the Aristotelian tradition, and his consequent avoidance of "explanatory analogies," as distinguished from "expository and mathematical" ones (302). I take explanatory analogies to be undemonstrative or improperly causal, expository analogies to be merely illustrative or instructional, and mathematical analogies to be idealized, isotropic, or in Daston's word, "regular" (303), rather than uniquely or individually phenomenal and occasional.[19] Daston further divides Galileo's mathematical analogies into three types, of which the last, which interprets mathematical entities in terms of physical ones, occurs with particular import when Galileo considers "the continuum [in Aristotle, the infinite divisibility of matter], for it was at once key to his understanding of both accelerated motion and the internal cohesion of matter" (305). Necessarily, this consideration depends on imagination. Employing analogy, as well as a particular conception of nature, Daston herself explains that

> Mathematics and the imagination, which occupied parallel positions between sensation and pure intellect, were . . . naturally paired in both [Aristotelian and Neoplatonic] theories [of psychology]. Geometric figures, being extended and therefore divisible, properly belonged neither to the intellect . . . nor to imperfect, mutable sensation. Rather they were, according to [the Neoplatonist] Proclus's prologue to his commentary on Euclid [in the fifth century CE], "projections" of the ideas of the intellect into sensory forms purified of imprecision and exhibiting "many likenesses of divine things and also many paradigms of physical relations." That is, mathematics, through the agency of the imagination, could render sensible objects intelligible and pure ideas visible. (305–6)

Daston concludes that the continuum, involving consideration of the subsensible and the infinite, is what it takes to make Galileo employ the imagination, "however cautiously, in order to render the invisible and unintelligible visible and comprehensible" (307). Daston's Galileo sounds more than a little like Aristotle himself, not to mention the physicist Max Planck, the founder of modern quantum theory.[20] Although, in the modern philosopher Malcolm Schofield's words, Aristotle is "tempted to view *phantasia* [phantasy, imagination] as a form of thinking, or at least as a thought-like component of thinking" in *De Anima*, Aristotle also acknowledges in the same treatise, as earlier noted, that imagination is likely to be deceptive: whereas "sensations are always true, imaginations are for the most part false."[21] The qualification "for the most part" is important since imaginations (*phantasmata*, representations, appearances, sensory images) are necessary for thought. As Schofield resolves this tension, Aristotle expresses "scepticism, caution, or non-commit[ment] about the veridical character of sensory or quasi-sensory experiences . . . [that is,] 'it *looks* thus and so [—but is it really?]'" (251, cf. 253, 259). Aristotle's view looks to be, at the very least, protoscientific.

Galison's essay is titled "Descartes's Comparisons: From the Invisible to the Visible" (311–26). Aside from specialized analyses of terms, in Descartes's early work the understanding (or intellect) requires the sensory imagination to function, much as it does in Aristotle (321). Citing Rule XIV of Descartes's *Regulae*, Galison emphasizes the essential role of imaginative reasoning to attain truth: as Descartes puts it,

> Other than that which consists in the simple and naked intuition of single independent objects, [all knowledge] is a matter of comparison of two things or more, with each other. . . . The chief part of our human industry consists merely in so transmuting these ratios as to show clearly an equivalence [*aequalitas*, having an "equation" among its meanings] between the matter sought for and something else already known.[22]

Descartes famously uses tennis balls and barriers to illustrate refraction and reflection, the omnidirectional pressure in a barrel of grapes to illustrate the propagation of light from a candle, and the use of a blind man's cane to indicate the instantaneous transmission of light: "'light passes before our eyes, through the medium of air and other transparent bodies, in the same way that the movement or the resistance of bodies, which meet this blind man, pass[es] across his hand through the medium of his staff.'"[23] Notably, French *comparaison*, Descartes's preferred term for such similitude, derives from the Latin words *compar*, "like, equal to another" (itself a combination of *cum*, "with," and *par*, "equal"), and *ratio*, "reason," the progenitor of French *raison*. A point Galison does not make is that *ratio*, the Latin version of Greek *logos*, also implies order, system, and even English *ratio*, or proportion, the last also in Latin the synonym for analogy (*proportio*).[24] Something else Galison does not observe is that analogy need not signify equivalence per se (as Descartes's *aequalitas* is translated above) but only a significant, proportional relationship: A:B :: C:D, not necessarily A:B = C:D; the latter, which is the equation for a quotient, is likely to be assumed by modern readers (2/4 = 5/10 [2]). In the citation above, Descartes may not have had a mathematical quotient in mind (especially since the learned translator's "equivalence," accompanied by the indefinite "an," is ambiguous on this point), and it is surely hard to imagine a universally regular quotient for the transmission of luminosity and either grapes in a barrel or a blind man's cane.

At this point, I want to interject a brief comparison that comments on the issue of shifting terminology with respect to analogy, which is already evident and also looks forward to a later stage of my argument that engages the relevance of Scholasticism, which adapted Aristotelian analogy to Christianity. Often this relevance is overlooked or ignored in a surprising denial of history, including scientific, artistic, and other kinds of intellectual roots and early modern networks. In a book entitled *The Semantics of Analogy: Rereading Cajetan's "De Nominum Analogia,"* Joshua P. Hochschild stresses the instability of the concept of analogy, which is inseparable from the entangled terminology expressing it, as it passed from the Greeks and Romans through late antiquity, to the Middle Ages, and thence to the effort of the Thomist philosopher Thomas de Vio Cajetan (1469–1534) to clarify it in the Renaissance period.[25] In a broad sense, Cajetan's effort, like Descartes's, Bacon's, and Galileo's, would systematize, or rationalize (Latin *ratio*), the use of analogy but within a differently conceived system. A proportional relationship, albeit not an equivalence, between the unknown and the known—in Descartes's phrasing, "the matter [that is, knowledge] sought for and something else already known"—is also a characteristic of analogy in Cajetan's logical treatise (Hochschild, *The Semantics of Analogy*, 156, 170).

Galison repeatedly emphasizes Descartes's opposition to the use of comparison in the old Scholastic philosophy, where, in Descartes's view, it was applied indiscriminately to unlikes, whether unlike relations, objects, operations, or categories.[26] In logical terms, Descartes was accusing Scholasticism of fallacious equivocation. According to Galison, Descartes requires similarity "in a very strong sense"—a litmus test of degree, albeit notably by fallible, culturally inflected human judgment, *not* by litmus paper, which is imbued with the reliable sensitivity to color of powdered lichen (323). Similarity—more exactly a mean between arbitrary equivocation and univocation—is what Cajetan also sought in analogy. Univocation indicates conceptual, rather than proportional, unity, and it initially appears to be similar to what Descartes had in mind in criticizing the Scholastics: figures are to be "explained in terms of figures and motion in terms of motion," as Galison explains, virtually quoting Descartes. "But," Galison continues, "even if the *quaesitum* is subvisible, a geometrically similar object, the *datum*, is presentable to the imagination" (323). Two points about this alternative situation, which posits analogy: if what is sought is invisible, its participation in univocation, or identity, sameness, is indefinite and elusive, as even is its likeness to some extent. Moreover, the *datum*, or given object with a proportional relation to what is sought or in question, the *quaesitum*, does not in itself ensure valid, scientific demonstration. It could, like Black's suspect mathematical model, easily confuse "accuracy of the mathematics with strength of empirical verification." Perhaps a risk worth taking but nonetheless a risk.

Here, *datum*, with its reassuring aura, and elsewhere Descartes's imputation to the Scholastics of haphazard category mistakes surely have a double edge. Category mistakes are not always to be seen only as blunders; sometimes, like mixed metaphors, they are creative, signs both of a mind and of history in process, and their interdiction can reveal to a cultural anthropologist that the interdictor simply has alien expectations of reality.[27] "A geometrically similar *datum*" also implies a regular, an isotropic, and, to an appreciable extent, an idealized (abstracted) relationship. However partially, it thus rationalizes the phenomenal, occasional world: crudely put, a rotten apple and a sound one are still mathematically two, but not "really." As Galison indicates, in Descartes's early work, cognition is closely tied to the external world through the sensory, indeed the pictorial, Cartesian imagination. In Descartes's later work, however, the understanding is "given a much more autonomous role" (326). Galison even suggests at one point that the late thought of Descartes inclines toward a new essentialism.[28] Descartes's thinking had Platonic as well as Aristotelian roots, and the history of mathematics is conspicuously compatible with Platonism and Neoplatonism, which mixed historically with Aristotelianism, both early and late. For my purpose, however, the most salient feature to this point of my argument is the elusiveness of definitive difference between a rhetorical, psychological, and epistemological technique of analogy within one system of thought in this period and what looks like the same technique, with fluctuations of degree or extent, in another one considered more "scientific" in the modern sense of this word.

While earlier than the essays by Gentner and Jeziorski, Douglass, and those in *Isis*, the intellectual and literary historian Brian Vickers's capacious chapter on "Analogy Versus Identity: The Rejection of Occult Symbolism, 1580–1680" has deservedly become a much-cited touchstone for the historical consideration of analogy in the early modern period.[29] Gentner and Jeziorski draw on it in explicit and implicit ways, and although Park does not in her *Isis* essay, she references a number of Vickers's related writings.[30] Essentially, Vickers examines the difference between science and magic. He treats the use of analogy by those practicing or proposing a new empirical science, such as Bacon, Kepler, Daniel Sennert (1572–1637), Van Helmont, Descartes, Boyle, and John Locke (1632–1704). He opposes such analogy to the conflation of the Saussurean signifier and signified, as well as of word and thing, in the mysticism and magic of mostly earlier Neoplatonic philosophers, alchemists, and kabbalists, such as Marsilio Ficino (1433–99), Cornelius Agrippa von Nettesheim (1486–1535), and, above all, Paracelsus. Jakob Böhme (1575–1624), to whom he also refers, is the single, chronologically later exception among the mystics. For Paracelsus, tropic relations have simply become literal and physical or chemical ones.

Even while associating analogy with metaphor, as does Aristotle, Vickers also opposes the new science to what he calls "Aristotelian philological science," by which he means a knowledge based on words, tropes, and logic rather than on numbers, observation, and experiment, and he charges its practitioners, like the occultists, with the manipulation of language in an effort to "describe or control reality" (103).[31] Rereading Vickers's chapter in a time when the assumption that language mediates, filters, and shapes our perception of reality is commonplace, I wonder what is so remarkable about the use of language to "describe or control reality" if this use has a rational basis, as in Aristotle, rather than an occult or magical one. When and why does a difference in degree become one in kind? Vickers's chapter serves to sharpen this question.

One of Vickers's strengths is his repeated acknowledgment that the rhetorical and logical traditions that derive from Aristotle, with whose writings those formally educated were likely to be familiar in the early modern period, recognized the difference between language and what he again calls "reality," as well as the difference between metaphorical and literal dimensions of meaning. Moreover, they did so in ways that the likes of Paracelsus did not. At the same time, however, Vickers's acknowledgment reports rather than engages this evident and inescapable tension in his argument. In it, from the vantage point of allegory, a continued, syntagmatic form of metaphor, oversimplification occurs when Vickers regards as "religious or moral allegory" Paracelsus's reading of God's curse in Genesis 3:17–18 that Earth should bring forth thistles and thorns. Paracelsus interprets the weeds as Tartar, the source of all human diseases within his system of equivalences (145). Properly speaking, this Paracelsian imposition of a specialized meaning onto the biblical text is allegoresis rather than allegory; it imposes metonymic code instead of working through a metaphoric process and substitutes virtual identity for the play of similarity with difference. Whereas metonymy, properly understood, is substitutive, coded, and ideological, metaphor, of which allegory is the continued, narrative form, is deviant, creative, constructive, and code breaking.[32]

Although Vickers's Paracelsian example is a brief moment in a long essay, both the connections *and* the distinctions among metaphor, metonymy, analogy, and allegory are important for careful, critical thinking. The same not irrelevantly pertains to the relation of (shaping) thought to (real) things, which Vickers muddles in his discussion of reference and the Saussurean sign. He wrongly identifies the signifier and signified within the Saussurean sign with the word (or sign, in popular usage) and its referent outside the Saussurean sign, that is, with what the word refers to in an act of speech. (The relation of a mathematical entity to the physical world offers a provocative parallel to that of the sign to its referent.) This is precisely the problematic relation of words to things increasingly evident in the history of analogy, not least in Vickers's derogation of "Aristotelian philological science" and Gentner and Jeziorski's reliance on Vickers. This relation includes metaphorical words, whether translative by etymological derivation or rendered translative by use. Metaphorical language, in fact, enhances the visibility of the familiar issues more broadly involving the relation of words, thought, and things that are at stake.

Besides tending to exaggerate contrasts, Vickers's pioneering effort does not consider the multifaceted complexity of Aristotelian analogy itself, the salience of analogy in Roman language and grammar, and the development of analogical thought in the Middle Ages, especially in the philosophy of Thomas Aquinas (1225–74) and later in that of Cajetan, the Renaissance Thomist earlier mentioned.[33] All these historical roots (and others) feed into early modern thought, and they represent the development of an imaginative thinking—indeed, of a poetic, or making (Greek *poiein*, "produce, invent, create, make")—that, as noted, shares more with modern physics than with the mechanized positivism toward which a good deal of the early new science was reaching. Such a poetic is variously linguistic, rhetorical, subjective, and speculative, but its aim is to express and in this way to substantiate the ever-elusive real. This poetic, whether in Kepler, Donne, or Milton, is finally my concern.

In the impressive trilogy entitled *Time and Narrative*, the modern philosopher Paul Ricoeur asserts the significance of Aristotle's having entitled his treatise on representation, the very treatise in which he addresses analogy most specifically and extensively, *Poetics*. This title is significant, insofar as "The act of emplotment [in drama and narrative] . . . extracts a configuration from a succession," an arrangement from chronology and a form from—or could it be a new or renewed form for—time.[34] The role of metaphor and specifically of analogy in science, whose autonomous and representational nature is itself recurrently in question both within and without its walls, has deep roots reaching back millennia, and among these roots, the *Poetics*. Consider this conclusion of a bold argument by Daniel Tiffany for the analogy of modern science with modern lyric: "Only if scientific materialism were reduced to a purely mathematical science—a development that would deprive physics of any claim to realism—could we say that poetics and physics are antithetical phenomena. The leading theorists of quantum physics, when confronted with the prospect of imageless substance, [have] found it impossible to renounce the regime of pictures in materialism."[35] The comparative, analogous, metaphorical imagination might be indispensable, after all.

Classical Roots: Aristotle's Philosophy, Varro's Language, Quintilian's Rhetoric

In the *Poetics*, when Aristotle defines a type of metaphor of the A:B :: C:D variety that is proportional, or analogical, his formulaic representation of the structure of analogy provides the sort of abstraction that enables broad application, as does a mathematical formula: if apples are not oranges, both are generically, conceptually, and hence more abstractly fruit.[36] Abstraction enables a bridge between the two cultures, as manifest in our own time, at the base of one of which is a system of verbal signs and, at the base of the other, of mathematical signs. At some point, purely material things must yield to signs pure and simple or nearly so, since these remain human constructs, before they are reconnected, we trust, to their origin in what actually exists. And so back to Aristotle's theory of analogy, the formulation of which is a version of metaphor that has a technical relation to, and a basis in, mathematics.

The Greek word *analogia*, "proportion," combines *ana*, "according to," with *logos*, English "ratio" or, more broadly, "order." Accordingly, the proportional metaphor Aristotle describes is relational, and certain of its terms might even be substituted for one another: for example, B and D can be exchanged. As he explains, transference (that is, metaphor) through

> analogy is possible whenever there are four terms so related that the second is to the first, as the fourth to the third, for one may then put the fourth in place of the second, and the second in place of the fourth. Now and then, too, they [users] qualify the metaphor by adding on to it that to which the word it supplants is relative. Thus a cup is in relation to Dionysus [god of wine] what a shield is to Ares [god of war]. The cup accordingly will be described as the "shield of Dionysus" and the shield as the "cup of Ares." (2:1457b16–22)

If the shield were to be substituted for the cup, however, the trope would become metonymic in my view, albeit in a metonymy deriving from the original proportional/analogical metaphor. This metonymic substitution, or transaction, only reminds us again that in the Roman tradition of rhetoric, which descended from the Greek one and became definitive for early modern Western Europe, the umbrella term for all tropes is the same as that for the specific trope *metaphor*, namely, *translatio* (*trans*, "across," plus *fero/-latus*, "carry"). Stressing again that, seen historically, tropicality is a category based on something shared by its members, I do so knowing that in the early modern period rhetorics were written to distinguish the tropes, and blood was spilled both over them and over the relation of tropicality itself to what was thought to be real—in controversies about the Eucharist and religious vestments, for example.

As Aristotle's explanation of analogy continues, its association with poetic metaphor becomes even more specific:

> To take another instance [of analogy]: As old age is to life, so is evening to day. One will accordingly describe evening as the "old age of the day"—or by the Empedoclean equivalent;[37] and old age as the "evening" or "sunset of life." It may be that some of the terms thus related have no special name [that is, word] of their own, but . . . they will be described in just the same way. Thus to cast forth seed-corn is called "sowing"; but to cast forth its flame, as said of the sun, has no special name. This nameless act, however, stands in just the same relation to its object, sunlight, as sowing to the seed-corn. Hence the expression in the poet, "sowing around a god-created flame" (2:1457b22–30)

This last example is a particularly clear instance in which Aristotle relates what we recognize as acts, or operations, to one another. It further instances a catachrestic, as well as an analogous, metaphor, insofar as a proper word does not exist for the radiation of sunlight as the poet conceives it. Tropic identities can overlap; in practice, they are perceptual and dynamic. Notably, too, the first example in the inset, "old age of the day," compresses the underlying analogy and requires analytical expansion to bring its structure to light. Such compression is a characteristic of analogical thinking, and it will appear again.

Aristotle used the notion of proportion, or analogy, elsewhere in his philosophy in ways that further bear on its subsequent development, relevantly in the work of Aquinas and Cajetan. For example, he analogizes the underlying working of odors to that of savors and the perception and memory of magnitude to that of distance/time; he finds the heart in sanguineous creatures functionally analogous to the vital organ in nonsanguineous ones; he refuses explanatory power to numerical correspondences, such as seven vowels, seven strings on the scale, seven Pleiades, and so on (and on and on), considering them *merely*, incidentally, and insignificantly analogous; he also characterizes the notion of friendship to oneself as an analogous use of the word/concept *friendship* rather than an absolute, or proper, one, and he recurrently invokes the concept of proportion in relation to politics and justice, the latter a concern the Romans will further develop.[38]

Most pertinently, in the *Metaphysics*, Aristotle distinguishes numerical, specific, and generic unity (oneness) and then adds to these another sort of unity by analogy, which pertains to "those . . . related as a third thing is to a fourth," or as the number 3 is to 4 (2:1016b34–35). Notably, analogical unity directly pertains neither to genus nor sensible being here; to judge by the example offered, it is at once mathematical and in accordance with "the old Greek principle of the interaction of the like with the like,"[39] a principle that distantly resonates with the Cartesian requirement of similarity "in a very strong sense." Elsewhere in the *Metaphysics*, Aristotle distinguishes systemic otherness, sameness, likeness, and difference and specifies use of the term *different* when things "though other are the same in some respect, only not in number but either in species or in genus or by analogy," that is, the same in one of these three broad ways and only in some respect the same, and therefore without implying that analogy is identity or unrestricted univocity (2:1018a13–14). In a final example, he discusses analogous relations of actual and potential existence, observing that "all things are not said in the *same sense* to exist actually, but only by analogy—as *A* is in *B* or to *B*, *C* is in *D* or to *D*; for some [analogies] are as movement to potentiality [for example, as a man seeing to a sighted man with his eyes shut], and the others as substance to some sort of matter [for example, a statue of Hermes in a block of wood]." Notably, the basis of analogy in such instances is the systemic relation of actuality to potentiality, *not* simply, in the last example, the confusion of a block of wood with the statue that could be made from it (2:1048a30–b8).[40] Analogy in this final example is also a semantic relation with a basis in human perception, insofar as it is said to exist in some sense rather than to exist absolutely. Even though metaphysics pertains to the order of the real, here the analogy of potency to act belongs in some way to the order of saying rather than simply to that of existence. In sum, Aristotle's use of the term analogy is at once systemic and careful yet still various—subjective or objective, conceptual or empirical, semantic or existential, depending on its context. While I do not suggest that his use of analogy cannot be rationalized by specialists, the challenges to consensus and the opportunities for diversity in interpretive reception that it offers seem obvious.

The very variety—arguably, the instability or even the play—of Aristotle's statements about metaphor, unity, difference, and relationship, as instanced in the preceding paragraphs, underlies the problem of analogy in medieval semantics, a fundamental branch of logic. This problem concerns the use of analogous terms to make valid, as opposed to fallaciously equivocal, inferences. Far from being only academic (or Scholastic), it is at the core of the history and significance of analogy, involving what is traditionally known as the Aristotelian semantic triangle of word, concept, and thing. This is the very relation that Vickers invokes in an effort to explain the cognitive errors of magicians, mystics, alchemists, and the like, albeit with some confusion, as earlier noted, regarding Saussure's theory of language.

Among the ancient Romans as well, use of the term *analogia*, alternatively called *proportio*, occurs saliently in discussions of language and grammar that also prove influential in later centuries. Varro's *De lingua Latina* repeatedly considers *analogia*, as does Quintilian's *Institutio Oratoria*. Both these works indicate the centrality to Roman culture and society of appropriate language for thought and communication. Grammar, rhetoric, and etymology, the last understood as linguistic history, or grounding, were recognized to contribute both locally and systemically to meaning, as deconstructionists and other genealogists of cultural word-concepts have, in an understatement, noticed more recently. Varro (116–27 BCE) treats analogy in the context of linguistic derivation, especially with respect to case forms. Irregular forms are considered *anomalia*, "anomaly," and regular, or paradigmatic, forms, *analogia*, "analogy."[41] Varro also addresses the relative merits of *consuetudo*, "usage," and *ratio*, "theory," as Roland Kent translates these words, and then, retaining *consuetudo* but replacing *ratio* with *analogia* ("regularity," in Kent's rendering), declares them more closely connected than earlier authorities have supposed (IX.i.2). Concentrating for the moment on *consuetudo*—"usage, custom, similarity" (also termed *analogia* at times)—Varro compares the inflecting of nouns and verbs to the arranging of dining couches, among which one is of a different size or is misplaced, either too far forward or not far enough so. These anomalous forms are corrected according to *consuetudo* and *analogiae*: "common usage and . . . the analogies of other dining-rooms" (IX.iv.9). In passing, Varro's own analogy of dining couches is notable. It is certainly relational, and while hardly equating couches with words, it invokes a customary conception of order (*ratio*). Moreover, it is spatial: not inappropriately we might be reminded of other instances of spatial metaphors in connection with language in Roman writings, for example, *locus* (place), a name for the rhetorical commonplace, or the strange referential room (*ali-*

enus locus) that a word enters when used metaphorically, or transferentially, moving from one place to another.[42]

Varro's tenth book, which is addressed to Cicero, enumerates "four factors which limit the inflections of words: what likeness and unlikeness (*simile ac dissimile*) are; what the relationship (*ratio*) is which they [grammarians] call *logos*; what 'by comparative likeness' (*pro portione*) is, which they call 'according to *logos*' (ἀνὰ λόγον [*analogon*]); what usage (*consuetudo*) is" (X.i.2). By explaining these factors, Varro intends to clarify the origin and character of anomaly and paradigmatic analogy in language. Here, as elsewhere, the interlinking and the implications of his Latin terms, which translation tends to weaken, are suggestive in ways that subsequent passages clarify. Addressing *ratio*, or relation, with respect to new words, whether newly made by a poet or just newly encountered, Varro, sounding a little like a logician, announces that "each and every relation (*ratio*)" is like or unlike another, either the words being different but the relation the same, or vice versa (X.ii.36). For the same relation, he instances *amor/amori* and *dolor/dolori*, paired nominative/dative inflections, and he cites the nominative/accusative combination *dolor/dolorem* as an unlike relation. He makes another negative example of a nominative/genitive singular and a nominative/genitive plural; although the cases are analogous, the difference in matter (*materia*) and specifically in number—the *vocis figurarum*, "the forms of the spoken word," that is, the forms of *vox*, or "significant sound"—means that the relation of these pairs "cannot [of itself] effect Regularities (*analogias*)." He further emphasizes that it must be "by a proportionate likeness (*pro portione*) that the word has the same relation"; "then and not until then does this relation achieve what is demanded by *Analogia* of regularity" (X.ii.36). Varro's concerns with terminological usage, relation, likeness, and difference pertain to later Scholastic theorizing about the register, intention, and context in which analogous terms are used. The likeness he requires in analogy looks to be downright rigorous.

Varro next asks what a relation by proportionate likeness (*ratio pro portione*, or Greek *analogon*, "according to *logos*") might be and explains,

> If there are two things of the same class (*ex eodem genere*) which belong to the same relation though in some respect unlike each other, and if alongside [them] . . . two other things which have the same relation are placed, then because the two sets of words belong to the same *logos* (λόγον) each one is said separately to be an analogue and the comparison of the four constitutes an *Analogia* (ἀναλόγ[ί]α). (X.iii.37)

Kent, our translator, compares the same relation (*eadem ratio*) between the sets Varro describes to that in mathematics between two ratios of equal value that make a proportion—A:B :: C:D again, although it is not clear whether Kent does or does not have a quotient in mind (2:562n37a).

Clarifying examples of proportional relation now follow Varro's initial exposition: when we say (*dicimus*) that one twin is like the other, we are speaking of one twin only, "but when we say that a likeness is present in them, we are speaking of both." In passing, note that Varro's argument belongs to the order of speaking, of language use, rather than to being as such: *dicimus*, "we say." The relation of the two orders is neither simple nor simplistic in earlier periods, another point, I would emphasize, that is important for later centuries. To the example of the twins, Varro adds that of the ratio of one coin to half itself to the ratio of another coin to half itself and declares the result "an analogue" (*ἀνὰ λόγον* [*analogon*]). He concludes, "when we say that both in copper and in silver there is the same relation, then we are speaking of *Analogia*," which Kent glosses as "Regular Relation" (X.iii.38, cf. 39).

Somewhat later, Varro observes a variety of analogies: in "unlike things," such as numbers (2:1 :: 20:10); in "like things," such as coins; in all other things "said to be in a state of comparative likeness, wherein there is a fourfold nature" (daughter:mother :: son:father, midnight:night :: midday:day) (X.iii.41). There is obviously a certain categorical logic (*logos*) to Varro's examples, or, as he remarks, poets make much use of such proportions in their similitudes, and the geometricians use it *acutissime*, "most sharply, subtly, pointedly"; Kent translates *acutissime* "keenly" (X.iii.41–42). That the difference Varro recognizes between poets and mathematicians is not one of methodological kind but of qualitative practice is both notable and notably Aristotelian. Varro sounds even more like Aristotle when he starts describing "two crossed relationships" (*rationes implicatas*: "involved, interwoven relations"), of which the two types are *derecta*, "direct, straight," whether "vertical, or horizontal," and *transversa*, "transverse, crosswise." As examples he offers tables of numerals and of verbal inflections and indicates a variety of sets based on them (X.iii.43–44).

Quintilian, writing in the first century CE, has additional nuances to offer regarding analogy, and his enormous influence in subsequent centuries further argues for his inclusion here. In the *Institutio Oratoria*, he begins a discussion of guidelines to be observed by speakers and writers with an assertion about fundamentals: "Language (*sermo*: 'conversation') is based on reason (*ratione*), antiquity, authority, and usage (*consuetudine*). Linguistic reason (*ratio*, *logos*) finds its chief support in analogy and sometimes in etymology."[43] Not much later, Quintilian notes the need of critical judgment (*iudicium*) in all cases, "especially as regards *analogy* (a Greek term for which a Latin equivalent has been found in *proportion*). The essence (*vis*: 'force, power, meaning') of analogy is the testing of all subjects of doubt by the application of some standard of comparison (*ad aliquid simile*) about which there is no question, the proof that is to say of the uncertain by reference to the certain," or of the unknown by the known, to recall terms used earlier with Descartes (Quintilian, I.vi.3–4). It would also be hard to find a closer analogue to the rhetorician Augustine's enunciation of the rule of faith—for Augustine more specifically of the central tenet of faith, charity—that, together with the analogy of faith, was so important for hermeneutics in early modern Protestantism.[44] Of course the examples Quintilian offers have nothing to do with faith, or more accurately, nothing immediate to do with it.

The other passage in Quintilian's *Institutio* of particular relevance occurs in a discussion of legal arguments. Quintilian observes that some distinguish between *analogy* (*αναλογίαν*) and *similarity* (*simile*), but he expresses his opinion that the former is included under the latter, in legal as well as in other kinds of usage (V.xi.32–34). He proceeds to a mix of examples:

> For the statement that the relation of 1 to 10 is the same as that of 10 to 100 certainly involves similarity, just as does the statement that a bad citizen may be compared to an actual enemy. But arguments of this kind are carried still further: "If connexion with a male slave is disgraceful to the mistress of the house, so is the connexion of the master with a female slave. If pleasure is an end sought by dumb animals, so also must it be with men." But these arguments may readily be met by arguments from dissimilars: "It is not the same thing for the master of the house to have intercourse with a female slave as for the mistress to have intercourse with a male slave; nor does it follow that because dumb animals pursue pleasure, reasoning beings should do likewise." Or they may even be met by arguments from opposites; as for instance, "Because pleasure is an end sought by dumb animals, it should not be sought by reasoning beings." (V.xi.34–35)

Quintilian both expresses the view that analogy is a subcategory of something more inclusive and, through his examples, suggests the perspectival origin of analogy, including its potential for social manipulation. Remembering Aristotle, I will recall the potentially suspect relation of analogy to the perspectives and likely deceptions of imagination and his similarly theorized awareness (Greek *theōria*) that analogies are "said," or humanly constructed by speech. Taken together, the considerations of analogy by Quintilian and Varro, as earlier by Aristotle, exhibit an approach to it that is anything but haphazard. They also attest to considerable awareness of significant construction, whether the signs involved are verbal or mathematical. The prominence of logic in the later Middle Ages, while also rooted in Aristotle, is plausibly a response to this awareness and an attempt to regularize it further.

Late Antique, Medieval, and Early Modern Translations: Boethius, Aquinas, Cajetan

A good reason for attending to the analogical tradition in Roman language and rhetoric is that the philosophers and theologians of the Middle Ages and early modern period, like virtually all other educated thinkers—natural philosophers and new scientists included—were schooled, indeed drilled, in Latin itself and in the Latin rhetorical tradition.[45] Historians of intellectual culture neglect this profoundly formative line of transmission when they move from Aristotle to Aquinas to Cajetan in considering analogy. The trajectory Hochschild outlines at least includes Boethius (480–524 CE), however, whose practice looks to me to have been influenced by the linguistic/rhetorical conception of analogy, which in turn traces back to the metaphorical/mathematical ratios in Aristotle's *Poetics*.[46]

Boethius, an accomplished Hellenist who rendered into Latin several Alexandrian, Neoplatonic, and Neopythagorean texts on mathematics as well as Aristotle's *Organon* (logical treatises), translates the Greek term *analogia* as Latin *proportio* ("proportion") and relates it to varieties of equivocation. But Boethius also employs the more abstract term *proportionalitas* ("proportionality") to characterize what Hochschild describes as "the technical mathematical sense of Greek *analogia*" and as an expression of "any kind of common relationship": more exactly, in a translation of Boethius's words in his *De Arithmetica*, "'Proportionality is a similar relationship of two or more ratios'" (Hochschild, 7–8). Proportionality, as Boethius conceived of it, clearly included more than specifically arithmetical signs. According to Hochschild, "for Boethius a geometric proportion is [also] most properly called proportionality (*De Arithmetica* 2.44), and it seems to be the best model for nonquantitative proportionalities or analogies, such as *sole* is to *foot* as *palm* is to *hand*." Boethius's effort to further terminological precision by employing *proportionality* also had the curious effect in later Latin philosophy of turning the term *analogia*—when simply transliterated from the Greek instead of being translated as *proportio*—into merely a "mean between univocation and pure [i.e., arbitrary] equivocation," that is, between semantic and logical sameness and difference. In an inversion of meaning, Latin *analogia* (transliterated) thus becomes what Aristotle designated *pros hen* equivocation, as distinct from what he designated *analogia* in his native Greek (Hochschild, 8). Note especially two points, the word "model" in the transfer of proportionalities from mathematics to human anatomy above (foot, hand) and the logical designation "*pros hen* equivocation," which I define immediately below, since it will prove important for discussion of Thomist analogy.

In Boethius's Latin, Greek *pros hen* equivocation is rendered *ad unum* (lexically, variant meaning "with respect to one, in relation to one, in consequence of one," and in my expanded gloss of the term *one*, "in relation to one primary meaning"). In modernity, its sometime equivalent would be "focal meaning" or, for Hochschild, "associated meaning" (2–4).[47] A classic example involves the word "healthy," whose meaning shifts when applied, say, to a person, to a diet or an exercise, and to a sample of blood or urine. Etymology, morphology, polysemy, inflection, cognation, and even tropicality, with all their ancient and modern controversies, variously converge on such linguistic shifts and extensions.

For Boethius and later for the Scholastic philosophers, transliterated Latin *analogia*, or *proportio*, therefore becomes mainly a linguistic phenomenon, Hochschild explains, while Boethian *proportionalitas* (the true Greek *analogia*), which affords "a fundamental insight about relationships between things, retains a wide significance . . . in mathematics, music, astronomy, architecture, and the physical sciences," all having the first of these, mathematics, in common. To my mind, the distinction between language and things that Hochschild sees in Boethius casts a long shadow forward to modernity—another shade of C. P. Snow's two cultures, the one scientific, the other verbal, metaphorical, and poetic. But deducing this shadow in Boethius, whose usage is less neatly and consistently split, Hochschild is anticipating efforts in the Thomist tradition of the Middle Ages and Renaissance to distinguish mere metaphor from "proper" analogy. A sobering reminder of the difficulty of maintaining this distinction and of the reach of mathematics beyond the mathematical disciplines comes in Hochschild's account of a medieval board game widely used to teach Boethian *proportionalitas* in the *De Arithmetica* and its "implicit links to the [thoroughly Aristotelian] idea of virtue as a (proportional) mean."[48] Thus understood, proportional relationship might still be based on a mathematical relation, but it nonetheless looks to modern eyes, as it did to Aristotle's in the *Poetics*, like a metaphorical transfer from one field of experience to another. The measuring of an intangible—virtue, in this instance—doesn't work very well in practice, as Spenser's giant discovered when he actually, physically tried to weigh truth and falsehood, as well as justice, in *The Faerie Queene*.[49] If human beings really do have analogue brains, which modern culture increasingly subjects to digitization, such measurement carries additional, proleptic irony.

The undiscussed elephant in my chapter at this point, as well as in the tradition of commentary on Aristotelian analogy, is Aquinas. History has centrally associated the concept and technique of analogy with his philosophy, especially as found in the so-called analogy of being (*analogia entis*), which involves different meanings of the term "being" and bears on the predicative knowledge of God.[50] The Aristotelian-Thomist tradition is significant to my argument insofar as it serves to theorize the connection of classical analogy to Christian faith. Surprisingly, Aquinas wrote no sustained treatment of analogy or proportionality itself, but over centuries Thomists have developed pertinent remarks in his writings into positions regarding the nature and use of analogy in metaphysics, epistemology, and logic. Cajetan in particular, although himself working in the Thomist tradition, departed from Aquinas's views on analogy, modifying them significantly. In what follows, I rely on Ralph McInerny's third and most recent volume on Thomist analogy to summarize the views of Aquinas before I return to the indispensable Hochschild, this time on Cajetan himself.[51]

An overview of the differing orientations of McInerny and Hochschild regarding the treatment of analogy in the Aristotelian and Thomist traditions provides a quick introduction to the issues they engage. Whereas McInerny considers Cajetan's *De Nominum Analogia* (1498) to be an erroneous reading of Aquinas, Hochschild finds in it Cajetan's effort to answer the objections to analogy of John Duns Scotus (ca. 1265?–1308) and therefore a very different project that has mistakenly been received as *the* authoritative reading of Aquinas rather than as a more original work in its own right.[52] According to Hochschild, Cajetan's treatise was not only innovative and influential in itself, but it also set the terms of debate about analogy for succeeding generations, indeed, for centuries of them (xiv). McInerny's view of this same treatise would protect Aquinas's Latin from reduction to Aristotle's Greek, a language Aquinas did not read, whereas Hochschild's view would free Cajetan's return to the meaning of *analogia* in the Greek tradition from dependence on Aquinas. This return suggests Cajetan's participation in the new humanism of his time, dedicated to the revival of Greek and Roman letters and learning. Taken together, McInerny and Hochschild are especially helpful for clarifying the logical, semantic challenges of analogy, and both are attuned to relevant modern questions, including those about metaphor and the implication of analogy in it, while not engaging these directly. The ancient Greek and Latin connection of metaphor with poetics, with making, inventing, and creating (*poiein*), somewhat threateningly underlies their concerns, as I read them.

McInerny, drawing on a range of Aquinas's writings to explicate the saint's thinking on analogy, minimizes reference to what he terms "transcendental analogy"—to being, as distinguished from "being"; McInerny's quotation marks around "being" are meant to signal his (and Aquinas's) focus on semantics as an aspect of logic (111, 161–62). He asserts that what Aquinas regards as analogous words ("names" used analogously) has no counterpart in Aristotle's use of the Greek term *analogia* and that there is no analogy of being, as distinct from "being," in Aquinas, thus none that is real, existent, and properly metaphysical, as distinct from logical, nominal, and conceptual—in short, mind made (124, 152). Yet McInerny does acknowledge Aquinas's engagement of analogous thinking about being/Being (humans/God), without which, as he recognizes, the epistemological alternatives would be pantheism or agnosticism, these aligning respectively with univocation and pure equivocation in semantics.[53] (Atheism is not at issue for him.) Once again, this acknowledged engagement pertains to the existential order rather than to the order of semantics—to the order of things rather than words. McInerny holds that analogy, as treated specifically in the writings of Aquinas, belongs to the order of words.

Aristotle's *pros hen* equivocals are, for McInerny, "exactly what Thomas means when he speaks of analogous names." These equivocals are various, deliberate expressions "with reference to one among them" (*ad unum*), within a priority of posterior to prior and of derivative to proper (46, 96). Of course, Aquinas does address proportionality, the Boethian word for Aristotle's *analogia*, but McInerny understands even this Thomist version of proportional unity as a type of *pros hen* equivocation, this time not *ad unum*, with respect to one, but *ad alterum* ("with respect to another," or, indeed, an Other). If various expressions have a proportional relation to another, they still remain focally oriented in meaning (102, 113–14). Crucially, such proportionality *ad alterum* pertains not only to words analogously common to God and to creatures, such as the word "knowledge" or "goodness," but also to verbal analogies between the sensible and the nonsensible—for example, between physical sight and intellectual comprehension: sight:imagination :: reason:intellect.[54]

McInerny's discussion becomes particularly relevant for my purpose when he confronts the relation of analogy to metaphor. I want to highlight the slipperiness of the distinction, and lack of distinction, that he recognizes between the two before considering crucial passages in Aquinas that indicate how subjective and intentional such a distinction must be. McInerny acknowledges conflicting statements and in fact an impasse ("aporia") regarding metaphor and its relation to analogous naming in Aquinas before he reaches for a solution. As it happens, he could easily have transported his solution from Varro, namely, usage, or *consuetudo*, which Varro associates with *analogia*. But Varro does not figure in the resolution McInerny reaches.

Approaching this solution, McInerny first relates the *extension* of a word's meaning(s) from its material roots through cultural polysemy over time, to what is now considered its proper, true meaning—a meaning more abstract, conceptual, ideal, or spiritual, since this is the purpose of analogy that concerns him, as it did Aquinas (114). Such extension involves the problem also known as linguistic sublation, "raising" or "supersession," of which examples are *idea* (from Greek *eido*, "to see") or *concept* itself (from Latin *com* + *capere*, "to seize together") or, indeed, even *issues* (Latin *ex* + *ire*, "to go out, flow out").[55] The analogy sight:body :: intellect:soul is still another traditional example of such raising. Material roots threaten the transcendence of thought built on the abstractions or elevations deriving from them. They expose it as metaphor—as artful, human construction. If the use of analogous naming is not likewise to be considered *mere* metaphor, the extension it involves must be semantically and logically distinguished from this trope. The alternative would be to consider human construction true or at least potentially true, and this would require some kind of faith. Somewhere between these alternatives lie hypothesis, theoretical modeling, and fiction.

In order to contrast "metaphorical or [other] equivocal modes [of signifying] to *analogical community*," McInerny defines the proper meaning of a given word as its most familiar one, that is, the one defined by current usage (129–30; my emphasis). In our modern age of monolingual dictionaries, Ricoeur, in good part a neo-Aristotelian, will identify such familiarity with the standard lexical meaning (*Rule of Metaphor*, 290–91). Varro's dining couches, arranged according to *consuetudo* and *analogiae*, "common usage and the analogies of other dining-rooms," again come to mind, bringing with them all the timeless questions about who or whose community determines what usage is and which analogies are appropriate, or "proper": who decides which arrangement of dining rooms is *right*? McInerny next invokes Aquinas's ranking of meaning from more to less complete expressions of the exemplar, for instance, from the exemplary lion to a merely lion-like, courageous man. He apparently conceives of this ranking as a form of associated meanings or, more exactly, as degrees of relation *ad unum* or *ab uno* ("to or from one," hence focal). But he also notes an instance in which Aquinas seems to think metaphor simply expresses a gradual subsiding of proper meaning into impropriety. Subsidence would put metaphor, too, on a continuum with the familiar or communal meaning, but McInerny hastens to neutralize this possibility (130–31). A view of metaphor as subsidence, even aside from McInerny's neutralizing, strikes me as reductive, contrasting with such persuasive arguments for the creativity of metaphor as Black's and Ricoeur's. Their arguments effectually connect metaphor with making, *poiein*.

Finally, with a reminder about usage, particularly as it pertains to the difference between physical and intellectual sight, McInerny turns to passages in Aquinas that ask whether, or in what verbal sense, light is to be found in spiritual things. In the *Summa Theologica*, Aquinas notes the physical origin of the word "light" (*lumen*), signifying "'that which makes things manifest to the sense of sight,'" and the extension (including sublation) of this word "'to signify anything that [has] made something manifest in whatever kind of knowledge.'" That last phrase, "whatever kind of knowledge," crucially opens the door to a controlling context. Aquinas concludes that if light is taken in its physical sense, it applies metaphorically to spiritual things, but that "'if it is taken (*accipiatur*: "received, willingly listened to") insofar as usage has extended it to any manifestation, then it is used (*dicitur*: "said") properly of spiritual things.'"[56] In this passage, with the verb *accipiatur*, "willingly listened to," we are not far from the biblical injunction to see if we have eyes to see and to hear if we have ears to hear (Matthew 13:16). We are crossing into faith, which, it must be noted, may legitimately be considered a context of usage and, in Aquinas's instance, one that is partly specified and partly just assumed. This sort of sociohistorical contextualization might fairly be said to haunt analogy, having its own temporal extension from the religious faith of the high Middle Ages, through the historical Enlightenment, to the greatly modified credence that pertains in modernity to scientific hypothesis, for example, about invisible phenomena in wave-particle physics (cf. Gingerich, "Kepler").

Before leaving Aquinas, I want to look at two additional passages about metaphor, both from his *Commentary on the Sentences* of Peter Lombard (*Scriptum super libros Sententiarum*). In the first he adjudicates the differing judgments, on the one hand, of Ambrose and Denis that "light" is said only metaphorically of spiritual things, and, on the other hand, of Augustine that "light" pertains properly to spiritual things. In short, he simultaneously engages the issues of equivocals and sublation/supersession. The whole passage in Aquinas bears quoting:

> Note that the bodily is transferred (*transferuntur*) to the spiritual on the basis of a similitude of proportionality (*similitudo proportionabilitatis*), but this similitude must be reduced to one of either univocation or of analogy (*analogiae*); so it is in the present instance (*et sic est in proposito*). In spiritual things, that is called light (*lux*) which is related to intellectual manifestation as corporeal light (*lux corporalis* [*lumen*]) is to sensitive manifestation. But manifestation is found more truly (*verius*) in spiritual things, and because of this Augustine is right in saying that light (*lux*) is more truly found in spiritual than in corporeal things—not according to the proper [sensible, physical] meaning (*propriam rationem*) of "light," but according to the meaning of manifestation (*manifestationis*).[57]

As McInerny surely realizes in commenting on this inset, the word "transferred" in its first line is the sign of metaphor—that is, the sign of *translation* < *transfero/-latus*. In order to avoid this implication, however, he again relies on controlling context and extended meaning to argue that the mode of manifestation "can be named properly (though analogically) by the term *lux* ['light']" (133). Qualifying the hypothetical proposition in the first sentence of the inset that the *similitudo proportionabilitatis* must be reduced either to univocation or to analogical metaphor (*similitudo analogiae*), McInerny concludes that the passage exemplifies precisely the difference between these (133). It further establishes that Thomist analogy is to be understood as *pros hen* equivocation. Thomist analogy is therefore *ad unum*, extended, polysemous, and in this instance, also raised, or sublated and superseded, *ad alterum*. Thus McInerny. To my eyes, however, Aquinas's twice-repeated phrase "more truly" indicates translation into another register of meaning, and it implies a leap, that of faith, into another order of being. Otherwise put, the translation is metaphorical, and the leap metaphysical, the one figurative and the other real and veridical—that is, *verius*, "more truly." Effectually, a third possibility, one between univocation and equivocation, is created.

The second passage also engages the *similitudo proportionabilitatis*, or proportional likeness, similarity, with respect to God:

> What is properly said of Him (*de ipso*) is truly (*vere*) in Him, but the things that are said metaphorically of Him are said because of a similarity of proportionality to some effect (*ad effectum aliquem*), like fire,

> in Deuteronomy, because as fire is to the doing away of its contrary, so God is to doing away with iniquity.[58]

"Here," McInerny points out, "there is no similarity of nature" or of the proper, conceptual meaning of the term "fire," only a similarity of effect or property, and, in short, a metaphorical relationship (133). His distinction resembles the scientific uses of analogy with which I began insofar as it rejects a basis in nature but accepts one in effect or function. Notably, modern historians are often similarly at pains to observe the avoidance by early scientists of a basis in nature for their analogies.[59] They, too, seem to want to distinguish something the mind constructs from a metaphysical claim about existence, or nature. Not engaging the distinction directly, they miss an opportunity to reflect on the nature of analogical method.

Recognition that the use of analogy by early modern scientists avoids a transcendent metaphysics does *not* establish the absence of metaphor from their method and logic or the incomparability, even the basic dissimilarity, of these to earlier methods and semantics of analogy, its structure, or its transferential, translative working.[60] Bacon's and Descartes's objections to the use of analogy in Scholasticism and Galileo's reluctance to embrace it tell an important cultural story, but this story is not about the method or structure of analogy per se. Peter Dear, a historian of science and technology with a keen eye for the historico-cultural implications of mathematics, makes the broader point that pertains here: "the conventional image of the Scientific Revolution as a new beginning that swept clean the philosophies of the past," allowing room neither for tradition nor for those using it, "is well on its way towards historiographical rejection."[61]

In the two centuries between Aquinas and Cajetan, the use of analogy became a more conspicuous issue in philosophy, as well as in related disciplines, one of which was science, traditionally designated natural philosophy, a branch of metaphysics. Only a generation after Aquinas's lifetime, Scotus rejected analogy as what he understood to be merely equivocation in Aquinas, insisting instead on its univocation (Hochschild, 43). He considered the Thomist claim that analogy is a mean between univocation and equivocation, sameness and doubleness—effectually, monism and dualism—semantically impossible (60, 143). Cajetan set out to avoid the sameness of the analogical concept relating two things and to ensure its sufficient unity to enable valid syllogisms, while avoiding the fallacy of equivocation (44). A very tall order.

Cajetan sidelines *pros hen* equivocation, the form of analogy that McInerny's Aquinas endorsed, and argues instead that the only true form of analogy is *proportionality* as such—what by Aristotle is termed *analogia* and by Boethius *proportionalitas*: A:B :: C:D.[62] As earlier with Aquinas, however, metaphor again proves a major challenge for him and, to my mind, in the end a stumbling block. In good humanist fashion, Cajetan employs the Latin rhetorical term *abusio*—in English "abuse," in Greek *catachresis*—to characterize the extension of the word "analogy" by Aquinas and his commentators to versions of analogy other than the true Greek one, Aristotelian *analogia* (Hochschild, 123). (*Abusio* is an inflated metaphor or the more pedestrian use of a word to refer to something similar that lacks a proper word.) Cajetan recognizes that *analogia*, even as used in his own lexicon, is at times metaphorical, but, like many a modern historian, he also tries to distinguish metaphorical analogy from "proper analogy," whereas Aristotle does not (11). His distinction resembles that of Aquinas in trying to differentiate between a proper analogy and a metaphorical one, but it differs from it in not turning proper analogy into *pros hen* naming. For the more classically oriented Cajetan, analogy can be either proper or improper—the latter (merely) metaphorical.

According to Cajetan, metaphorical analogy commits what we moderns call a category mistake, as when a field in bloom is said to smile; as Hochschild puts it, such an analogy crosses from one "context" to another, human to floral in the Cajetanian example he provides. In contrast, a properly analogous term truly belongs both to the original context and to the one "to which it has been transferred" (Hochschild, 11). Once again the word "transfer" indicates the ghostly, structural presence of metaphor in proper analogy, while the emphasis on continuity recalls that Cartesian proviso earlier cited, namely, that figures are to be "explained in terms of figures and motion in terms of motion." This is contextual, not to say categorical, agreement: likes to likes, as in the old Greek principle, too. The example of this agreement that Hochschild offers involves the term "vision," a variant of the familiar "sight": as he explains, "*vision* is really in the body, and *understanding* is really in the intellect, [and] so the intellect can be said properly to *see*, by analogy with the sense in which the eye sees" (11–12, 125). But I would ask whether the intellect sees as the eye sees if the intellect is understood to be nonsensible and impassible, as it regularly was in medieval and early modern Christian philosophies. Does the analogy between physical and intellectual sight afford a viable mean between univocation and equivocation, or does it merely express a fallacy of relationship? Does it mask what is really a gap? The "mind's eye" in which Shakespeare's Hamlet sees and idealizes his dead father comes to mind, along with the many questions it raises about what he sees and why he sees it.[63] Certainly for a modern reader, the "mind's eye" of Hamlet is metaphorical, and, for a single-minded reader of Cajetan, Shakespeare's play could look at times like a staging of the controversial status of analogy as a form of thought. In the two centuries of reformation immediately after *De Nominum Analogia*, the issues engaged by Cajetan's treatise were increasingly in the cultural air of Europe. Faith, vision, and real presence are just three of the terms for relevant issues that would roil England and the Continent.

In a discussion resonant at once with late twentieth-century debates about metaphor and with Sidney's *Defence of Poesy*, Hochschild seeks both to minimize Cajetan's appeal to metaphysical considerations and to excuse the circularity of his view that an analogy is metaphorical when it employs metaphor. He explains that a metaphorical predication is spoken intentionally only *as if* it were true. Thus "the intention of a metaphorical predication is not to say what is literally true, or to say what is false, but to express some truth by way of an improper terminology" (Hochschild, 126–27).[64] Situated between truth and falsehood, such metaphor would appear to resemble what we variously term hypothesis, theoretical modeling, or fiction. The metaphorical analogy therefore uses a pair of words as if they signified another pair proportionally. In contrast, a proper analogy uses a pair of words simply to signify another pair proportionally but without fictional intent (128). Intention, apparently, is all. Distantly, we might be reminded of Sidney's argument as to why Poets don't lie, namely, because they don't assert truth: the poet "nothing affirms, and therefore never lieth"; of course, Sidney also refers to himself as "a piece [that is, bit] of a logician," a designation that applies to Cajetan, as well (Sidney, 95, 123).

Hochschild eventually acknowledges that Cajetan's "account of a mean between univocation and equivocation . . . require[d] him to invoke the metaphysical notion of proportional unity." Where Scotus had denied "the semantic possibility of analogical signification," he continues, Cajetan found it possible precisely "because analogical relationships are metaphysically real"—and, in another word, *proper* (138). If the relationship is real in spiritual usage, we have again crossed into faith, as happened when Aquinas spoke of a truer use of the word *light* than a merely sensible, physical one.

Another distinction Cajetan offers between proper analogy, which is metaphysical, and tropic metaphor is that one analogue in proper analogy can be known without the other, but in metaphorical analogy, both must be known (Hochschild, 156). This negative distinction granted to proper analogy is made for divinity, as well as for speculation, exploration, and hypothesis, whether physical or transcendent: according to it, man:human being :: God:divine being, an analogy in which the latter analogue exceeds human knowing—human grasp. In the analogy blooms:field :: smile:face, or "the field smiles," we recognize a metaphor when we see the second set (smile:face).[65] But in proper analogy, the second set is something infinite or invisible—in short, unknown—that does not signal a metaphor to us. Its unknowability, like divinity or infinity, is a given, a matter of judgment, faith, or both. At the same time, the availability of proper analogy to speculation and hypothesis is productive, provoking and nourishing to intellection. One might object, of course, that this state of affairs fails to tell us what proper analogy really is or what sort of true knowledge it offers. It seems to imply correspondence without asserting it—to operate as a kind of working hypothesis or else a belief. In other words, it seems to have it both ways, at once to be a kind of univocation and equivocation in disguise. But isn't this metaphor again, the trope that embraces predication and its denial, that simultaneously is and is not (A=B, A≠B)? Notably, the smiling field, the example of metaphorical analogy that contrasts with proper analogy, is strongly poetic, even catachrestic, in its projection of a human physiognomy and response onto nature. It exemplifies Cajetan's ranking of kinds of analogy—indeed, kinds of metaphor—on the basis of their subject matter rather than their structure and, in short, on the basis of what is considered real. In short, my argument has circled back to its beginning, the discussion of Hallyn, Gentner and Jeziorski, and Black.

Before leaving the smiling field, I want to summon into consideration a brief Miltonic description of heaven's landscape in *Paradise Lost*, at once to suggest its relation to Cajetan's discussion and, with a chapter concerning Keplerian science next on the horizon, to offer a foretaste of specifically literary connections. Milton's description represents heaven metaphorically without making a literal claim on truth. In Book III, the angels in adoration cast their amaranthine garlands to the "bright / Pavement that like a Sea of Jasper shon / Impurpl'd with celestial Roses smil'd."[66] Strictly speaking, it is the bright pavement that smiles, but it does so because it has become purple-red with the flowers covering it, distantly recalling the fictive metaphoricity of Cajetan's smiling fields in bloom. Shining simultaneously like a sea and like jasper, a green gemstone—at once undulant, immense, opaque, yet also translucent—the smiling pavement is figuratively infused with otherworldly life. Although the jasper and flowers derive from the imagery of Revelation, and this derivation imparts some authority to them, the piece of description I have cited is fictive, "though what if Earth / Be but the shadow of Heav'n, and things therein / Each to other like, more then on earth is thought?" (V.574–76); Raphael's question in a later book gestures at something more than merely human invention, as do the poet's earlier allusions to Revelation. In Cajetanian terms, the landscape described nonetheless consists mainly of brilliant category mistakes and metaphorical predications spoken only *as if* they were true. In relatively more Sidneian terms and otherwise in modern ones, the description mainly would be an objective correlative for an imagined idea and, in short, a fiction. Mainly.

Closing the Frame

When Giorgio Agamben, a modern historian of intellectual culture, follows analogy back to its roots, he, too, locates them in ancient Greece. He frankly *radicalizes* (his word) Aristotle, extending hints or developing possibilities in Aristotle's thought in order to push against the dichotomous logic that for centuries has been dominant in the West.[67] Agamben finds an *analogon*—unnamed but present—in Aristotle's explanation of a paradigm in the *Prior Analytics* (69a13–15), in which a singularity is irreducible to either the particular or the universal, the example or the law, the sensible or the conceptual, induction or deduction, and so on, and he reads this *analogon* back into Plato's dialectics as well (18–19, 22–26).[68] Agamben offers as his example of a paradigm the Latin noun *rosa*, whose case ending is at once specific to it and general for the first declension: "Through its paradigmatic exhibition (*rosa*, *ros-ae*, *ros-ae*, *ros-am* . . .), the normal use as well as the denotative character of the term 'rose' is suspended," thereby enabling "the constitution and intelligibility of the group 'feminine noun of the first declension,'" of which it is one case, one member, one instance (24). Varro would surely have agreed, a coincidence that again enforces a suggestive continuity within a historical difference, a movement back to the future in a process that is innovative rather than merely repetitive. In this process itself we might recognize the working of constructive analogy.[69] Simultaneously, however, we must recognize its historical grounding.

I began my discussion of analogy in this chapter by questioning whether its modern and early modern scientific exemplars break definitively with metaphor and thence with Aristotelian analogy, including what Vickers terms "Aristotelian philological science." Having looked at the variegated history of analogy from Aristotle to early modernism, with glances at alternative statements about modern analogy in philosophy and physics, I can reaffirm that Aristotle's view of analogy as a distinguishable kind of relational metaphor was operative from his own time through the early modern period and that in modernity it continues to be so productively.

CHAPTER 5

Proportional Thinking in Kepler's Science of Light

Light is at the center of Kepler's optics, astronomy, and cosmology. Verbal and mathematical analogy, whether as concept, proportion, or both, is crucial to his methodology and, indeed, to his habits of thought. In addition to these thematic and methodological reasons, Kepler plays an exemplary role in my study because he is an intellectual hybrid who combines ideas about light, deriving especially from Neoplatonist and perspectivist traditions, with mathematical and physical discoveries anticipating those of Descartes and Newton, both of whom number among his debtors. The longstanding debate as to whether Kepler's work is the culmination of medieval perspectivism or the emergence of mechanization, the climax of Neoplatonic emanationism or the beginning of celestial mechanics and modern geometric optics, has gradually given way to the recognition that it is both. In this regard, he is typical of much that was happening in the first half of the seventeenth century.[1]

More generally, however, an exclusive emphasis on newness in the science of the seventeenth century continues to privilege progress as it is measured from and by what succeeded it—by what looks more like us. Even our customary, laudatory label "The New Science" exhibits this built-in bias if it is used as a shorthand for the period. Failure to acknowledge what really is new and different in the period surely risks historical distortion as well, but at least in Keplerian scholarship, it has been less entrenched. Assessing Kepler's work from any single vantage point conjures a partial picture, one based either on cherry-picking his writings or on overemphasizing parts of them. Kepler is better seen whole, simultaneously as a bridge builder from past to present *and* as an original thinker. Even seeing him as a contradictory or transitional figure, although he is arguably both in part, overweights what is gone and what is to come at the expense of Kepler's complex and fascinating present. This is a present in which everything is interconnected rather than subject to the hard boundaries of subsequent periods or to the greater fragmentation associated with modernity.

A striking feature of Kepler's writings is his interest in, and engagement of, the interface of the immaterial with the material that light effects, as well as with correspondences and other connections between the celestial and terrestrial realms. Not surprisingly, analogy, the salient means of linking the known with the unknown, the abstract with the sensible, and the mathematical with all of these, is conspicuous and pervasive throughout his work. Kepler's *Ad Vitellionem Paralipomena*, concerning light, optics, and their bearing on the observation of astronomical phenomena, is the treatise on which I particularly focus, although my discussion of Keplerian analogies will extend to his other writings, the bulk of which are astronomical and cosmological. Kepler's *Paralipomena*, also known as his *Optics*, is fundamental to the rest of his thought.

Keplerian Light (and Analogy)

In the first chapter of the *Paralipomena*, Kepler treats the nature and origin of light, from which all else might be said to proceed. He locates these in the triune God, who created our world "like as possible to himself," giving it "the most excellent figure of all, the spherical surface."[2] In Kepler's later *Harmonice mundi*, God will be both a geometer and a celestial musician, indeed, "the superexistent geyser of geometry and harmony."[3] But at the outset of the *Paralipomena*, Kepler is focused more intently on geometry, as we might expect with optics. Once having located light and sphericity in the Godhead, he notes that God's act of cosmic creation has also established mathematical quantities and geometrical distinctions between the curved and the straight. He then develops an analogy between the Trinity and the originary point at the center of the spherical solid and the surface to which this point spreads out, communicating itself: the center, the spherical surface, and the interval between them, he declares "clearly three" yet "nonetheless one" (19). The spherical is also the geometrical archetype of light, as well as of the world, although light, lacking corporeal matter, is the purer expression of this form. Significantly, the spherical surface that Kepler envisions expresses emanation from a point-source, and, to insist on the obvious, it is a surface. Conceptually, intervening radii span what Kepler calls the interval, the space between; they radiate from the center and could be represented by geometrically straight lines, although Kepler does not explicitly assert that they are.

On such a Neoplatonized Christian and geometrical foundation, Kepler turns next to the creation of cosmic light, which gives "form and growth to everything" on the first day in Genesis. This principle, light, excels all else in the corporeal world: it is the "matrix of the animated faculties," the origin (implicitly a womb: *matrix*) of all those expressing a soul, and the "chain linking the corporeal and spiritual world" (19). As link, light "has *passed over into*," crossed over into, and thus observes the laws that pertain to the operation of this world: accordingly, in the body of the sun "nests" the "faculty of communicating itself to all things," namely light, and it is for this reason that the sun sits in the center of the universe, "pour[ing] itself forth equably into the whole orb," much as might a divinity (19–20; my emphasis).[4] Not surprisingly, Kepler was an early and ardent proponent of a fully heliocentric universe, which was an advance on the heliostatic Copernican one and better suited his metaphysics of light. Moreover, the linkage of the immaterial to the material world that light enables—its "passing over into," crossing into, or translation (*translation* < *transfero*)—is a vital feature of Kepler's science. As the relationship, in fact the equivalence, of these terms suggests, it shares ground with metaphor and analogy.

In the *Paralipomena*, thirty-eight propositions about the physical properties of light now follow, but as W. Pauli and Gerd Buchdahl in particular have stressed, Kepler's initial, prior assumptions about the created cosmos are operative, not perfunctory, simply pedagogical, or merely ornamental. These metaphysical and theological assumptions operate along with, rather than despite, his equally pronounced commitment to mathematical precision, observation, and experience.[5] Moreover, they motivate and inspire his conceptual creativity and might be said at once to require and to cooperate with his data. His running argument with the English Rosicrucian Robert Fludd significantly centered on the difference between freewheeling mystical symbolism and accurate, mathematically based diagramming.[6] Kepler had no interest in the mysticism of numbers per se, however, as distinct from harmonic and geometric proportions.[7] Of these proportions, "inasmuch as life is more excellent than the body, the form than the material, by so much does harmonic adornment excel the geometrical": to restate this analogy formulaically,

> (A) life : (B) body :: (C) form : (D) material :: (E) harmony : (F) geometry

Harmony is clearly superior to geometry. Harmonic form expresses animate life; geometrical form inheres more tangibly in the measurable world of matter (*Harmonice*, 80). The spherical representation of God is a "figure," or figuration, not the Godhead as such. It moves closer to matter, beginning to partake of it in some iconic way.[8]

According to Kepler, his propositions about light in the *Paralipomena* are based on "Euclid, Witelo, and others" (20). Witelo, a Polish scholar of the thirteenth century, wrote the reigning optical treatise prior to Kepler's. It was published in three editions in the sixteenth century, one of which, used by Kepler, included Alhazen's earlier treatise, which William H. Donahue describes as its only rival (*Paralipomena*, xiin3) and David C. Lindberg describes as the most important optical treatise between antiquity and the seventeenth century (*Theories of Vision*, 57–58). Alhazen (Ibn al-Haytham), whose seminal treatise on optics was not translated from Arabic into Latin until the late twelfth century, was the first to offer a punctiform analysis of the visible object into point sources that emit rays, a conception that Kepler embraces, albeit with a distinction between light and its motion that is his own and affords a typical example of the Keplerian modification of inherited theory. Alhazen also integrated the mathematical with the anatomical and physical, as did Kepler, again making highly significant changes in this model and thus both utilizing and reaching beyond Alhazen's achievement.[9]

Kepler conceives of light as a projection from its source "towards a distant place"; that is, another place some distance from it. It is subject to geometrical laws and "considered in place as a geometrical body," a proposition critical to Kepler's mathematical analyses (20). In the second citation, "place" has become the point on the surface of the illuminated object, the destination of light, so to speak. This meaning is unavailable in the first citation, which is momentarily mysterious (what place?) on first reading. In this little instance of the reflection of content by verbal form, light has, as it were, leaped from the first citation just above to the second, even as light leaps from "place" to "place," source to (incidental) destination, without the use of any medium. But the speed of light begins to anticipate my explanation.

Since Keplerian light "partakes of size and density," it is measurable, yet it "has no matter, weight, or resistance" of its own and therefore no mass (20); it thus participates in measurable quantities while lacking them in itself. Again, Kepler further distinguishes between light and its lines of flow, which he calls rays, also stressing that a ray is not light itself but the motion of light and, when diagrammed geometrically, its iconic representation. He refers this difference to the physical analogy that undergirds his conception: thus "a ray is nothing but the motion itself of light . . . as in physical motion, where . . . [the] motion is a straight line, while the physical movable thing is a body," or, more formulaically restated, (A) light:(B) ray :: (C) body:(D) straight line (22). A great number of Kepler's analogies are explicit; perhaps even more are implicit and clearly a pervasive habit of mind. The explicit analogies usually combine demonstration with proof, which should not be confused with certainty. The failure to respect the difference between analogy and identity has led to some of the more dubious claims about Keplerian epistemology.

In Kepler's view, light not only moves in rays (geometrically straight lines) without utilizing a medium but also does so instantaneously; it leaps over what lies between, as if this were no more (in my own term) than a vacuum. Its speed is infinite, as is the force projecting it (22–23). Light is also a kind of surface that is affected (for example, tinted or deflected) by the surfaces of the bodies it encounters and can affect them in turn, as when it warms them or bleaches their color.[10] Here, in effect, is the interface of immaterial light with the surface of a material body, and it enables participation and interaction. Each light ray is bounded, and thus acted upon, by its point of contact on the encountered surface that is illuminated: "the boundaries of the infinite rays of light are the infinite points; that is, the [illuminated] surface which is, *as it were* [*quasi*], composed of them" (23; my emphasis).[11] Here, "as it were" and elsewhere an occasional "as if" are deliberate Keplerian signs of counterfactuality, or metaphoricity, and creative theorizing. Along with Kepler's frequent, pointed recourse to analogy, they testify to his awareness of the extent to which his science crosses into translation—*translatio*, the Latin name for metaphor, the trope of which analogy is a version; a translation carries (*fero/-latus*) something, anything, across (*trans*) from one place, one mode, one realm, one system of signs, to another. In this respect, it operates like light, which "passes" from immaterial purity "over into" the laws pertaining to this world. Kepler's surface is not *really* composed of bounded light rays, or points, except as it is seen and understood through the lens of the metaphorizing geometer.

An instance of the phrase "as if" that is further telling by comparison occurs in another Keplerian treatise, *Astronomia nova*, when Kepler explains the *species* (forms, emanations) of power that proceed from the sun to move the planets and in numerous, but not all, ways resemble the emanations of solar light. These two emanations (or forms, *species*) are in Lindberg's witty wordplay "different species of the same genus" ("Genesis," 40): "they [the *species* of power] proceed according to that very [same] law of emission [as light]: they do not possess their own boundaries, but just as the surfaces of illuminated things cause light to be *considered as surfaces in certain respects* [*consideretur quasi* (as if) *quaedam superficies*], because they receive and terminate its emission, so the bodies of things that are moved [planets] suggest that this moving power be considered *as if* [*quasi*] a sort of geometrical body . . ." (*Astronomia nova*, 382; my emphases on English). Along with the counterfactual "as if," the phrase "considered as surfaces in certain respects" clarifies Kepler's simpler but equally metaphorical assertion in the *Paralipomena* that light is a geometrical surface. Both phrases indicate a mental act that is hypothetical in nature. Kepler's analogy between light and the motive power of the sun also testifies to the extent to which light (and thereby the *Paralipomena*) pervades and underwrites his cosmology. It does so as much as does mathematics and, if anything, more so. Yet Lindberg's judgment in his masterful essay on "The Genesis of Kepler's Theory of Light" is finally too absolute regarding Kepler's realities:

> When Kepler . . . probe[s] the nature of light, he finds it to be a mathematical surface—not some corporeal thing possessing a mathematical surface, but a mathematical surface in and of itself, an immaterial, self-subsistent, geometrical entity. The ultimate realities, here as in the search for cosmic order, prove to be mathematical. (42)[12]

The claim that the ultimate realities, indeed light itself, are simply mathematical and more explicitly geometrical overlooks, or perhaps discounts, Kepler's reservations, qualifications, and, critically, his awareness of the extent to which his theory is analogical. It also falls short of his location of ultimate reality in a Godhead that contains mathematical archetypes—mathematical ideas—but is not *simply* the same as them.[13] For all Kepler's enthusiasm about the sun as a reflection, instrument, or vehicle of divinity, he insists that the orbital motion of the planets is not the work of a mind except insofar as the "harmonic attunement" of the entire universe "is the work of the highest and most adored creative mind or wisdom," in a word, the biblical God (*Epitome*, 87–88).

Immediately after the Keplerian hypotheses ("as if") just above, in a significantly reflexive response to them, Kepler shifts into a traditional, dialogic mode of objection and response that effectually dramatizes thoughtful debate. He considers a possible objection to his claims on three counts: the variable extension of the sun's motive power, the immateriality of this power, and its interaction exclusively "with a mobile subject, such as the [material] body of a planet" (382). "To lack matter and yet to be subject to geometrical dimensions" as in the instance of the sun's emitted power, he continues, "appear to be contradictory." He then replies that the sun's motive power, while immaterial, is "destined to carry matter," namely, a planet, and is therefore subject to geometrical laws (383). To recall a parallel statement in the *Paralipomena*, upon the creation in Genesis, immaterial solar light "passed over into" the laws of this world; this, too, was a translation, a crossing from one place, one realm, to another, much as in a metaphor or analogy. Kepler's God, like Milton's, is an analogizer. Continuing to rebut the possibility of contradiction, Kepler notes as well that the planetary motions effected by immaterial solar power occur in space and time, which are geometrical entities. The argument he considers strongest, however, relies on the "authentic" (reliable, foundational, natural: *plane genuinum*) example of light, "companion" to the sun's motive power and, like it, "nesting" in the sun, not entirely unlike the Miltonic Spirit that "Dove-like satst brooding," punningly nesting, on and at the creation of the world (383).[14]

Accordingly, Kepler quickly summarizes the characteristics of light pertinent to the sun's motive power that he has earlier and more painstakingly established in the *Paralipomena*, referring explicitly to them with the confidence of a man who is now presenting established facts. Although immaterial, light, too, operates "with respect to place, suffers alteration, is reflected and refracted, and assumes quantities so as to be dense or rare and to be capable of being taken as a surface wherever it falls upon something illuminable" (383). While neither light nor the sun's motive power exists in the space between source and object, however, they differ insofar as "light . . . flow[s] forth in no time," but "this [motive] power creates motion in time." Light is instantaneous, whereas the motive power is not. Yet the ways in which they accomplish these actions, he adds, are the same. Here again, his explanation addresses the vital interface of immateriality with materiality, and it is accompanied by the qualifying recurrence of counterfactuality, an underlying analogy, and a shift in "places" (registers, realms, modalities, applications):

> Light manifests those things which are proper to it instantaneously, but requires time to effect those which are associated with matter. It illuminates a surface in a moment, because here matter need not undergo any alteration, for all illumination takes place according to surfaces, or at least *as if* a property of surfaces and not as a property of corporeality as such. On the other hand, light bleaches colours in time, since here it acts upon matter *qua* matter, making it hot and expelling the contrary cold which is embedded in the body's matter and is not on its surface. In precisely the same manner, this moving power perpetually and without any interval of time is present from the sun wherever there is a suitable movable body, for it receives nothing from the movable body to cause it to be there. On the other hand, it causes motion in time, since the movable body is material. (383; my emphasis on "as if")

Buchdahl has described Kepler's theory of the surface and more exactly of its density as, in effect, Kepler's "fatal doctrine," but this is also a fatal Cleopatra he can't do without (271, 282). This theory, like that of ray motion, is another that is peculiar to him, and it is what enabled him to come close to the sine law (that is, the Snell-Descartes law) of refraction, even if not quite reaching it. It is present in the inset just cited, along with qualitative analysis, physical observation, causal explanation, and, perhaps I should add, much analogy: for one out of numerous examples, (A) solar light:(B) bleaching colors :: (C) solar motive power:(D) orbiting planets. These characteristics make the inset feel Aristotelian, and the following stage of argument, which is briefly teleological, does nothing to dispel this impression. Yet it is Kepler's very next chapter that focuses on an analogy between the sun's motive power and that of the magnet, with an up-to-date reference to William Gilbert's novel theory that the earth is an ensouled magnet, which Kepler first welcomed and then embraced. Without Gilbert's theory, whose status shifts from an alternative explanation in *Astronomia nova* to the theory of choice in the *Epitome*, the sun would require some sort of mind, soul, or "divinity" to distinguish—or by another Keplerian analogy, to glance at (385–87)—its movable, but otherwise "inert," planetary targets.[15] Of course the mathematical evidence that supports or disproves all this analogical theorizing and yields Kepler's first two valid laws of planetary motion, concerning orbital ellipses and areas, will be based on what light makes visible for observation, testing, and measurement. By measuring such *illuminata*, which Kepler believed available for human enlightenment, he sought to bring the distant heavens ever closer. Defining all three of Kepler's laws in my introduction, I referred to his third or harmonic law in *Harmonice mundi* as itself "fundamentally analogical": if according to this law the cubes of the mean distances of any two planets from the sun are proportional to the squares of their periods of revolution, then, by the analogy underlying application of the law to multiple planets, A:B :: C:D :: E:F :: G:H, and so on. Analogy, with a basis in mathematics, is not simply rhetorical.

Since Kepler revised his work over decades, holding onto his core principles and beliefs even while making significant revisions, his theories sometimes present moving targets. Before proceeding, I want to consider one of them that is important to the interface of the immaterial with the material. Much has been made of Kepler's note in the second edition of *Mysterium Cosmographicum* regarding his change from the attribution of planetary movement to a solar soul (*anima*) in 1596 to a solar force (*vis*) in 1621, elsewhere a solar virtue (*virtus*), at least in Keplerian treatises from *Astronomia nova* to the *Epitome* (*Mysterium*, 203n3).[16] This change is invariably, perhaps too easily, cited as evidence of Kepler's increasing adherence to the mechanization of science. In the intervening years between Kepler's editions of *Mysterium*, as his note reports, he has pondered the fact that the motive power of the sun weakens and its light "grows thinner" with distance, and he has concluded that this power is corporeal, proceeding, as it does, from the body of the sun: "this force is something corporeal, that is, an emanation which a body emits, but an immaterial one" (203n3). "But an immaterial one" is the rub. The sun in later Keplerian writings still has some sort of mind or soul, even if this is the ultimate, not the immediate, source of emitted power: here, an analogy true to Kepler's thought might be ventured between the physical source of motion and what underlies it, as when an animated (ensouled) body throws an object. This solar emanation of power, like light, would remain an immaterial one, leaving the interface of the immaterial with the material a continuing Keplerian principle, if also an issue recurrently asking for more explanation.

Kepler approaches a notion of physical energy but continues to circle around it, not least because of his basic commitment to circular form. Comparing Galileo to Kepler, E. J. Dijksterhuis notes the dominance of "perpetual circular motion" in Galileo's cosmological thinking, as in Kepler's, despite Kepler's discovery of elliptical planetary movement, which in Kepler's view is inferior and more materialized than circular movement.[17] But Dijksterhuis also notes Galileo's greater interest in the kinematical than in the dynamical, in the how than in the why, and in description than in explanation (338). In contrast, the more religious and arguably more philosophical Kepler looks for an explanation and ultimately a cause. Mathematics in Kepler is likely to be a check on theory—a confirmation, invalidation, or extension—rather than its origin, although there are arguably significant qualifications of this likelihood, for example, regarding the camera obscura, aerial illusions in a museum, eclipses, and celestial bodies at dawn or dusk, all of which are subject to measurement. In such instances, however, Kepler's theory of light nonetheless remains the underlying constant.

If Kepler's sun possesses some sort of soul, albeit with immaterial emanations from its material body, his Earth has a soul as well. For example, when faced with the problem of the Earth soul's distinguishing geometric configurations (aspects) that influence the weather from those that do not, Kepler first traces light rays reflected from the surfaces of planets to points along the Earth's equatorial perimeter; then he sections the circular Earth by extending these points inward and from the sections derives polygons of various dimensions; finally, he relates the distinction in the Earth's responses that he seeks to numerical harmonies. Patrick J. Boner explains further, citing Kepler's *Gesammelte Werke*:

> Not every [aspect] . . . comprises corresponding movements of the air, but only that which in geometry is found to be harmonic. *Yet geometry is a thing of reason, which on its own has no efficacy.* It is therefore necessary that the said geometry acts objectively, contained in [the form of] the aspects. But it is not able to act objectively in anything whatsoever, save only in the animate faculties, just as music upon being heard inspires the farmer to dance. Thus, that which triggers the motion of the air according to the ordinance of the aspects formed in the Earth is the partaker of an animate faculty present in the Earth; and since it stirs the conditions of the weather clearly by its own jurisdiction . . . it is therefore made up of those things rational, that is to say, it is a certain type of soul.[18]

In short, Kepler's geometric and harmonic realities are here dependent on some form of living Earth soul with the capacities of recognition, instinctive response, and the further generation of archetypal shapes, such as snowflakes, not to mention honeycombs and fossils. While thus capable of perception and responsive action, such a nonhuman soul is incapable of discursive thought. Serenely, compared to many of his contemporaries—the speaker of John Donne's *First Anniversarie*, for one—Kepler also understood the phenomenon of new stars or comets in connection with the creativity of the same, underlying, archetypal force.[19] A devout Lutheran, he was no skeptic.

Returning to the *Paralipomena* with a broader sense of the implications of Kepler's metaphysics and physics of light, I want only to pause over several of his statements about color before engaging his theory of vision. Traditionally, color belongs to the discussion of light, and for Aristotle, it constitutes the subject of vision and the cause of the visible, a position that Kepler rejects (*Paralipomena*, 43). Although Keplerian, like Aristotelian, vision is intromissive, for Kepler light is not the activity of a transparent medium, as in Aristotle, but instead "the activity of colors, insofar as they are seen, or radiate": light, rather than color, is the cause of the visible (46). Color itself, considered apart from vision, Kepler defines as "light in potentiality, light entombed in a pellucid material." Distinctions among colors result from variations in the density and rarity of matter, in its pellucidity and darkness, or in "the spark of light [*lucula*], which is condensed into matter" (24). White and black are exceptions to Kepler's definition of color as light buried in the pellucid—buried as if awaiting redemption by light—insofar as these two colors are opaque in the highest degree, in his view—white resembling the shining body of the sun, and black a dark body. "In a certain way," Kepler explains, they are "material or corporeal colors" (24, margin). Their opacity hinders light, and therefore "light practices deadly enmity with all matter," especially with blacks, "which are as it were (*quasi*) materialized darkness" (51). Pure black lacks all potential light, Donahue comments editorially, and its purity is evident only by contrast since it is nearly devoid of radiation. Donahue also wonders aloud "that matter has so great a power, that through it light produces its enemy, the color black—that is, nothing but rays of darkness, painting themselves to some extent on the surface presented" (24n25). With Milton in a chapter ahead (and one behind), the power of matter is suggestive, as is Kepler's description of black as "materialized darkness"—*quasi*, "as if" or "as it were."

The interaction of color and light is yet another encounter between materiality and immateriality that is mutually affected and affecting. "Light strikes upon color . . . from all directions," illuminating it "through [by means of or else by permeating] the solidity" of the pellucid object with the result that "the illuminated colors radiate in an orb," as does light itself inherently. At the same time that the light is spherically "reflected as if [*quasi*] from surfaces," it is tinted and thus assumes the color of the medium or object. Color has therefore become "a correlate of light for mutual action" (*Paralipomena*, 34–35). Kepler summons this same argument in *Astronomia nova* to explain the periodic time of different planets' orbits by means of an analogy. Although light is constituted for illumination, color is only illuminated to a relative, various degree by it, "for colour intermingles (*confundit*) its own peculiar *species* [form, likeness] with the illumination of light, thus forming (*efficit*) some third entity (*tertium quippiam*)."[20] Likewise, although power is constituted for motion, the speed of the immaterial, motive power of the sun is hindered by "some sort of matter possessed by the surrounding aether, or the disposition of the movable planet to rest," as well as by its own weakening over distance (383–84). Moreover, if the light of the sun is truly, properly analogous to the sun's moving power, it cannot truly, properly be the same as it. As a version of metaphor, analogy will always include difference as well as likeness. Kepler, as if to mark the limits of his analogy, carefully notes that light radiates spherically in straight lines while the moving power, although also spreading out in straight lines, does so circularly; he notes as well that, because light is impeded by opacity, if it were the same as the motive power, darkness would result in the cessation of planetary movement, as is patently not the case.

Keplerian Vision (and Analogy)

Kepler's study of optics combines punctiform analysis with the geometry of refraction and contemporary anatomical work on the structure and operation of the eye in order to establish what Kepler characterizes in the *Paralipomena* as "the true opinion" based on "irrefutable examples [*experimenta*]"—in short, a theory confirmed by the evidence. According to the précis that he offers as a preview early in this treatise,

> From the sun, and from the colors illuminated by the sun, there flow out forms [*species*] similar to each other; and . . . in this flow itself they are diluted, until they strike upon a medium that is in some proportion opaque, and there they represent their source; and . . . vision occurs . . . when the opaque wall of the eye is colored in this manner. (*Paralipomena*, 48)

More exactly, the opaque wall of the eye is the retina, and the projected *species*, or forms, depict the visible object on the retinal screen. Kepler adds that the vision achieved can be either blurred or distinct, depending on the conditions of reception, that is, on whether the "images of different colors are intermingled," or not. Kepler's terms in describing the process of vision generally participate in those of his sources and more generally in the lexicon of his culture, evident here in the word *species*. Elsewhere, however, he is at pains to define terms he considers particularly important or to invent new ones, but his practice in the inset at hand is not an isolated instance. Not surprisingly, the epistemological implications of his optics and the question of their relation to the past and the future are entangled in his word-concepts.

The term *species* in the inset above has a range of possible meanings that is notoriously broad and potentially carries a good deal of theoretical baggage, particularly as pertains to the perspectivist tradition, the leading influence in Kepler's inheritance of optical theory.[21] Yet it is sometimes the inclusiveness, flexibility, and, indeed, the necessary breadth and indistinction of Kepler's terminology that enables him to link, or translate, different realms, modalities, applications, and the like, as well as to speak across disciplines. The loose word "place," discussed early in the preceding section, is a functional case in point. In the immediate inset, the term *species*, understood by Donahue in his translation as "forms" here, links sun and earth, light and vision, solar source and retinal picture, in a Keplerian system that most other words, even Keplerian *ray*, or light-motion, might not have done as obviously and effectively for his contemporaries. To disqualify Kepler's terminology, whether traditional or innovative, without trying to analyze its use risks missing what he has actually written.

Lindberg's gloss on *species* is further suggestive here. Lindberg cites a passage from Roger Bacon, an influential perspectivist of the thirteenth century, which enumerates the various meanings of *species* in different contexts: namely, "form, image, similitude, species [*sic*], idol, phantasm, simulacrum, intention, virtue, passion and impression." In passing, the fact that context determines meaning for the medieval Bacon is worth notice. Bacon also explains that the constant in each of these various uses is for *species* to be "'the first effect of an agent; for all judge that through species [all] other effects are produced'" (Lindberg, "Genesis," 19, 35–36). As the first effect of the rays illuminating the visible object, Kepler's designation *species* in the preceding inset would make further sense in describing what is, for optics, the (incidental) destination of this effect, or what is depicted on the Keplerian retina.

But perhaps Kepler himself provides another, more specific gloss for the *species* of light and color that together flow into the eye from visible objects. In a passage from *Astronomia nova* earlier cited, he explains that these two kinds of *species*, "intermingling," have formed "some third entity" (384). Simultaneously, the *species* of color has been illuminated on the surface of the visible object, and the immaterial *species* of light has been tinted by color. This action is "mutual," and in the process of vision both *species*, or the third entity, will also be bent by ocular refraction. In all these ways, the inflowing *species* instantiate the interaction of the immaterial and material that is a Keplerian hallmark.

At the beginning of the fifth chapter of the *Paralipomena*, which analyzes the process of vision, Kepler carefully limits the concern of optical studies to the eye itself, excluding what occurs in the psychological faculties beyond the retinal wall. Having traced the emanated *species* (effects, forms, third entity) of light and color entering from the outside world and striking this opaque retinal surface, where outside meets sensitive, animated inside, he stops, or more exactly, almost stops:

> vision occurs when an image [*idolum*] of the whole hemisphere of the world that is before the eye, and a little more, is set up at the white wall, tinged with red, of the concave surface of the retina. How this image or picture [*idolum seu pictura*] is joined together with *the visual spirits that reside in the retina and in the nerve*, and whether it is arraigned within by the spirits into the caverns of the cerebrum to the tribunal of the soul or of the visual faculty; whether the visual faculty, like a magistrate given by the soul, descending from the headquarters of the cerebrum outside *to the visual nerve itself and the retina*, as to lower courts, might go forth to meet this image [*idolo*]—this, I say I leave to the natural philosophers [*physici*] to argue about. (180; my emphasis on English)

Earlier noting that *idolum* (now reappearing in this passage) is a relatively unusual term in classical Latin, Donahue annotates it as the Latinized form of Greek *eidolon* and, while acknowledging that it can mean "ghost, apparition" (hence Keplerian *imago*), aligns it instead with its use "by the Epicureans to denote the small images given off by illuminated bodies" (155n160). Initially his gloss might suggest the Keplerian *species* of light and color that flow into the eye, particularly insofar as *idolum* is one of the many synonyms of *species* enumerated by Roger Bacon. But such Keplerian *species* are at most an intermingling of materiality with immateriality (a "third entity"), whereas an Epicurean image, like most perspectivist ones, would presumably be corporeal. In the next sentence of the inset, however, Kepler addresses the juncture in vision of what is depicted on the retina with the spirits of the visual sense, which are corporeal since they belong to sensation or to "the lower courts." The fact that the neural spirits are not simply behind the retina but touching and participating in it bears emphasis because it is regularly overlooked. In traditional faculty psychology, sensation ascends to imagination, which is also sensitive, and thence in successive steps to the higher faculties. If Donahue's Epicurean gloss is right,

Kepler's doubling of terms in the inset, using "image [*idolum*] or picture" to refer to the product on the retina, attempts to look two ways at once, simultaneously toward the visual sense and toward what is received by the retina, and it affords a further instance in which, at and on another Keplerian surface, that of the retina, intermingling variously occurs: interior with exterior and corporeal with incorporeal. Of course, glossing hardly makes the difficulties of Kepler's terminology disappear, but at least it attempts to engage and explain it. His use of terminology appears to be deliberate, and it asks for attention.

Momentarily, in connection with Kepler's analogical speculation about the visual process beyond the retinal wall, two earlier discussions in this volume that have broad cultural relevance come to mind: in Chapter 2, Lodowyk Bryskett and his character Spenser's attempt to reconcile inside and outside, mind and matter, and, in Chapter 4, Aquinas's and Cajetan's efforts to address analogical metaphor, including the relation of physical to spiritual light and sense to intellect. Such connections will also pertain to the chapters that follow, for instance, to the analogy of the two-sided "written Rols"—inner mind and outer skin—near the end of Donne's *Second Anniversarie* and to Raphael's analogy of being in *Paradise Lost*, Book V.

As noted, Kepler *almost* stops at the juncture of the retinal picture with what is behind it. Yet his little allegory of the court of judgment that operates in the animated interior of the cerebrum plays indulgently with the process that lies beyond ocular geometry and anatomy.[22] Significantly, this psychological process within is imagined at once to be rational and judicious. It is also dramatized as a vignette and therefore presented as an expression of the imagination.[23] Kepler subscribes to the dominantly Aristotelian faculty psychology that he shares with most of his contemporaries, and what lies beyond his retinal wall accords with it. Anyone tempted to mystify a supposition of darkness beyond the retinal wall should reread Kepler's "praises of shadow!" in the preface to the *Paralipomena* (16). It is only by contrast that we earthlings can approach the knowledge of light. Contrast is to be observed, measured, and used in the service of reason, hypothesis, and revelation rather than feared.

Before proceeding to a fuller presentation of the process of vision, Kepler hedges his theory again. He reverts to the problem of the afterimage, whose intermingling of its color with the present object pertains to the reliability of Keplerian visualization, which has been much debated in recent years. At the conclusion of the first chapter of the *Paralipomena* (before its appendix), Kepler worried this problem, namely, that a "surviving image is blended with the colors of other things to the viewing of which the vision, so permeated, is transferred" (42). Eliminating a number of erroneous explanations, he settles on "the remaining possibility . . . that what inheres in the eye is an image of the action and effect, not of light but of illumination"—that is, of having been illuminated. A brief flurry of analogy based on the physics of motion—more specifically on the impetus theory, which is roughly similar to the modern concept of momentum—quickly follows an analogy based on sense experience:[24]

> In like manner, the sensation of pain from a blow persists, and this is understood as a kind of image [*species*] of that effect. In like manner also, in violent motion, a kind of image [*species*] of it carries over into the projectile, and carries it forth to some place even after the one who had given the motion removed his hand. (42)

Kepler's investment in the afterimage, as he explains, is that its lingering should belong to the loci of sensation, the network of nerves in the retina and the optical nerve and "the [visual] spirits . . . borne by them," and not to the transparent ocular humors, lest the larger principles pertaining to transparency, light, and vision be violated (42, cf. 180). He makes his investment clear from the start, and when he reintroduces the phenomenon of the afterimage in Chapter 5 of the *Paralipomena*, he recalls his earlier explanation, adding that the strength of the color (that is, the great impetus of its striking the retina) accounts for its lingering: "This image [*species*], which has an existence separable from the presence of the object seen, is . . . in the spirits, and vision occurs though this impression of images [*specierum*] on the [neural] spirits. However, the impression itself is not optical, but physical and mysterious" (181). It is sensational, the physiological result of impetus, or force, on the nerves and spirits (and presumably thence on the imagination), and does not belong to the optics of geometry and physics.

In Kepler's fullest single statement of his optical theory, which follows, he no longer refers to the image *or* picture (*idolum seu pictura*) that joins, also mysteriously, with the interior of the cerebrum, as he does in introducing his little allegory of a psychological tribunal, but now he refers to the picture alone. The origin and existence of this picture, it becomes clear, he needs to distinguish from the nerves and spirits, even though he is well aware that it must in some way be accessed by them and conveyed to the brain. His explanation of the means—indeed the mechanism—of optical picturing derives largely from his geometrical studies of ocular refraction and includes the reversal from left to right, as well as the inversion, of what is seen. It is most efficiently quoted in full:

> Vision thus occurs through a picture of the visible object at the white of the retina and the concave wall; and those things that are on the right outside, are depicted at the left side of the wall, the left at the right, the top at the bottom, the bottom at the top. Further, green things are depicted in the color green, and in general any object whatever is pictured in its own color within. The result of this is that if it were possible [*si possibile esset*] for this picture on the retina to remain while the retina was taken out into the light, while those things out in front that were giving it form were removed, and if some person were to possess sufficient keenness of vision, that person would recognize the exact configuration of the hemisphere in the compass of the retina, small as it is. Moreover, that proportion is preserved, such that when straight lines are drawn from the individual points of objects that can be viewed to some particular point within the compass of the eye, the individual parts within are depicted at very nearly the same angle at which the lines will have come together, so much so that not even the smallest points are left out. And to such an extent that the fineness of this picture within the eye of any person you please is as great as the acuteness of vision in that person. (181–82)

The final sentences of Kepler's description could almost suggest the accuracy of a photocopy, coincidentally another process involving the action of light (Greek *phōs*) on a surface.[25] Kepler's assertion of "the fineness of this picture" is unqualified apart from an individual's ocular deficiency, such as myopia. This accuracy of depiction does not shield visual perception from illusion and error, however, a subject to which I'll return.

Kepler famously reports that he tied himself "in knots for the longest time" in attempts to avoid the conclusion that the retinal picture is inverted and laterally reversed (221). Finally, the precision, consistency, and rationality of the correspondence between picture and external object persuaded him that what he had discovered was not absurd or incredible. His interpretation of this correspondence is that "Nature has found an excellent proportion," indeed, has produced an iconic analogy mathematically scaled in every detail (184). If this discovery has "horrendous epistemological implications," as A. Mark Smith has proposed, Kepler is unruffled by them. Smith later concludes that Kepler was "not fully aware—perhaps not even aware at all—of" the entailments of his theory, as is evident "from his continued invocation of visual spirits and species in explaining visual perception."[26] Smith, reading backward from "the next generation of optical theorists, who attempted to mechanise the visual process to the point of dispensing almost entirely with perceptual intentionality," faults

Kepler for not having been one of them even while not considering him such (42).

That the terminology Kepler employs is simply accidental, ornamental, unaware, or undeliberated is unlikely, as I have earlier argued. Kepler's first chapter begins with a discussion of etymology and proper terminology; his chapter on vision begins with the etymology of *oculus*, "eye." He pauses to define terms, his distinction between his use of "image," *imago*, and "picture," *pictura*, being the most familiar instance. Might it be that the Keplerian entailments Smith envisions—a blind eye, no meaningful place for perceptual intentionality, the shift of analysis wholly to the object, the absence of a comprehensible link between eye and brain—are simply not there in Kepler's writing, whereas the visual spirits, species, and their likes effectually are, and they ask for comprehension? Read carefully, Kepler's use of these traditional terms is generally controlled or clarified by the context he provides, one that is sometimes verbally and syntactically so hypercorrect as to be circuitously wordy: his explanations of the afterimage and his use of *idolum* have afforded brief examples. In a perceptive study of optical terminology, Isabelle Pantin summarily indicates that Kepler employs *species* in a careful, technical sense in his original demonstrations, notably in the instance of the afterimage and in other demonstrations in which sensation is involved (255–56). Several instances I have cited in the present section support this view, perhaps the occurrence of *species* in the first inset and later of *idolum* being problematical exceptions.

A blind eye or a dead eye is not what Kepler describes with wonder just above as he imagines what would happen if the retinal picture could be examined by an external viewer—turned, in effect, into a snapshot. Instead, such an externalized retinal picture is contrary to fact ("if it were possible") and purely hypothetical—the record of a moment's actual sight. Perceptual intentionality, moreover, is simply not Kepler's subject; he explicitly limits his concern to the science of optics, based on light and geometry. Mere paragraphs before the citation above, he has also implied in his allegorical tribunal something of what happens in perceptual intentionality within the subject, a bow, as it were, to what does not immediately concern him. The Keplerian observer may use instruments and "see" the heavens scientifically and truly only with measurements, as Kepler tells us, but he has no more disappeared than has the broader context of Kepler's writings.[27] In *Harmonice mundi*, Kepler will remind his readers that "the eye has been made as it is because the mind is as it is, and not the other way round," or in a pithier version, "the eye is so, because the mind is so, not vice-versa."[28] The same passage of *Harmonice* adds that geometry, which Kepler considers innate in the mind, is "certainly not received within through the eyes" but upon creation has "passed into" (*transivit*) God's human image (*imago Dei*). Kepler has not changed his view from the initial chapter of the *Paralipomena*: its "passing over" or "into" mirrors that of light, divine and celestial, into the natural laws that operate in this world. The anticipation of a Cartesian separation of subject from object is available to readers of Kepler who seek it, but it is by no means *necessarily* there. The eye of the beholder is decidedly in play, then and now, and this eye, modern as it may be, is not just an objective one. This is again my point, and it is a historical one.

Kepler's distinction between an image and the retinal picture has also received considerable attention, and it bears on the implications of his optics. Kepler is a careful and deliberate organizer, summarizing or otherwise introducing each chapter, calling attention to various stages of his argument, and explaining each step of his procedures. Given these methods, his explicitly defining an *image* in his third chapter, on catoptrics, but not explaining his contrasting use of *image* with *picture* until more than halfway through Chapter 5, on the visual process, is noticeable. It might be strategic or another result of Kepler's giving an account of his research that avowedly reflects its actual course, notoriously at times including dead ends after pages and pages of frustratingly detailed, mathematical investigation. When he does make the contrast between *image/imago* and *picture/pictura* clear, he signals that this is the meaning of each word in the present context and presumably going forward.

Kepler's full title for Chapter 3, "On the Foundations of Catoptrics and the Place of the Image [*Imaginis*]," indicates the context of his immediate concern, which pertains to the location of a reflected image as specified by the ancient *cathetus* rule, the received versions of which Kepler considers a "disgraceful stain in a most beautiful science" (76). According to this rule, as explained by Olivier Darrigol, "the image should be at the intersection of the prolongation of the visual ray (that is, the ray reaching the center of the eye) and the perpendicular drawn from the object to the [reflective] surface" (26). For me, the positing of such an image might also be conceived as passing through the looking glass to locate the image on the other side of it. So conceived, questions of virtuality and reality clearly pertain to the image.[29] To a modern, in fact, this image looks like an optical and mathematical fiction. How fitting that the author of *Through the Looking Glass*, namely, the Victorian Lewis Carroll, should also have been, like Kepler, a skilled mathematician.

Kepler, having summarized his own view of the location of the image, begins what he terms his "demonstration" as follows:

> First, I place at the entrance the definition of the Image, taken from catoptrics, into which we are entering. *The Optical writers say it is an image, when the object itself is indeed perceived along with its colors and the parts of its figure, but in a position not its own, and occasionally endowed with quantities not its own, and with an inappropriate ratio* [*proportione*] *of parts of its figure.* Briefly, [then,] an image is the vision of some object conjoined with an error of the faculties contributing to the sense of vision. Thus, the image is practically nothing in itself, and should rather be called imagination. The object [itself] is composed of the real form [*species*] of color or light and of intensional quantities [measurable intensities].[30]

Much later, in Chapter 5, Kepler recalls this definition of an image in the context of treating visual problematics, such as looking through an aqueous globe set before a window, viewing objects that are too close to the eye to be seen clearly, and simultaneously seeing different intensities of light together. His fourth proposition in Chapter 5 reads, "Darkness accommodates an image, but when a stronger light arises from the direction of the image, the image passes away. For an image is in part a creature of intention [*intentionale ens*], the work of the sense of vision" (193). Some form of illusion operates here and partly involves subjective perception and presumably also in part what I have referred to as the mechanics of vision.[31] Kepler's note on the passage confirms this impression. It explains why "a bluish and nearly colorless material is laid beneath glass mirrors" and why "when darkness outside surrounds glass windows, i.e., at night by candlelight, the windowpanes serve us as mirrors" (193n38).

Subsequently in Chapter 5, still in the context of visual problematics and beneath the heading "Definition," which is positioned conspicuously between two propositions, Kepler explains, "Since hitherto an Image has been a Being of the reason [*Ens rationale*], now let the figures of objects that really exist on paper or upon another surface be called pictures" (210). Real rather than imaginary, intentional, or virtual, such figures have objective existence, in contrast to an entity of imagination, reason, or, more generally, of intention or, in short, something psychological. They are comparable to the early modern view that words, whether written or audible, have objective existence as things.[32] They include reflections on the wall of a camera obscura, on the retinal surface, or on paper held to capture the light of a solar eclipse. They answer to the dependable, authenticating laws of geometry.

Alan E. Shapiro treats the difference and relationship between Keplerian picture and image in ways that are simultaneously illuminating and, for the hasty reader, potentially misleading. He refers first to the definition of *image* cited in my preceding paragraph, then to the earlier use of *image/imago* by Kepler and others, and finally rationalizes a relationship between this term and *picture/pictura*:

> Nonetheless, in general the distinction that Kepler drew between *imago* and *pictura* was not as sharply demarcated as he makes it seem here. Earlier, in chapter 2, where he explains image formation in a camera obscura, he refers to the image cast on the wall mostly as an *imago* but also as a *pictura*. Kepler's successors certainly understood the two to somehow be related and, like Kepler, they freely interchanged the word *imago* and *pictura* for the projected image. [Here, Shapiro's note: "In contemporary scientific usage and illustration, picture, or image in a printed book was called, indifferently, *imago* and *pictura* (and also *icon*)."] A *picture* was evidently some kind of image—that is, it belonged to the genus of image—because it was seen by refracted or reflected rays rather than by straight or direct rays, like an object. Seeking a common basis for the two would have been a meaningless task had they thought that the two were utterly distinct kinds of entity. (286–87)

Despite Shapiro's inclusion of the word "here" at the end of the first sentence, his noticing that Kepler appears *earlier* to use *imago* and *pictura* interchangeably and his registering this usage as a deficiency neglects three facts. First, Kepler carefully includes reference to earlier use, including his own, in his current definition: "hitherto an Image . . . now let the figures of objects . . ."; that is, what he did earlier, as well as what others have earlier done, is distinguished from, and potentially off-limits for, his current usage. Second, Kepler tells readers repeatedly that he is presenting his findings in such a way as to reflect the process of his own discoveries, including false starts, in order to achieve greater authenticity. Third, both scientific discovery in this period and humanist procedure often self-consciously reflect process, provisionality, hypothesis, conjecture, and "play."[33] If a treatise is committed to these in the interest of enhanced credibility, it is misreading to adopt a "gotcha" approach and ignore them. Shapiro's remarks are very helpful insofar as they characterize the concept of *picture* as a subcategory of the larger term *imago*. With this sensible revelation, much—not all—fretting about Kepler's terminology is further put into a properly diminished perspective.

In a discussion now deservedly classic, Svetlana Alpers has explored and analyzed the relevance of Kepler's ocular picturing as a cultural model for

Dutch painting of the seventeenth century. Although she, too, if reluctantly in passing, labels the Keplerian eye dead and passive, she also argues that the Keplerian model entails an embrace of the optical artifice of lenses and the pinhole camera and, more generally, of the juncture of "finding and making, of nature and art" (33). The Keplerian eye itself effectually has a lens within it, which refracts and focuses light, and its function is otherwise analogous to the pinhole camera (*Paralipomena*, 184, 191–218). Alpers also maintains that Kepler is "Far from raising doubts about the value or use of the . . . instrument[s] of sight," whether ocular, mechanical (or both), even when he uncovers their errors and deceptions (35). His investigations account for such flaws and control for them, enabling more accurate astronomy, as in Kepler's *Rudolphine Tables*. Aligning the Keplerian model with the trust of the lens and of "picturing" among the early modern Dutch, Alpers then remarks Christiaan Huygens's conclusion that "nature is art and art is nature" and Jan Vermeer's *View of Delft* as a claim about "the meeting-place of the world seen and the world pictured," which "Kepler defined mathematically, [and] the Dutch made a matter of paint." Ultimately, Alpers shows "the recognition [by the Dutch] that there is no escape from representation" (35).

Alpers's argument raises two questions for me: how different were the implications of these developments from those of the dominant view when Kepler wrote, and how does analogy bear on them? First the question of representation: I want to recall, if only in passing, the related problems of representation in language and in painting that go back to the ancient Greeks, for example, to the layered ironies of Plato's *Phaedrus* and *Cratylus*. If the Platonic worldview had a warrant, at least for a while, in the realm of Ideas, or Archetypes, Kepler and most others in his period had a comparable one in the biblical God of creation. The epistemological status of mathematics, privileged in Platonism, was demoted in Aristotelianism and recognized as merely rational—hardly a negligible value yet still a far cry from Platonic Idealism. In short, representation had been inescapable and problematic for a very long time. Elizabeth Spiller observes that the notion of visual distortion was nothing new in this period but that its extension to all acts of perception was (103). Whether it was extended to all acts of perception sorts oddly with Kepler's claims for the "fineness" of the retinal picture, and this is again a question to which I will return. There is surely some truth in Spiller's view but likely also some overstatement.

Second, the question of analogy. Alpers's move from Kepler's to Huygens's conclusion that nature is art, art nature is too easily mistaken as identity, sameness. It is instead metaphor, as is Kepler's dubbing what is on the retina a picture. Alpers acknowledges this fact when she describes Huygens's conclusion as a *near* collapse of the distinction between art and nature and Kepler's adoption of the term *picture* as a simile (35, 37). Whether the latter is a simile or a metaphor (the better historical term in this instance), the relation of art to nature is essentially analogous in Kepler's writings. Analogy, a form of metaphor, partakes of difference as well as of likeness. The same observations apply to the modern description of the Keplerian eye as a dead eye or a blind eye (at least in a live, sighted person). Both are metaphors, and even calling the crystalline humor of the eye a lens looks like a metaphor as well, an instrumental back-formation: after all, why not call a glass lens a crystalline humor? The metaphor of a dead or blind eye accompanies and correlates with the denial of sentiency to the Keplerian eye, which, on consideration, is another dubious modern assertion. The pupil of this eye moves and changes, dilating and contracting in response to light, and, of course, the surface of the eye is sensitive to other kinds of contact. As Kepler remarks, moreover, eyes age, change, and grow weaker. They do so because they are mortal and living. Perhaps Kepler has not erected a barrier between the eye anterior to the retina and the brain/soul behind it so much as he has greatly clarified their relation insofar as it pertains to visual reception of the outside world. Perhaps the real truth could be that there is no escape from metaphor or from the need to recognize and account for it.

A third question, also present in Alpers's argument, unavoidably arises from the *Paralipomena* itself as well as from a number of more recent commentaries on it. This is the question of the accuracy and veridicality of Keplerian vision and its entailments. Again, analogy bears on this question. A strong, if unwitting, example of this bearing is present in Ofer Gal and Raz Chen-Morris's essay tellingly titled "Baroque Optics and the Disappearance of the Observer: From Kepler's Optics to Descartes' Doubt."[34] Summarizing the consequences of "the significance of Kepler's formulation of optics," these coauthors explain that it "had no place for forms and visual rays [corporeal *species* and extramissions], and without them the teleology of the optical process was lost, together with the essential veridicality of vision and the importance of optics as an epistemological anchor for all other sciences" (199). They go on to acknowledge that Kepler's optical writing was, in fact, epistemologically oriented, "but instead of guaranteeing the authenticity of human visual knowledge in general, it was aimed at supporting the empirical underpinning of Kepler's new astronomy." True enough, as is their next point, that "we can trust images [retinal pictures] because they are outcomes of a purely natural, causal process" subject to experiment and theorizing. As they continue, however, their conclusive assertions become less persuasive precisely because they have dissolved, not respected, Kepler's analogies:

> Yet this trust [in eye and instrument] comes with a steep epistemological price tag: if the instrument is not prone to error more than the eye, then the eye is as vulnerable to error as the instrument. Passively receiving "illumination" like any instrument, the eye is not merely comparable to "a closed chamber" [camera obscura]—it is one: "the pupil takes the place of the window." The cornea is truly nothing but a lens; the retina nothing but a screen, essentially the same as the paper or the pavement. The picture on the retina, like the one on paper or the pavement, is *not* an accurate reflection of the object. It is a fuzzy "row of infinite figures," the shape of each caused by the accidental shape of the aperture; "the pupil" or "window." (199)

The quotations in this inset come from the *Paralipomena*. The first of them is the only instance I noticed in which Kepler likens the eye to the pinhole camera or camera obscura; it is a Keplerian metaphor or, if you wish, metonymy. The extent to which Kepler relied on the camera in developing his theory of vision is debated. Be that as it may, this moment in the discussion of vision is the closest Kepler comes in his treatise to identifying the two. What immediately precedes and follows the fragment quoted, however, strongly indicates that Kepler's phrase "takes the place of" actually refers to another Keplerian translation from one place to another and, in short, a metaphor. Immediately before the fragment cited, Kepler writes, "nearly the same thing happens that we proved in chap. 2 above happens in a closed chamber"—again, *nearly* the same [*penè idem*] (*Paralipomena*, 184). Immediately after the fragment, he writes, "the crystalline takes the place of the panel opposite, except that [*nisi quòd*], because of the proximity of the pupil and the crystalline, the complete intersection has not yet been brought about here, with the result that everything is still confused"—*except that*. Here is difference, however partial, as well as likeness. Here as well is analogy, not identity, and it is used skillfully by Kepler, if not always as reliably by his modern readers.

The quotations at the end of the preceding, coauthored inset are drawn from another passage, not even about the eye but about reflection in a camera obscura, the familiar passage in Chapter 2 in which Kepler uses threads depended from the four corners of a book and drawn through a polygonal aperture to trace with each thread "a narrow row of infinite figures of the hole [that] outlined the large quadrangular figure of the book on the pavement" (56). The point of this experiment, which was evidently inspired by the writings of Albrecht Dürer and Leon Battista Alberti, is to answer the question "Why a [light] ray that has entered through four cornered figures does not make four-cornered shapes, but circular ones."[35] The book in the experiment stands for the sun, notwithstanding their different shapes, one rectangular and the other circular. Kepler's experiment confirms that the round shape reflected through a four-cornered aperture is "not that of the visual ray [or of bundled rays, as earlier authorities had suggested] but of the round sun itself" (56). The fuzziness of the retinal picture supposed by our coauthors is not found on the page invoking the analogy of the camera obscura with the eye, but it can be illustrated by the meticulous experiments, replicating and expanding Kepler's with the threads, pinhole apertures, and the images reflected through them that Giora Hon and Yaakov Zik describe and illustrate in their article on "Kepler's Optical Part of Astronomy (1604)."[36] Again, experience shows, to summon a phrase of which Kepler is fond, that reflected images on walls are often fuzzy, but if this fuzziness is what our eyes see all the time, we need an appointment with the ophthalmologist. The Keplerian analogy of the eye with the camera obscura is part likeness, part difference. The collapse of this duality into identity, as in the Gal/Chen-Morris inset, gives us either Kepler the magician or Kepler the empiricist, not the real, compound Kepler.[37]

In the interest of efficiency, I want to continue with Gal and Chen-Morris's conclusions since I agree with much of what they next argue, dissenting only from their tendency to overstate the case, which is a representative, not an isolated, example. In fairness, I should mention that their conclusions, or "consequences," are paradigmatic soundbites compared to the more extensive arguments that precede them. At the same time, however, their claims are put forth as the essence of these arguments. They are well aware that "Kepler was no skeptic"; nonetheless, they abstract the supposed kernel of his optics from the shell of language, rhetoric, founding assumptions, and the broader culture in which it is embedded, and from all of which, to my mind, it is inseparable without misrepresentation:

> Kepler's optics abandoned the epistemological assuredness that traditional optics provided by relinquishing the assumption that "the arrangement of the species [is] *exactly* as the object." For Kepler, "it has been demonstrated most clearly, *from the very structure of vision*, that it frequently happens, that an error befalls the sense of vision." Visual errors were of course nothing new, but Kepler's was a new concept of error. In the Aristotelian paradigm, errors were created by the

> intervention of the human imagination; the visual data were indubitable. With the new optics the doubt was directed at the very images perceived. (200; italics of Gal/Chen-Morris)

The first quotation in this inset comes from John Pecham, a medieval perspectivist, here summoned to represent the Aristotelian element in perspectivism (Gal/Chen-Morris, 198). The second quotation occurs at the conclusion of Kepler's section addressing specific instances of visual error and includes the adverbial qualification "frequently," as cited, yet without the immediately following, final phrase of Kepler's same sentence, "in estimating the size of bright things" (*Paralipomena*, 236). In other words, the Gal/Chen-Morris generalization in the inset is inadvertently misleading.

But it is the rest of this inset that most interests me. Its first sentence ("Visual errors . . . new concept of error") is followed by a footnote to Giora Hon's essays on error, the more recent of which, "Putting Error to (Historical) Work," finds in Kepler's practice the acceptance of a degree of play (*Spiel*), by which Hon intends an "unavoidable ignorance" associated with "inherent gaps" and, I would add more freely, with some degree of leeway.[38] I wonder whether Hon's provocative *Spiel* could even be extended to Kepler's little psychological allegory; to his *Somnium* (Dream), a piece of early modern science fiction; to his playful treatise on the snowflake; or more broadly to the humanist concept of *serio ludere* (serious play). If so, its broader cultural context is a little less apocalyptic than Gal and Chen-Morris imagine and somewhat less unavoidably ignorant than Hon suggests.

The last two sentences of the inset, concerning the reigning Aristotelian paradigm, as distinguished from Keplerian optics, are certainly sound, but they still make the contrast between the two so stark as to risk oversimplification and misunderstanding. The Aristotelian paradigm combines sensation and psychology; the Keplerian one does not, even though, as I have argued, Kepler's position regarding the relation of sense perception to optics is not as absolute in separating them as his commentators'. In *De Anima* (*On the Soul*), Aristotle holds that though "sensations are always true, imaginations are for the most part false," even while adding that imaginations (*phantasmata*: representations, appearances, sensory images) are necessary for thought.[39] As proffered in my previous chapter, however, Malcolm Schofield suggests that Aristotle expresses "skepticism, caution, or noncommit[ment] about the veridical character of sensory or quasi-sensory experiences . . . [that is,] 'it looks thus and so [—but is it really?]'" (252; cf. 253, 259). The Aristotelian tradition, like the Platonic one, witnessed many a variation on its central themes over the centuries, but the necessity of the senses as a basis for any mental process (possibly and variously excepting intellection) remained a constant, even though the senses were known to provide data that had already been mediated and filtered by them.[40] Moreover, if the Aristotelian eye observed an optical illusion or the merely apparent existence of an image in a mirror, it truly reported this deceptive sight, and it was up to the various faculties such as judgment and reason to adjudicate its reality, as in Kepler's allegorical tribunal. A tribunal renders judgments.

Kepler also traces the awareness of a number of optical problems, enigmas, and corrective devices back to Aristotle, Ptolemy, and other predecessors: in the *Paralipomena*, for example, the problem of a rectangular aperture but a round reflection of the sun, as recently discussed, and further in several instances recounted in Kepler's Chapter 5, Section 5, titled "Those things that recoil upon Astronomy from the means of vision; or, on flawed vision," the very chapter cited by Gal and Chen-Morris in the most recent inset above. If the Aristotelians thought "the *arrangement* of the species [is] exactly as the object," as Gal and Chen-Morris put it (my emphasis on "arrangement" replacing theirs on "exactly"), Kepler thought "that *proportion* is preserved, such that when straight lines are drawn from the individual points of objects that can be viewed to some particular point within the compass of the eye, the individual parts within are depicted at very nearly the same angle at which the lines will have come together, so much so that not even the smallest points are left out" (181–82; my emphasis). Neither the Aristotelian nor the Keplerian paradigm, both of which are intromissive, proposed that the reality of the object was identical to what the eye received; had it been so, the tree, or whatever else, would have sprouted in the eye. In both paradigms, there was some degree of mediation and representation. What Kepler did, and significantly did, in the *Paralipomena*, crucially having begun with his metaphysics and physics of light in Chapter 1, was to explain the causes of many optical flaws and illusions and to account rationally for visual reception, primarily by means of anatomy and mathematics. *Primarily* leaves room for observation and experiment and, above all, for analogy, which is implicit in mathematical, as well as in verbal, language applied structurally to the real world, anyway.

Keplerian Analogy

A section on analogy might at this point seem surprising since the preceding sections on light and vision have hardly avoided this topic, which

Kepler's writing urges. They have not made it their focus, however, the first having focused on light and the second on vision. In this section, I want briefly to include a number of Kepler's specific statements about analogy in the *Paralipomena* and to focus on four instances of it that involve his arguments, experiments, and demonstrations, in short, his methods and evidence. The first of these comes in Kepler's Appendix to Chapter 1, in which he rejects Aristotle's theory that vision requires the use of a pellucid medium between the object and the eye, whereas his own theory proposes that light simply and instantaneously leaps over the intervening distance. In a relevant passage, Kepler wonders aloud whether Aristotle might have "proposed this motion or affect of the pellucid [air] analogically and not entirely physically" (53). This possibility would make Aristotle's theory more acceptable to Kepler, but it only leads to another question as to why Aristotle did not "rather imitate the nature of things and set up an analogical motion in light," implying, incidentally, that analogy itself is natural. Kepler next declares his key problem with the Aristotelian position, namely, that "light or form, a mere accident [in Aristotle], cannot support any motion." Motion, as we have repeatedly witnessed, is more important to Keplerian theory than is often recognized. The Aristotelian position on light or form is directly opposed to Kepler's principles, and to it Kepler objects that an Aristotelian body, which is material, as is Aristotle's pellucid medium, cannot be moved in an instant. In Kepler's reassertion of his own, opposing view, "the motion of light (without time) is such as to square with the quality of its body (without matter), the two being analogical" (53). Additionally striking throughout this argument is the extent to which analogy and physics are themselves correlates. An underlying metaphysics is operative as well in "the nature of things" and more specifically in the immateriality of light. Donahue's figurative use of the verb "square" in the last quotation, meaning "accords, conforms" but also signaling the potential for geometrical representation, is surely justifiable in the larger context of Kepler's analogical thinking.[41]

Another instance of Kepler's methodical use of analogy occurs in his "Demonstration of those things that have been said about the crystalline [humor of the eye] in regard to the means of vision" (191). It starts empirically, with what can be derived from "common experiences." This derivation involves a crystalline globe filled with clear water and placed before a window and a piece of white paper on the other side of the globe, distant from it by the radius of the globe. When these arrangements are in place, "the glazed window with the channels overlaid with wood and lead, enclosing the edges of the windows, are depicted with perfect clarity upon the paper, but in an inverted position." Additional physical demonstration follows with the room darker and the aperture smaller (effectually a pinhole camera), which produces a distinct picture on the paper of what is outside the room. But when the eye is applied at the same distance from the crystalline globe as was the paper, there appears a "confusion of the objects represented through the glass," the glass itself appearing entirely bright, red, dark, or the like. If the eye moves nearer or farther from the globe, the objects become either larger or else smaller, distinct, and inverted on the nearest surface of the globe. Similarly placed, the sheet of paper depicts nothing. Kepler explains that

> All these things happen with regard to an aqueous globe, because of the refractions and the shape, as a result of there being some convexity in the shape. And since the crystalline [humor of the eye] is made of convex surfaces, and it is also denser than the surrounding humors, just as water in the glass [globe] is denser than air, therefore, whatever we shall have demonstrated concerning the aqueous globe in this way, and using these media, have also been proved concerning the crystalline, with privileges reserved to it because of the particular convexity of shape, inconsistent with the convexity of the globe. Let us proceed, then, to the demonstration of those things that happen in relation to a crystalline or glass globe. (191–92)

What follows is a geometrical demonstration regarding the place of the image [*imaginis*] "when the object is observed with both eyes through a globe of a denser medium" (192). Empiricism gives way to geometry, the higher form, which rationalizes and intellectualizes the observations based on sense. Before it does so, however, a revealing analogy has been established between the aqueous globe and the crystalline humor, as well as between the passage of light rays through each of these to the paper and retina as screens, and, finally, between experimental observation and ocular functioning. The analogy is quite convincing but nonetheless still hypothetical, and it remains so even when abstracted, theorized, and to this extent "confirmed," in Kepler's lexicon, by geometry. Relevantly, Darrigol enumerates "at least three reasons" why Kepler's analogy is "imperfect": "the incoming beams are narrowed by the pupil of the eye, the crystalline humor does not have a spherical shape, and a first refraction occurs at the cornea" (29). The Keplerian analogy between an aqueous globe and an ocular humor thus includes difference not only with respect to "the privileges reserved" to the crystalline humor "inconsistent with the convexity of the globe," as Kepler himself has acknowledged, but also because any analogy, if it really is an analogy, would be "imperfect." Perfection would be identity, sameness: the aqueous globe "literally" (as we like to say) in the eye.

In preceding sections of this chapter, especially the last, Kepler's analogies between light and the physics of motion, such as the impetus theory, several times have attracted my attention. I want to look more closely at a particular cluster of these in his Propositions 18 to 20 of Chapter 1, on light. Frequently Kepler describes light as "striking" an object and, as the notion of striking might suggest, doing so violently or forcefully. The cluster to be considered illuminates his reason for choosing such a descriptive word. Proposition 18 concerns reflection: "Light that has fallen upon a surface is made to rebound in the direction opposite to that whence it approached" (26). Since "motion is an attribute of light," he adds, so is the particular kind of motion called "impulse" (*impulsus* < *pello*: "beat, strike"; "impel, propel"). First analogy, then *causal* identity follow: "what in physical motion is hardness of the colliding bodies . . . in light is the bare surface, or the bounding and shaping of bodies. . . . But the cause of rebounding's occurring in both physical motion and [immaterial] light is in the violence of the motion." Next follows an explanation of rebounding, or, with light, of reflecting. Throughout it, the analogy between body and light is sustained:

> since the motive force cannot all be destroyed by the collision, the motion will accordingly continue beyond the end of its line, which is the surface [that is struck]. But it cannot do this straight forward; for in the former case, body is in the way of body; and in the latter, surface is in the way of surface; in the former the motion would go into the solid, in the latter partly so. . . . What is left, therefore, is that it move in the opposite direction. (26)

Kepler concludes that the power that drives a movable toward something and the power that drives it back from something in the way are the same "because the collision is considered to occur at a *point*" (26; my emphasis). The mechanical analogy, with its concluding "point," thus anticipates translation to geometry. This translation has also been anticipated by the phrase "the end of its line" in the inset, which is initially puzzling and therefore conspicuous without a geometric diagram. Kepler's argument in this proposition seems to mirror his process of thought. It also serves as yet another illustration of why some claim his work for the onset of mechanism. One verbal detail in this proposition and the next, the Latin *mobile*, "movable" or "that which is moved," is further telling in this connection, however: Buchdahl takes *mobile* to mean "mobile body," and he remarks that materiality is "subtly smuggled in" here by Kepler (279); Donahue translates *mobile* more abstractly as "a movable" (pluralized as "movables" in Proposition 19) and thereby avoids a simple materialization of the object. Donahue's choice is truer to Kepler's text, which summarizes an analogy between the action of a material body and that of immaterial light. It hovers in a conceptual and causal dimension other than either alone and accounts for their movement. This dimension is at once theoretical and actual.

Proposition 19, which could be illustrated perfectly by Descartes's rebounding tennis balls, an illustration actually derived from Kepler's analogies, further treats reflection and establishes that the angles of incidence equal those of reflection.[42] The much longer Proposition 20 then addresses the more complicated issue of refraction, which critically pertains to Kepler's explanation of vision. The proposition itself is simple enough: "Light that has approached the surface of a denser medium obliquely, is refracted towards the perpendicular to the surface" (27). This time, Kepler starts with geometry, diagrammed, and he satisfactorily demonstrates what he means. But a question of a sort I have tracked nonetheless arises for him: "by what faculty could it happen that the influence of a pellucid surface [could] be imposed upon light?" (28). He responds that motion and the varieties of motion belong to light: namely, "impact upon a denser surface, and the overcoming of it, and a certain amount of resistance of the medium overcome. Moreover, that this also happens necessarily in physical movables, whenever a globe is spun into the water, provided it goes beneath the water, is shown thus." What follows (at great length) is entirely lucid, continues the comparison to physical projectiles, and concludes that here is a "kind of bending back," or refraction, of light "entirely similar to those which occur in projected natural bodies," or missiles (29). Analogy again offers a reasonable explanation of the interaction of the immaterial with the material at a surface. Once again, moreover, constructive, analogous metaphor has respected difference and avoided the false assertion of identity.

With an apology for digressing, on the grounds that a better place to correct error is not available, Kepler proceeds next to statics, or "the principle [*ratio*] of the balance" (29). His digression continues for pages, but I will just select a couple of examples to my purpose near his conclusion: "as C is to D, so is FG to GB. . . . Therefore, as C is to D, so is DK to KC" (32). Analogy is obviously present in Kepler's argument here, which is geometrical, and it is followed next by a return from what he now calls a preliminary consideration to "physical motion, which is common to light," as it is to the matter of balances about which he has been digressing (33). Although there are times when Kepler distinguishes between analogy and geometry, often the one continues into the other, complements the other, or makes the other possible. Moreover, this movement works in both directions and makes tight distinctions between Kepler's methods misleading. It also refuses to separate mathematics from its less pure applications and, indeed, from the imagination.

Kepler, still with an eye on balances, now considers the projection of a globe or missile against a panel or the current of a river against oars—more analogizing. He concludes that there is the "same ratio [*proportio*] of resistance of equal powers here in violent motion, as there was of unequal weights above in motion [that is, in balances] that is in accord with nature: that is, the lines, not the angles measure it" (33). Eventually, after further demonstrations, he concludes that "this behavior of violent physical motion also flows back to what is analogous in light" (34). Again, Kepler invokes analogy and thinks analogously on virtually every page of his treatise, and everything flows back to light, where it started. His digressing to consider statics, moreover, turns out to be relevant, not digressive after all.

The final example I want to examine includes Kepler's best-known statements about analogy, which occur in the *Paralipomena*, as it happens. Although Kepler's practice better exhibits the working of analogy in his thought, lacking these statements, this chapter would be incomplete. Over the course of about six pages in Kepler's fourth chapter, explicit reference to analogy occurs a dozen times, and it further pervades his practice (105–11). The fourth chapter is titled "On the Measure of Refractions," and the analogical concerns of its third and fourth sections, "Preparation of the true measurement of refractions" and "On the sections of a cone," spill into the fifth section, which treats the quantitative measurement of refractions directly.[43] Kepler begins the third section with his experience observing images as they appear in mirrors and in water and his doubts about the optical theories available to explain them. Restricting his reasoning to mirrors, he reports that he finally cut the "Gordian knot of catoptrics . . . by analogy alone" (105). He realizes that it is only because of reflection from a polished surface that images occur in a place other than the object seen and, by analogy, "that in water, too, images ascend, and approach the surface . . . only because of the refraction of the spark of light that has flowed from the object into the eye." Having pursued this first analogy to a second, the analogy of an image in a convex or concave mirror to an image in rarer or denser media, he now assumes that the plane surface of water has "a certain type of curvature," as do curved mirrors. He therefore turns next to conics in geometry since they curve as well, knowing that his study will be fruitful for optics per se and for the optical part of astronomy (110). This is the point at which, because conic sections are difficult to grasp, he chooses to treat them first "in a mechanical, analogical, and popular vein," playfully cautioning pure geometers to "be indulgent" (106). If obvious, it is worth noting again that play is a constructive element in Kepler's writing and thinking and further noting that mechanics, analogy, and popularity are here lumped together. In the present context, only geometry is superior, at once purer and simpler—in short, more abstract and conceptual—although, as we have recurrently seen, analogy can and frequently does occur in geometry, as well.

"Speaking analogically rather than geometrically" in order to explain conic sections, Kepler next proceeds to relate various geometrical figures to one another until he reaches the conclusion that

> the opposite limits are the circle and the straight line: the former is pure curvedness, the latter pure straightness. The hyperbola, parabola, and ellipse are placed in between, and participate in the straight and the curved, the parabola equally, the hyperbola in more of the straightness, and the ellipse in more of the curvedness. (107)

He turns next to foci, terming them such "For the sake of light, and with an eye turned towards mechanics," but also adding that he might have called them "centers" since they occur on the axes of conic sections, if only the latter term had not already been preempted for use with the hyperbola and ellipse (107–8). Having explained the foci of conic sections for the hyperbola, ellipse, circle, and parabola, he adds, "It therefore follows by analogy that in a straight line the pair of foci (we speak thus of the straight line, contrary to custom, only to fill out the analogy) coincides with the straight line itself, and is single, as in the circle" (108). We might wonder why Kepler would bother only to fill out his analogy, if it were not for what comes next. In the circle the focus is at the center, "receding as far as possible from the circumference." In each of the other figures, it recedes less and less, until the straight line "falls upon it." He fills out the analogy in the interest of encompassing relationships that are proportionally symmetrical and of the archetypal geometry of the curved and the straight. He has shown how the circle and the straight line are the limiting cases, other figures fitting in ordered steps between them. As Gérard Simon has written, what Kepler "is actually looking for, is . . . [the] immanent order" of the world, and here "the search for analogies is taking place . . . inside the realm of mathematics" (72, 81). With this recurrence of the curved and the straight, the circle and the line, we are back to the first principles of God's creation, with which Kepler's *Paralipomena* and my account of it began.

More in the same vein follows with respect to the diameter (Kepler's "chord") and to the axis (Kepler's "sagitta") of the geometrical figures concerned, starting with the circle and with each figure in order descending "in all ratios [*proportiones*: proportions], until [the sagitta] . . . vanishes in the straight line" and the diameter "is made infinite since it coincides with its own arc (to speak thus improperly, since the arc is a straight line)."[44] At this point, Kepler famously explains, not to say enthuses,

> For geometrical terms ought to be at our service for analogy. I love analogies most of all: they are my most faithful teachers, aware of all the hidden secrets of nature. In geometry in particular they are to be taken up, since they restrict the infinity of cases between their respective extremes and the mean with however many absurd phrases [*locutionibus*: "modes of expression"],[45] and place the whole essence of any subject vividly before the eyes. (109)

Geometry now serves analogy rather than the reverse, and analogy knows the created secrets of nature, the principles at its core. Geometrical terms, which exist to enable quantitative accuracy, are the servants of something greater than they, but used with some insight and imaginative freedom, they still facilitate the beauty of mathematical simplicity.

Kepler moves from his enthusiastic declaration of the virtues of analogy to the mechanical portion of this section, already anticipated in his gloss on the term "focus," which he has employed "with an eye turned towards mechanics." Geometry has given way to analogy, which, with all the explaining of hyperbolas, parabolas, and the like, has still centrally engaged geometry, and now analogy gives way to mechanics, which will turn out to afford—indeed, to be—another analogy, but now a physical one. Once again, Kepler's methods do not neatly split off from one another, one of my arguments from the outset of this chapter. Kepler, now entering his mechanical introduction to conics, adds to his praise of analogy in the inset above, "Furthermore, in the description of the [conic] sections also, analogy has been of the greatest help to me" (109). The description, or rather the *analogical demonstration*, of the sections he intends is accomplished with threads. He explains, along with an illustration, how to draw a hyperbola, an ellipse, and a parabola by using threads, fingers, and pegs: thus "the analogy [has] showed (and geometry confirms) how to draw [these]" (110). This time analogy, geometric confirmation, and mechanics are all wrapped up together. In the next section, he is finally ready to return, "at analogy's urging," he informs us, to the "angles of reflections through various angles of incidence of light upon concave mirrors" and to employ a good deal of geometry (111). But he continues to invoke the extended geometrical analogies of the preceding sections of preparation, in which we might naïvely have supposed him only to be indulging us and, in playful part, also himself. I want my chapter to trail off into Kepler's own, exemplary words, which, I have also argued, should be heeded more attentively:

> You could now guess immediately that because the hyperbola does the opposite of the refractions, the ellipse, being the hyperbola's opposite, is going to do the same as the refractions, and will accommodate itself to the measure. This guess is made the more probable in that on this basis the analogy gives the concave spherical mirror to the medium lacking refraction, and in this mirror the rays going out from the center and those bent back coincide, and do not enclose any angles. Let B be the focus of the ellipse, A the opposite focus. . . . (112)

And so on.

CHAPTER 6

Analogy, Proportion, and Death in Donne's *Anniversaries*

Ample evidence exists that Donne read at least one, and possibly two, of Kepler's earlier writings on astronomy, that he actually met Kepler, and that he probably knew, or heard, at some point about Kepler's experiments concerning vision. Eventually, he was at least aware of Kepler's *Harmonice mundi*. My Introduction has treated these observations in greater detail. I reemphasize here, however, that my present concern is with the larger cultural pertinence for Donne's *Anniversaries* of Kepler's thinking, which is broadly representative of a large body of thought "at the dawn of modern science," rather than with the direct influence of Kepler's writings on Donne's.[1] Approaching Donne's *Anniversaries* with these givens, I want especially to recall the kind of "reasoning by which [Kepler] . . . fuses geometry, astronomy, and [eventually] music theory into an integrated account of proportion."[2] To this summary by Albert Van de Schoot, I would add that Kepler's fusion participates in a worldview, indeed, a cosmos, that is also religious. His pronounced interest in a demonstrable, active interface between immaterial light and material surfaces is a significant aspect of his cosmic view, which expresses a qualified continuity rather than unity. Variously mathematical, physical, metaphysical, and harmonic, *proportion* is clearly a complex, multidimensional term in his thinking, and more so when its complicated historical background from the ancient Greeks and Romans, to Boethius, the Schoolmen, and the Renaissance humanists is factored into consideration—the background that I addressed in Chapter 4. Proportion is variously a translation of, a synonym for, or a specialized version of analogy. It, too, bears on Donne's *Anniversaries*; as a word (or cognate), it occurs over a dozen times in them, and it is further pervasive as a concept. Yet its occurrence comes with a difference that reference to Kepler intensifies. With two exceptions, explicit occurrences of the word *proportion* and its cognates are all in *The First Anniversarie*, and these are overwhelmingly associated with loss. In *The Second*, it recurs in the context of a human perfection exceeding all measurement and later in the context of redemptive grace.[3] It thus reflects the larger movement and relation between these two complementary poems. The extent to which these major poems by Donne actually respond to the new science is less salient and significant in them than what they discover about death. The relation of death to life (and light) is finally what they trumpet. Although Donne employs proportion significantly, his radical answer is firmly anchored in the body, albeit also within the embrace of religious faith; in comparison, the deeply religious Kepler's answer is more abstractly anchored in Neoplatonized proportion.

More immediately approaching Donne from the account of Kepler's writings in my last chapter, I should nonetheless be more specific about the strength of Kepler's religious commitment, which earlier has only been mentioned. Kepler, having trained for the ministry but having then been diverted from it by historical circumstances, remained a devout Lutheran all his life, as congregations, magistrates, and rulers with different religious affiliations, together with the religious turmoil of the Thirty Years War, repeatedly affected his employment and residence. At the same time, it is hardly surprising in view of Kepler's writings, dealing, as they do, with time, place, spirit, matter, and bodies and also manifesting revisionary theory, persistent experimenting, and new discoveries, that he should have encountered some problems with established religious thinking. His basic difficulties pertained to the Lutheran doctrine of ubiquity and the Eucharist in the Augsburg Confession of his youth.[4] Max Caspar, Kepler's biographer, memorably elaborates on these issues:

> [Kepler] inclined to the Calvinist conception in the doctrine of communion, followed the Catholics or, as he said, the Jesuits, as regards the doctrine of ubiquity, and on the other hand rejected Calvin's

> "barbarous" teaching of predestination and could not agree with Luther's book about the captive will. (217)

Kepler was excluded from communion by his pastor in Linz, where he lived from 1612 to 1626, neither the first time he had encountered trouble because of his beliefs nor the last. In this respect, he shares some provocative ground with Donne, who is almost exactly his contemporary, as well as with the more distant Milton, whose grand tour began eight years after Kepler's death and four years before Galileo's.

The First Anniversarie: Donne's Speaker and the Loss of Proportion

The first of Donne's *Anniversaries* was published in 1611 and then republished with the second in 1612. Exceeding a combined length of a thousand lines, both poems mourn and celebrate the death of Elizabeth Drury, the fourteen-year-old daughter of one of Donne's patrons, a young woman whom Donne had never met. Although both are thus occasioned by a specific death, their scope is much broader: in *The First*, comprehending not only the decay of the world but also the loss of meaning, or loss itself, and, in *The Second*, the relation of an immortal soul to mortal death. Wesley Milgate, introducing his edition of Donne's *Anniversaries*, equates T. S. Eliot's *Waste Land* "in content to an epic" and then adds that "it was left to Milton to write the full epic statement of the themes which Donne engages."[5] The consuming concern of Donne's *First Anniversarie* is nonetheless with death, and of his second with life, not so much in as through, and only through, death, more specifically the physicality of death. Milgate's perceptive comparison thus invites further reflection and expansion.

To turn to the two *Anniversaries* after chapters on analogy and Kepler is to unsettle seemingly obvious expectations still further. Despite the many repetitions of "proportion" and its variants, relatively few structured analogies—that is, A:B :: C:D—as distinct from simpler metaphors and other forms of similitude, are to be found in them. At a cursory glance, the high proportion of satire in *The First Anniversarie. An Anatomie of the World* and the proximity to sermon of *The Second Anniversarie. Of the Progres of the Soule* might bear on the *relative* brevity, understatement, and paucity of properly structured analogy in them as a whole and, in *The First Anniversarie*, on its evident suppression, notwithstanding the salience of the macro/microcosm relationship, whose contribution to *The First* is unstable and negative.[6] But these generic affiliations hardly account for this relative lack overall, not only because we can find such analogies in Donne's satires and, conspicuously, in his sermons but also because they are not specific enough to the form and content of these extraordinary poems. When eventually, dramatically, and focally *The Second Anniversarie* rediscovers the use of properly Donnean analogy, its use is all the more salient, significant, and constructive.

Coming to Donne after Kepler also puts into perspective the kind of philosophy the speaker of *The First Anniversarie* is so downcast about. The speaker's view is much embedded in the old Ptolemaic philosophy.[7] Although he mentions new stars, worrisome meteoric phenomena, the relation of the sun to earth, and famously the discrediting of the fiery sphere, the concerns he most anxiously fixes upon and elaborates belong equally and sometimes exclusively to the old philosophy, notably the apparent irregularities in solar and planetary movement and the use of longitudes and latitudes roughly to locate and measure celestial movement. His evocation of the newly established principle of magnetism as only a binding force that has departed with the death of Elizabeth Drury treats magnetism as if it were only a restatement of the cosmic principle of love, attraction, and friendship found in the ancient myths and their Renaissance retellings and as if it were only a loss rather than, conversely, further evidence of systemic cohesion, as in Gilbert and Kepler (FA 221–22).[8] He worries especially about violations of roundness—that is, of the traditional perfection of circular form—whether these familiar variances are evident in celestial movement or in the circumference of the earth, whose surface is blemished by mountainous peaks, hardly requiring a telescope to be noticed; he fails to mention Galileo's projection of the same blemish onto the moon, although Donne was acquainted with *Sidereus Nuncius* and possibly with Kepler's *Somnium*, a piece of science fiction, both of which treat lunar topography.[9] There is nothing new in much of the speaker's complaining, and there is also some incoherence about it: if the fiery sphere has no place in a heliostatic or heliocentric universe, neither does irrational solar movement.[10] The precession of the equinoxes and the apparently retrograde orbits of the planets, moreover, were phenomena that either the Ptolemaic or the Copernican hypothesis had long been able to "save," or account for adequately, even if Donne's speaker had not yet heard of Kepler's discovery, published in 1609, of orderly elliptical orbiting by the planets.[11] Aside from ancient reports of irregularities in the immutable heavens, other ones had also been observed in recent times, well before the *Anniversaries*. One new star discovered by Tycho Brahe and others in 1572, along with additional, worrisome heavenly phenomena, bears on the argument of

Spenser's Mutabilitie Cantos, written in the late sixteenth century but unpublished until 1609, two years before Donne's *First Anniversarie* and highly relevant to its themes of instability and decay.[12] Specifically, as Donne knew, Galileo and Kepler had identified more new stars in the decade preceding the first of his poetic *Anniversaries*. If an accumulating weight of evidence had the potential to be further disturbing, these most recent discoveries also lacked the shock value of something never heard before. Chicken Little had been clucking for quite some time, and more than one response to her clucks is both conceivable and evident in the scientific community as well as in the broader one.[13]

When Donne puts Copernicus, Kepler, and Galileo into the satire *Ignatius His Conclave* (1610, pub. 1611), the last two astronomers, both keen advocates of the new philosophy, are excluded from hell, although Copernicus gains entrance, along with Machiavelli, whose work even bishops read avidly; Columbus, whose discoveries memorably excited Donne's imagination, is also admitted, as is Paracelsus, whose writings engaged Donne's medical interests. Interrogating Copernicus in the *Conclave*, Ignatius, founder of the Jesuits, skeptically demands what he has discovered to advance Lucifer's kingdom and scoffs with ironic ambiguity, "Do not men believe? do they not live just, as they did before?"[14] If Donne's satire of the three best-known Renaissance astronomers is pungent, it is also playful, and it is an awkward fit for the gloomy excess of his speaker in *The First Anniversarie*. Donne, while subject to melancholy, shared many of the Neoplatonized interests and beliefs that buttressed Kepler's optimism and profound, if sui generis, Lutheranism.

In *The First Anniversarie*, Donne's obsessive speaker, his attention preoccupied by death, not only shares the concerns of Spenser's Mutabilitie Cantos but also the attitude of Spenser's disheartened speaker in the Proem to Book V of *The Faerie Queene*, the Book of Justice (1596, 1609). Considerably longer than any other of Spenser's Proems, the unusual fifth has a truly dramatized speaker, one whose distinct attitude and socially familiar voice claim attention for themselves, apart from the allegorical narrative, before abruptly, ambiguously, and only at the end ceding place to it. Spenser's speaker is unremittingly pessimistic, moving as if on script from one dispiriting observation to the next with hardly a pause for stanza breaks. Like Donne's speaker in *The First Anniversarie*, Spenser's compares an antique age of virtue with the lamentable present and complains about the apparently irrational movements in the heavens—once again, the precession of the equinoxes, the obliquity of the ecliptic, and the orbits of the planets, all of which could be fully rationalized and usually were, as Spenser well knew. He summons hearsay and wizards' tales to buttress his scientific argument, and he despairs about a loss of meaning, specifically a lack of correspondence between word and thing:

> For that which all men then did vertue call,
> Is now cald vice; and that which vice was hight,
> Is now hight vertue, and so vs'd of all:
> Right now is wrong, and wrong that was is right,
> As all things else in time are chaunged quight.
> Ne wonder; for the heauens reuolution
> Is wandred farre from, where it first was pight,
> And so doe make contrarie constitution
> Of all this lower world, toward his dissolution.[15]

Evoking the relationship of macrocosm to microcosm, the final lines of this stanza assume the flow of influence from higher to lower, greater to lesser, and at least to this extent they retain a traditional order amid decay.[16] Although pervasive irony in this Proem supports a reading of Spenser's speaker as a fin-de siècle worrywart, a case against it has a basis in the recurrence of this speaker's sense of degeneration within the narrative of Book V as well as in the Mutabilitie Cantos. Instead of a simple endorsement or rejection of the speaker's view, these larger contexts invite the reader's judicious assessment of it, as befits Spenser's Book of Justice.

The extent to which Donne's speaker in *The First Anniversarie*, not unlike Spenser's, participates in a staged performance can be underestimated. The facts that the decay of the world was a highly conventional topic both anciently and in the late sixteenth century, that Donne often exploits convention creatively, and that he repeatedly dramatizes a voice or persona in his poems and often assumes roles in his sermons enforce the likelihood of his doing so in both *Anniversaries*.[17] His openly assuming a role in *The First Anniversarie* might be said to lessen his need to do so as obviously in *The Second* or equally to suggest that his role in *The Second* is less fully and openly a mask. Likewise relevant to Donne's role, again especially in *The First*, is the conventional presence of paid mourners in the society of Donne's time. Mourning could be a public performance and still be meaningful as an expression of sympathy, respect, and community as well as an opportunity for mutual charity between mourner and bereaved patron.[18] Robert Drury's patronage of Donne, along with the fact that Donne had never met Drury's daughter, the dead woman he mourns so extravagantly, invites sympathetic, not simply cynical, evocation of such staged mourning.

Confronting "here and there . . . a sense of strain and a relative lack of clarity" in *The First Anniversarie*, Milgate implies that Donne might already have "conceived, or even partly written, a satiric 'anatomy', for the completion and publication of which Mistress Drury's death provided the occasion" (xxxiv). Earlier, Louis L. Martz, likewise having complained about a lack of control and about clumsy transitions in this poem, had remarked that if only the parts of it that he had identified as meditations were taken together and the rest of the poem (eulogies and moralizations) were omitted, the result, if given a conclusion, "would form a complete—and a rather good—poem."[19] Milgate's and Martz's observations, which are typical, underscore the extent to which the parts of the poem that decry the world seem to have a separate, prescripted status, one that could openly signal its staging of, not simply its participation in, a conventional topos.

Similarly critical of *The First Anniversarie*, Frank Manley, like Milgate an editor of the *Anniversaries*, occasionally includes the acknowledgment in his notes on *The First* that the passages in question are "difficult," "compressed," or "obscured."[20] For example, in one complaint about the disproportioned cosmos, Donne's speaker objects to the human desire to tame and possess the visible heavens instead of expending the effort required on earth to find a higher truth:

For of Meridians, and Parallels,
Man hath weau'd out a net, and this net throwne
Vpon the Heauens, and now they are his owne.
Loth to goe vp the hill or labor thus
To goe to heauen, we make heauen come to vs.
We spur, we raine the stars, and in their race
They're diuersly content t'obey our pace.

(FA 278–82)

The development of these lines is energetic but also restless and disorderly—in short, somewhat irrational. The speaker's metaphors begin with the fabrication and casting of a net for fish or game, then shift to the climbing of a hilltop in search of truth, apparently a recollection of Donne's "Satire 3," before finally turning into the forceful riding and controlling of astral horses, possibly with an ironic and contrasting memory of the most visionary moment of Plato's *Phaedrus*.[21] The overall effect of these successive metaphors is to devalue the scientific investigation of the visible heavens in favor of heaven itself.

Ben Jonson is notoriously reported to have remarked that "Donne, for not keeping of accent, deserved hanging," thus calling attention to Donne's occasional disregard for an idealized formality.[22] Embracing a concept of imitative form that need not be perfectly regular or, indeed, always proportional in a single, certain way—as Donne's metrics are not—I wonder whether a poem that decries the loss of all coherence might stage a dramatic speaker who is somewhat lacking in this feature—not enough to ruin the poem but enough to invite thoughtful assessment ("What's wrong here?") rather than merely sympathy ("Poor John"). There are plenty of Shakespearean examples of this sort of negative, disorderly, irrational decorum: Lear, Othello, Leontes, and Hamlet, to name some. The admirable efforts that have been made to outline the structure of Donne's poem might in their very variance be to some extent their own undoing. If more than one structure is reasonable and credible, the possibility of the others persists as a nagging doubt.[23] The situation is not wholly unlike the availability of various astronomical and cosmic theories that worries Donne's speaker. Multiplicity undermines surety. More meaningful, perhaps, is the fact that in actually reading (and repeatedly rereading) at least *The First Anniversarie*, awareness of much more than recurrence and therefore of structure is limited and only inescapable in the refrains about the lost *She*. Insofar as structure is perceptible in reading, moreover, it provides an ironic undercurrent of order. Again, insofar.

Donne's presence in the *Anniversaries*, most obviously in *The First*, is mediated by a persona, indeed, to a significant extent a conventional mask, and as a result, it is less intellectually and emotionally autobiographical than is sometimes assumed. Although my stressing the role of a persona in the *Anniversaries* is not new, it bears anew on criticism in which this conception is inactive, and it comes in an argument couched in different terms.[24] A persona, a role, a mask, a ventriloquized voice—as you will—qualifies but does not cancel Donne's personal investment in these poems or the relevance of the prose texts written by him a little before or soon after them: in addition to *Ignatius His Conclave* (1610/11), the letters from Mitcham (1607–09), the prefaces to *Biathanatos* (1608) and *Pseudo-Martyr* (1610), *Essays in Divinity* (1614?), and the sermons, which begin in the spring of 1615.[25] Considering this chronology, the *Anniversaries* offer a tempting pivot from Donne's years of melancholic uncertainty to those of settled faith. A little too neat, perhaps, and too Waltonian. Neither *Anniversarie* is the transcription of the prose texts in the unmediated way the annotation of parallels sometimes implies. Many of Donne's other secular and religious poems strike resonant chords with the facts of his life even while they frustrate demonstrable autobiography, and famously at times, Donne puns on his own or his wife's name in his poems, thus signaling and inviting a connection to his life even while doing so ambiguously and elusively from within a poetic fiction. To argue for a persona or dramatic speaker in both *Anniversaries*, as in other poems by Donne that typify his self-presentation, is to recognize the significant differences, not just the similarities, between both these elaborate fictions and the prose texts surrounding them. The major difference, aside from the poetic medium, might well be the sustained, meaningful presence of a persona in both, although most crucially and conspicuously in *The First Anniversarie*.

"A Nocturnal upon S. Lucy's Day, being the shortest day," the best and most personal of Donne's short elegiac poems, affords an instructive point of reference for his persona in *The First Anniversarie* and serves to introduce it. In the opening stanza, his speaker's anguished response to a lover's death suffuses the world outside him. His speaker pictures a world seen through the lens of despair:

> 'Tis the year's midnight, and it is the day's,
> Lucy's, who scarce seven hours herself unmasks,
> The sun is spent, and now his flasks
> Send forth light squibs, no constant rays;
> The world's whole sap is sunk;
> The general balm th'hydroptique earth hath drunk,
> Whither, as to the bed's-feet, life is shrunk,
> Dead and interred; yet all these seem to laugh,
> Compared with me, who am their epitaph.[26]

The speaker's point of view moves from the darkness of midnight and spent sunlight to a sickroom and beyond, or rather, beneath it—from dying to gravesite, from the condition of earth to individual extinction, from outside to inside, and from general objectivity to specific subjectivity. In the last line, the ghostly speaker, as himself the epitaph, has become the last word on dying, death, and burial, the last utterance on them and of them. He is devoted, dedicated, to death. In the second stanza, the speaker offers himself for study, as if for anatomy, and notes his alchemical rebegetting "Of absence, darkness, death; things which are not." Only in the third stanza can he begin to differentiate the death of his love from his own, and only here do readers actually know that someone besides the speaker is involved or that the death which has occurred is physically real. In the fourth stanza, he names her death, finally acknowledging it directly, even while parenthetically still trying to deny it: "By her death (which word wrongs her)," he has now become the elixir of nothing (28).

The similarities of "A Nocturnal" to *The First Anniversarie*, or *Anatomie*, are evident: the despairing viewpoint, its enveloping the whole world, its fixated gaze, its avoidance of naming, its desire to dissect, to expose, to cut open and apart. These are appropriate to a perception confronting but not yet emotionally able to accept the death of a loved one, and they are often more realistic, we might suppose, than pious assurances about heaven. As Donne acknowledges in *An Elegie vpon the death of the Ladie Marckam* (1609), tears of earthly passion overflow our banks in the face of a beloved's death, and even the soul's tears "Take all a brackish tast, and funerall" (6–10). Nothing has worth or pleases. *A Fvnerall Elegie*, published with the *Anniversaries* and generally considered Donne's earliest poem about Mistress Drury's death, follows a suggestion of her final resurrection from the grave with a questioning of its present consolation:[27]

May't not be said, that her graue shall restore
Her, greater, purer, firmer, then before?
Heauen may say this, and joy in't; but can wee
Who liue, and lacke her, here this vantage see?
What is't to vs, alas, if there haue beene
An Angell made a Throne, or Cherubin?
We lose by't.

(45–51)

Referring to the lack of a woman he has never seen, this speaker openly and sympathetically performs the conventional role of a public mourner. He gives voice to the family's feeling of loss.

Differences between *The First Anniversarie* and "A Nocturnal," aside from those of genre and length, are also evident: the absence in the *Anniversarie* of the personal, erotic connection between the speaker and the dead woman and the corresponding absence of the progress during the course of the lyric from the anguished internalization of death to its gradual distancing and externalization. If there is some small degree of progress in *The First Anniversarie*, it lacks emotional conviction. The speaker's lack of intimate, personal bereavement and his expansive use of the conventional topos of worldly degeneration also help to account for the inescapable impression of excess, which is enforced by the occasional difficulties, compressions, and obscurities that Manley and others have observed. This impression simultaneously generates critical distance and offers a corrective. The poem tactfully enables those grieving, notably the Drurys, a way to a better perspective by offering as a mirror for grief a venting of irreplaceable loss and by generalizing this loss to the world, at the same time at least partly deflecting it from the young woman. Indulging despair, recognizing this excess, and at least beginning to see what is wrong with the picture it paints are a way forward.

Relevantly, George Puttenham observes of "Poeticall lamentations" that they are altogether contrary to rejoicing, as "euery man saith":

> and yet it is a peece of ioy to be able to lament with ease, and freely to poure forth a mans inward sorrowes and the greefs wherewith his minde is surcharged. This was a very necessary deuise of the Poet [both past and present] and a fine, besides his poetrie to play also the Phisitian, and not onely by applying a medicine to the ordinary sicknes of mankind, but by making the very greef it selfe (in part) cure of the disease.[28]

The second sentence indicates that Puttenham has in mind not only "a mans" own sorrows but also another man's. He later adds that the poets have sought "by their arte to remoue or appease [sorrows], not with any medicament of a contrary temper, as the Galenists . . . but as the *Paracelsians*, who cure . . . making one dolour to expel another" (63). Notably, Puttenham's analogy connects a physical theory of medicine to a psychic one, a textbook transference further analogous to that of physical to intellectual vision, recurrently examined in my earlier chapters. His observations are a good fit for Donne, whose interest in medicine was pronounced and lifelong.

At the end of *The First Anniversarie*, the speaker might not be quite where he began, but he still seems to be stuck in the past, the Old Testament, the body, and this ruined world. After more than four hundred lines, he even seems to be overhearing his own rant and tiring of it: "the worlds carcasse would not last, if I / Were punctuall in this Anatomy. / Nor smels it well to hearers" (FA 439–41). As occasionally before in the poem, however, there are also some squibs of light at the end, intimations of the promise to be found in the past, in the Old Testament, and in death itself: "the soule of man" is "borne but than / When man doth die. Our body's as the wombe, / And as a mid-wife death directs it home" (FA 451–54). The appearance of constructive analogy in these lines—midwife:womb :: death:soul—is promising in itself, and it looks forward to *The Second Anniversarie*, where a variant of it occurs with grace as a third, intervening term (SA 214–15). Yet the analogy also seems somewhat abrupt and perfunctory in its present context, a little too much like an add-on and thus like the rest of the conclusion. For me, it recalls the brief reminder in an earlier passage to "feed (not banquet) on / The supernaturall food, Religion," which incongruously arrives—noticeably interrupted by the speaker's parenthetical swipe at superfluity—in the midst of a cascading lamentation for the loss of the heart, the very center of vitality (FA 187–88). It is as if the speaker's own heart were not in his reminder of religion at this earlier moment. If so, then this moment, as well as the concluding analogy of the midwife, are further examples of a form whose elements, such as timing, tone, and even syntax, mark it as imitative and self-consciously reflexive at noticeable points. At the end of *The First Anniversarie*, the speaker, who overhears and finally tires of his own excess, has found a Mosaic sense of purpose stronger than his original one, which was overwhelmingly to chasten the world by its worthlessness. Like Moses, however, he has failed to reach the promised land. He only offers fame at the end, not redemption.

The First Anniversarie: Macro/Microcosms and a Drury

With the speaker's general trajectory in *The First Anniversarie* in mind, I want briefly to look at the analogy between the macro- and microcosms, a commonplace that conspicuously informs *The First* and, with a difference, is also found in *The Second*. Doing so asks for preliminary background both with respect to recent scholarly discussion and to rhetorical theory. The relation of macro- and microcosms has variously been described as an analogy, a metaphor, a metonymy, and a synecdoche and, in a subtle essay by Marshall Grossman on both *Anniversaries*, as all of these.[29] Each of these rhetorical figures might be considered a variant technique of expression, structure, and perspective. Each also participates in a basic similarity, a metaphoricity, the principle of figuration or troping itself—translation.

Recently, Elizabeth Harvey and Timothy Harrison have taken a rhetorical cue from Grossman to argue that Donne's *Anniversaries* "attempt to heal" the epistemological rupture between macro- and microcosms, or between object and subject, "by suturing together two tropes—analogy and metonymy."[30] Their essay is accomplished; my argument intersects with it recurrently, while also dissenting from their view; and the issues involved are important—enough so as to invite discussion. This starts for me with the terminology that grounds and informs their argument. What Harvey and Harrison call "resonance" is the result of the stitching, or suturing, that covers over a wound, as their figure of the suture indicates. To suture is to align surfaces and sew them together; to address or resolve what lies beneath is another matter. A surgical suture, moreover, is only required by a gaping wound in the flesh. *Resonance*, another figure, is a materially based principle of iteration that Harvey and Harrison link to Renaissance correspondence, magnetism, swerving atoms, signatures, words, and especially music, all to be found in the *Anniversaries* (982–83). They further relate the term *resonance*, which is theirs, not Donne's, to a variety of other Renaissance occurrences, most tellingly to the 1615 medical treatise of Helkiah Crooke, in which "'resonance or consonance . . . is nothing else but an ordinate or proportionable mixtion or temper of contraryes amongst themselves; . . . a kinde of military order or consent in disagreeing'" and thus involves proportion.[31] Crooke's context is musical inasmuch as it involves the soothing effect of music on crying children, and he evidently refers to harmonic proportion. Somewhat surprisingly, then, Harvey and Harrison identify analogy, or proportion, with "an equality of ratios," or a Euclidean quotient. They acknowledge its broader use by Aristotle but not its essential, constructive metaphoricity in the *Poetics* (985). But whatever the specific structure of analogy, Harvey and Harrison locate "the philosophical heart of the *Anniversaries*" in the "layered analogy of the soul's ascent," by which they intend the figures that succeed the ascent proper in *The Second Anniversarie* (982). They argue that these analogies lend body to an otherwise missing cosmic coherence and in this way become a metonymy, or substitute, for it. Herein lies the suture. For the speaker of *The Second Anniversarie*, whose soul is the imagined vehicle of the ascent proper, Mistress Drury thus embodies—physically—"the resonant principle" that provides "the world with an analogical medium through which distance can be bridged and a metonymic model through which all parts touch a common denominator" (993, cf. 996). So far so good, but not really.

Metonymy is rightly defined by Harvey and Harrison as substitution, yet it is also more narrowly described by them as the substitution of one category for another (986–87). In historical terms, this description displaces the past rather than simply connecting it to the present. Aristotle, for a classic, opposing example, gives the categorical transfer "from genus to species, or from species to genus, or species to species, or on grounds of analogy" as a primary example of metaphor.[32] Harvey and Harrison also attribute contiguity to metonymy as a prime, definitive distinction, invoking the authority of George Puttenham but actually assuming a dubious, modern alignment by Roman Jakobson.[33] For Harvey and Harrison, metonymy then becomes Luce Irigaray's trope of touch, physical touch, and thus the rematerialization of a Donnean turn, that is, a trope, away from materiality. In this way, they dissolve the trope, or figure of speech, back into its material and more specifically bodily origin.[34]

Admirable, complicated, and, for my purpose, representatively materialist, Harvey and Harrison's project expresses what they characterize as the embodied perspective of modernity in which, so to speak, neither faith nor intellectual vision has legs. From this perspective, as they necessarily conclude, "Donne's imaginative exercise is . . . doomed" to failure (1002). Their view raises questions about particular readings, which will occur in subsequent notes and involve the status of figuration, of difference, and of narrative movement, or progression. But the broader question is whether faith is a historical reality—again, historical—and what an answer means for the interpretation of Donne. Another basic, related question is whether imagination has any purchase on human reality, then or now, and if so, of what sort, and a third question, likewise related, concerns the capacities of human language and most especially of analogical language.

Metonymy, as indicated in Chapter 4, is not only substitution but also cultural code. To summarize: in the classical rhetorics on which those of the Renaissance rely, metonymy, like all the other tropes, is traditionally a subcategory of metaphor, also known as translation. Metaphor/translation is the umbrella term for all the tropes as well as the name of the specific trope metaphor: this is to say that all tropes are in some way metaphorical—translations or transfers. The specific trope metaphor is constructive and creative, not merely coded and substitutive, and analogy is a version of metaphor. Over time, however, metaphors (the specific trope) can be frozen into dead metaphors or become coded metonyms: the metaphorical analogies of the body politic and of the macro/microcosm afford pertinent Renaissance examples. These lack tropic life, either wholly or to a considerable extent by Donne's time; increasingly they are perceived as nonfigurative—"literal" is a popular word for this development nowadays; "natural" is another.[35] Yet dead metaphors can also be reawakened, revived, and modified, becoming again an active source of enlightenment. But history never repeats itself without difference; if it did, time will have stopped.

In short, hard borders between the tropes are deceptively taxonomic and to write of a need to suture them risks dramatic overstatement. With use and in use, tropes have fuzzy edges; they shift and readily merge in the process of thinking. Classical rhetoricians make the point that taxonomic definitions are mainly helpful for the education of those unskilled in rhetoric, such as students. It is not until after the Enlightenment that tropes are culturally straitjacketed, and similitude/*similitudo*, perhaps the broadest figure of all, morphs into the narrowly defined modern *simile*, a comparison using *like* or *as*. Moreover, the status of rhetoric as a kind of second nature fades first in, then further after, the nineteenth century, and as psychoanalysis becomes a disciplined practice, metonymy becomes more prominent and privileged—an ideological participant in much larger social and cultural changes. In short, as I have argued throughout this book, rhetoric has a significant history that participates in that of thought, both reflecting and shaping it in different periods.[36] Its use is not historically neutral. What looks like a suturing of tropes in the modern period might be something quite different in an earlier one.

In Donne's *First Anniversarie*, the dominance of a relationship between the macro- and microcosms is everywhere evident. This relationship is a cultural metonym as familiar to the seventeenth century as the metonyms "White House," "Downing Street," "the Vatican," or "the Kremlin" are to us. Donne has his speaker employ the macro/micro metonym with important differences, however, which might be said to stir, if not fully to awaken, its ancient, analogical metaphoricity. His speaker confuses and perverts the flow of influence between macrocosm and microcosm: "She to whom this world must it selfe refer, / As Suburbs, or the Microscosme of her, / Shee, shee is dead" (FA 235–37). The *She* referenced here, the macrocosm "whose rich eyes, and brest, / Guilt the West Indies," could hardly be imaged more bizarrely and materially; if she is greater than the Indies, she is materially so—not analogically higher or other but merely the same (FA 229–30).[37] Within a providential context, the flow of influence is normally positive, not grounded, as in this *Anniversarie*, in absence and death, "things which are not." The speaker's reversed perspective, effectually from the lesser to the greater, the nearer to the farther, the individual to the general, is defining, and if we stop with this reversal, it is not only fallen but also unredeemed and, from another perspective, quite modern.[38] In *The First Anniversarie*, however, such reversal suits and rises from a worldview overwhelmed by grief and aligns with the twilight existence the speaker acknowledges, as well as with the original, pre-Adamic corruption of the created world that he later imagines (74, 193–96).

Although Elizabeth Drury's name appears only on the title page of each *Anniversarie*, prior to each of Bishop Hall's prefatory poems, its absence from the text of the poems is itself notable, the conventionality of this absence notwithstanding.[39] Donne, creatively exploiting convention in the instance at hand, has made the absence of Mistress Drury's proper name from the body of the poem significant. The meaning of her name pertains to that of the much-debated but nameless *She* within the poems themselves. The given name *Elizabeth*, a Greek transliteration of Hebrew meaning "oath of God," becomes, in William Camden's *Remains concerning Britain*

(1605), "'Peace of the Lord, or Quiet Rest of the Lord,'" coincidentally a gloss relevant to Spenser's Mutabilitie Cantos.[40] Implicit in Camden's gloss is the peace that passes understanding, the Sabbath, or "rest," at the end of time. This is the Sabbath to which the end of *The Second Anniversarie* looks. In the Bible, Elizabeth is the name of the mother of John the Baptist, the prophet—figuratively, the messenger or trumpeter—of Christ, a rough parallel to the paternity of Donne's muse (Mistress Drury being too chastely virginal for maternity) and to his speaker's trumpeting at the end of *The Second Anniversarie* (SA 34–36, 527–28).

The surname *Drury* is further significant, connecting with the many references to riches, particularly in *The First Anniversarie*. This name, or word, in the Middle Ages signified a gift, beloved person, keepsake, or treasure. Early in William Langland's *Piers Plowman*, published four times between 1550 and 1561, the striking figure of Lady Holy Church tells the dreamer, "Whan al treasures are tried . . . truth is best . . . It is as dere worth a drury [treasure], as deare God him selfe."[41] The *OED* (2014) cites less immediately pertinent uses in 1575 and 1594 in Scottish texts, which appear to conflate *drury* and *dowery* (that is, *dowry*). Notably, Chaucer also uses *drury*, referring to sexual love, in *The Tale of Sir Thopas* and *The Romaunt of the Rose*. Donne's interest in the meaning of proper names, including his wife's and his own, has already been mentioned, and any reader of his *Essays in Divinity* or his sermons notices his pronounced interest in the significance of biblical names, such as those of God and Adam. With particular reference to the meaning of *drury* as "treasure," biblical verses from the Sermon on the Mount found in Matthew 6:20–21 and Luke 12:33–34 are remarkably suggestive: "lay up for yourselves treasures in heaven, where neither moth nor rust doth corrupt, and where thieves do not break through nor steal: *for where your treasure is, there will your heart be also*" (Matt. 6:20–21; my emphasis).[42] The end of this passage seems a good fit for Donne's *Drury* and particularly for the section of *The First Anniversarie* that focuses on humanity's loss of its heart with her death (FA 173–74ff.)

Within the text of *The First Anniversarie*, however, what is insistently, inescapably stressed is the loss of Mistress Drury's name, which is simultaneously the loss of the very word *drury*, since in Renaissance usage *name* and *word* are synonyms. Despite this loss, the *Variorum* edition of the *Anniversaries* has an exhaustive section on the nameless *She*, which results from the mass of scholarly attention to her significance (6:293–317). A very few scholars have argued that *She* is "a blank counter," later updated to an "empty signifier" or a blank screen, but most have wanted to fill the void that the speaker relentlessly laments, and one could even argue that the poem invites (and foils?) their efforts.[43] What has nonetheless emerged as a credible consensus regarding what *She* represents is, with varying twists and emphases, the image of God.[44] Yet it is loss, namelessness, the lack of a center and so of meaning, stability, and identity that the speaker insistently voices, as he addresses a sick and apparently dying world: "But this is worse, that thou art speechlesse growne. / Thou hast forgot thy name, thou hadst; thou wast / Nothing but she" (FA 30–32). He then elaborates, "Her name defin'd thee, gave thee forme, and frame . . . yet none / Offers to tell vs who it is that's gone" (FA 37, 41–42). The point is that Mistress Drury's loss represents something missing, gone, and not exactly determinate: a pure earthly origin, meaning, the cement or balm of all virtues, the heart of everything, a binding force, coherence, beauty, color, proportion, symmetry, harmony, correspondence, comprehension—Donne's speaker tries them all. Earthly existence has no value or meaning in her absence; its center is missing. This is the point! All that lingers is a twilit memory, the "glimmering light" of her unnamed ghost (FA 70, 74, 370).

Approaching the section of *The First Anniversarie* that focuses persistently on disproportion in the heavens, Donne's speaker advises his reader that the only way to avoid "The worlds infection . . . [is] to be none of it" (245–46). Observing that even "the worlds subtilst immateriall parts / Feele this consuming wound, and ages darts," he rejects the best the world has to offer (FA 247–48). These "subtilst parts" include light, considered immaterial in the Neoplatonic tradition and by Kepler, and perhaps also aether, subtler even than light in Robert Fludd's reading of Plotinus, as are the equally subtle aetherial rays of the stars, themselves of aetherial, quintessential substance that is often identified with *pneuma* (Greek for "breath, wind, vital spirit").[45] To explain the ravages of age, Donne's speaker turns next to the decay of beauty, defined both as color, whose visibility requires light, and as proportion, which, in the next fifty lines and more, largely means sphericity. In Kepler, light and sphericity are in the Godhead as archetypes. They radiate from the center of being into this world and share an interface with matter. In contrast, Donne's speaker, looking at the physical heavens from earth, is not interested in abstraction, archetypes, or geometry, unless in their visible absence. He sees only appearance and the "disproportion [of] that pure forme," which should be but is not "their Sphericall / Their round proportion embracing all" (FA 251–52, 257). As of the sun's imperfection, "So, of the stares [planets] which boast that they do runne / In Circle still, none ends where he begunne. / All their proportion's lame" (FA 275–77). Repeatedly, the speaker scoffs at efforts to measure movement in the heavens or to "save the appearances," eventually coming down to earth, whose surface is disproportioned, too, and he further scorns the absurdity of a physical hell within the center of earth and the disproportionality—again the halting, mismatched members, the lameness—of worldly justice (reward and punishment). There is even a touch of petulance in the speaker's tone, and, to put it colloquially, the wheels are beginning to fall off the wagon. Now coming as if ironically full circle, the speaker concludes "That beauties best, proportion, is dead" (FA 306). As if caught in a moment of self-reflection, he then adds that "euen griefe itselfe, which now alone / Is left vs, is without proportion" before throwing himself into another mournful tangent, to use a geometrical word suited to the form of this cosmological section (FA 307–8). Once again, Donnean form appears to imitate substance.

Before leaving *The First Anniversarie*, I want to linger briefly over two odd passages, one near its beginning, the other near its end. The first considers the beneficent effect of Mistress Drury's memory on those who knew and valued her, a memory inspiring in them a "faint weake loue of vertue and of good" (FA 71). Those who follow her example are proof against inbred corruption:

(For all assum'd vnto this Dignitee,
So many weedlesse Paradises bee,
Which of themselues produce no venomous sinne,
Except some forraine Serpent bring it in.)

(FA 81–84)

The notion of a weedless paradise in a fallen world anticipates Milton's "paradise within," an inner perfection. Yet its jingling presence (sin/bring it in) in a lengthy parenthesis between the conditional clause that precedes it and the apodisis that follows it in the surrounding sentence (not cited) is syntactically intrusive and distracting; in short, the parenthesis is conspicuous both for these reasons and because it offers a literal, or historical, impossibility: paradise has long since been lost, and as Donne's speaker later suggests, the whole world, like mankind, was "almost [already] created lame" anyway because the angels had already fallen (FA 191–95). This is a no-win condition for absolute, totalized innocence, as Milton, too, will realize: *innocence* means "not" (*in*) "nocent" (< *nocens*, "harm-full") and therefore negates something prior to it. The intrusive parenthesis about weedlessness in *The First Anniversarie* sounds merely hyperbolic or like a dissonant joke—the comment of a speaker who cannot waste the piece of wit that has just occurred to him, however weakly motivated. The tone of the parenthesis fits oddly with its immediate and broader contexts and participates in a larger feature of this poem: whenever religious belief is mentioned, or inserted, there seems to be something problematical about the utterance—perfunctory, abrupt, inadequate, dissonant. For example, Eden itself never really exists (finds expression) in the poem but decays from its first hour; therefore this paradise is present only to the extent that it is not (FA 192–202). In addition to the weedless and Edenic paradises, other examples so far in this chapter have included the speaker's injunction to "feed (not banquet) on . . . Religion" and the brief analogy of birth to rebirth in his conclusion. If Donne's speaker is really a Christian, he is an unconvincing one.

The second strange passage exhibits a word that our own age might call a hot button: "though she could not *transubstantiate* / All states to gold, yet [she] *gilded* euery state" (417–18; my emphasis). Gilding—earlier, more awkwardly, and in a more bizarre context, spelled "Guilt," that is, *gilt* (230)—suggests shiny superficiality, not real change; it is hardly an unqualified supplement. (Its unfortunate spelling in line 230, while not necessarily authorial, further qualifies the earlier compliment.)[46] Few words in the Reformation are more notoriously controversial than *transubstantiate* and its cognates. In connection with the general process of *transmuting* baser metals to gold, "transubstantiate" suggests alchemy here, but its more specific evocation of Eucharistic controversy is at once inescapable and relevant to the kinds of lost relationships Donne's speaker mourns.[47] The transubstantiation that *She* cannot effect indicates the limitation of human power specifically with respect to the matter of the physical world. Insofar as this allusion is inescapably religious, it specifies the Roman Catholic doctrine that the bread and wine of communion are changed (*transubstantiated*) into the body and blood of Christ through the priest's words of consecration, which have been empowered by God. This change is substantial—actual and real—and it pertains to the substances of the supper themselves, not simply to the faith or perception of the priest or communicant. In this way, it is considered objective—out there. Simultaneously, however, it defies physical perception—sight, touch, and taste—since the appearances, or formal surfaces, of bread and wine remain. Renaissance Catholics considered the Reformers carnal because they did not believe that God had such power over matter. The Reformers considered the Catholics carnal because they did not believe that faith by itself was enough but needed to objectify it.[48] To the Reformers, the Catholics' real presence requires substantive expression and is therefore impervious to the truth of the spirit. In contrast to the Catholics, the Reformers might be said to have granted more autonomy to faith but simultaneously also to matter; their perception of matter answers overwhelmingly to the testimony of their senses. This kind of perception was a sign of the times, and Donne's speaker is its voice in *The First Anniversarie*. Matter can only be gilded/guilted, not transvalued—that is, in some way moved across (*trans*) the barrier, the gap, the space between this sensory world and any another possibility. Once again, Donne's speaker is grounded only in the material world and the body of death in this *Anniversarie*.

The most promising intimation of all at the very end of *The First Anniversarie* lies in another word, which is the final characterization of *She* and in this sense her epitaph, at least for now. This word is "incomprehensibleness," and it has rightly been glossed as "illimitableness" and thus as finally beyond our rational understanding, or, figuratively, beyond our grasp (469).[49] Not coincidentally, "incomprehensibleness," like "infinity," is rooted at once in the physical world and in its negation, in this instance the Latin verb *prehendere*, "to grasp, seize, catch," and *com*, "together" (cf. "prehensile"), the combination of these negatived by *in*, or "not." Literally, that is, physically, it suggests the grasp of a hand. The significance of *She*, incomprehensibleness, is well beyond this physical grasp. In a word, then, we are back to the beginning of the poem, in which *She* is beyond definition, or limit, and so quite nameless since to name is to know, or to understand and comprehend—to grasp. Nearly five hundred lines have not changed this basic fact. Unlike the planets, of which "none ends where he begun," moreover, Donne's loquacious lamenter, in thus completing a circle, has at last come back to what he earlier said and in doing so has begun to grasp it anew. His understanding now allows for something infinite, since incomprehensible and illimitable, and therefore for something beyond the limits of reason. Momentarily, like a flash of light, this allowance comes in a word six lines from the end of the poem. Nonetheless, it finally comes.

The Second Anniversarie: Outset to Visionary Ascent

The Second Anniversarie, a companion to *The First*, is generally considered the superior poem. It is certainly more buoyant and coherent even while its speaker, operating within a looser, less monotonously iterative form, conveys greater spontaneity. If by its end he has assumed the public role of a prophet and preacher, for most of the poem he has exhorted his own soul, repeatedly calling on her to rouse herself, to think, imagine, and ponder.[50] Effectually he has preached a sermon to his soul, as the conclusion of Donne's "Hymn to God my God, in my Sickness" characterizes this practice. He has practiced what he will eventually preach. He has been in exemplary mode but also in exploratory mode, engaged in a sort of thought experiment analogous to those of contemporary cosmologists.[51] His experiment is intellectual and also metaphysical insofar as it exceeds, or passes beyond, the known world of sense.

Although the opening of *The Second Anniversarie* focuses again on a world that is effectually dead, a series of similes (in the modern sense) that illustrate the world's paradoxical state of inanimate motion exhibits a good deal of poetic life, or *energia*. This motion is natural, but it would make a good analogue for mechanical life. The ship that strikes sail but keeps moving, the beheaded man whose members twitch and contract, the audible crackling of ice as it melts, and the cracking sound of strings breaking on an otherwise silent lute: all these are comparable in the poem to the lively motion of rotting and thus like the life-in-death that paradoxically still remains in the carcass—the fallen flesh (Latin *caro*, "flesh," and *cadere/casum*, "fall")—of this world. With a bow to the dead woman's power to inspire poetry, specifically the present poem, the speaker now appears more openly from the outset as a poet, a true maker rather than merely a dissector. As such, his thought needs only a chance word, *Venite*, a welcoming summons, to turn to the Resurrection, the time when God will summon his faithful to him (44):[52]

> Thirst for that time, O my insatiate soule,
> And serue thy thirst, with Gods safe-sealing Bowle.
> Bee thirsty still, and drinke still till thou goe;
> T'is th'only Health, to be Hydropique so.
> Forget this rotten world.
>
> (SA 45–49)

Like the Latin word *Venite*, "Come," these lines are full of memories, for example of Joel, whose third chapter locates the traditional site of the last day in the Vale of Jehoshaphat. Joel's chapter focuses first on judgment rather than on the restoration of paradise, but Donne's use of it here emphasizes health and salvation. Joel 3:11–18 reads in relevant part, "Assemble yourselves, and come . . . for the harvest is ripe: come get you down; for the press is full, the fats overflow; . . . And it shall come to pass in that day, that the mountains shall drop down new wine, and the hills shall flow with milk, and all the rivers of Judah shall flow with waters, and a fountain shall come forth of the house of the Lord."

As an analogue, the following lines from Langland can represent interpretive tradition; in them, Christ is speaking:

For I that am lord of lyfe, loue is my drynke,
And for that drinke to day I dye apon earth
I faught so me thrusteth [I thirst] yet for mans soule sake,
May no drinke me moyst ne my thurst slake,
Tyll the vendage [vintage] fall in the vale of Jsaphat
That I drinke right rype must [very ripe wine] *Resurrectio mortuorum*.
....
And all that bene my hole brethren in bloud and bapti[s]m
Shall not be dampned to death that is without ende.

(Cc.ii^{v})[53]

Often in the sermons, Donne voices Christ or effectually blends into him, a practice sanctioned by the familiar exhortation to be Christlike, in some sense actually becoming him—bathing in his tears, drinking his blood, lying down in his grave, and rising again with him.[54]

The most salient line in the redemptive turn in the attention of Donne's speaker is "serue thy thirst, with Gods safe-sealing Bowle," an obvious reference to Eucharistic communion, a mutuality, a sharing, a communication, a coming together. Referring to the wine of the Eucharist, not the bread, he avoids the more usual locus of sectarian controversy while quietly making a Protestant gesture: only Protestants observed lay communion in both kinds, bread and wine, rather than bread alone. Moreover, while Donne's bowl lacks the excited sensuousness of Keats's grape bursting against the palate, it uses the Eucharistic element that has the greater appeal, whether pungent or sweet, to the sense of taste. In this way, the communion for which the speaker thirsts insatiably in this world joins sense to salvation within the context of faith. More emphatically within this context, by evoking the Supper the poetic speaker also moves into metaphor, even analogical metaphor. Excepting Luther himself, the leading Reformers agreed that the Supper *crucially* involved figurative language, and their various views gathered under the umbrella of translation, or metaphor.[55] Finally, Donne's speaker combines an *affirmation* of sense with an *affirmation* of spirit. That these affirmations occur early and affectively in *The Second Anniversarie* and do so in connection with faith offsets the paradoxes of merely lifelike death with which the poem begins and sets the context in which the poem and its soul will progress. The course, indeed the narrative, of this progress is essential to the poem's meaning. Conversely, the plucking out of its parts can produce an alternative and static poem, even more so when the significant difference between the two *Anniversaries* is neglected.

Coming right after the speaker's thirst for salvation, however, another of his addresses to his soul might seem ambivalent, and it has led to the assertion of his backsliding into doubt about the immortality promised by Christ and sealed (certified) by the wine of the Eucharist—"Gods safe-sealing Bowle."[56] The speaker, having observed that this world is no more than a carcass, warns his soul, "thou art fed / By it, but as a worme, that carcas bred": here "as" means "as if a worm" rather than merely "in the same way," unless it refers in the latter sense to the inferior, organic soul, which encompasses the vegetative and sensitive workings of the body (SA 55–56). Meaning "as if," the word "as" signals counterfactuality rather than the unlikely identity of the inorganic, intellective soul with a material worm.[57] In the next line, the speaker then asks,

> And why shouldst thou, poore worme, consider more,
> When this world will grow better then before,
> Then those thy fellow-wormes doe thinke vpone
> That carkasses last resurrectione.
>
> (SA 57–60)

Rather than thinking the soul merely a material worm in an unaccommodated literalism, the speaker addresses his soul affectionately, sympathetically, and also a little playfully, as if to a child, much as Donne's speaker addresses first a flower and then his own heart in "The Blossom": "Little think'st thou, poor heart." Intimate address ("thou"), tone ("poore worme"), and hypothetical situation (the subjunctive "shouldst"), following the counterfactual "as," make all the difference in the inset from *The Second Anniversarie* cited just above. Those "fellow-wormes" can as easily and figuratively be ensouled persons as merely worms and can as easily and ironically contain a rebuke to the soul's fellows. Neither Donne nor Donne's speaker supposes that worms "do thinke." Figurative language is crucial to *The Second Anniversarie* in a way, to an extent, and with a complexity that is not found in *The First*, which is stuck, excepting only squibs of light, in a twilit, physical world.

When next Donne's speaker invites the soul to consider at length a scene in which the body dies, which is a traditional subject for meditation on the last things, salient figures include, first, death as a groom with a lighted taper; second, a death rattle in the throat that is punningly a "Diuision," both a separation of body and soul and a welcome melody; and third, the rotting body. Once buried, this body becomes a prince whose "Wormes," now courtiers, "insensibly deuoure their state," the body politic (85–118). Each of these is an analogy, the first and last (groom, prince) most obviously, the one with the pun more intricately: a death rattle is to the body as music to the soul. The pun on "Diuision" compresses two dimensions of reality into a single word, comprehending both in the same time and place, as well as both in the imagined scene and on the printed page, thus uniting them temporally and spatially. The implicit analogy underlies and extends the insight encapsulated in the pun. Again, as before in *Venite*, a verbal dimension is conspicuous and operative.

Having imagined the body dead and buried, if only for "a saint Lucies night," in the next movement the speaker begins by praising the harmonious constitution of Mistress Drury while living and ends with a mnemonic variation on the figure of death as a groom (SA 119–21, 156). Death, still retaining the familiar role of a household servant, now becomes an usher with the key to unlock a door (SA 156). Geometrical analogies cluster in the middle of this movement, and for the first of two times in *The Second Anniversarie*, the word *proportion* appears (SA 141). It does so in relation to the perfectly balanced constitution of Mistress Drury while she lived, not to bemoan its absence so much as to enforce the point that no one, not even she, can enter heaven except through the door to which death holds the key. Here, in the living Mistress Drury, a "proportionable mixtion or temper of contraryes amongst themselves," like Helkiah Crooke's, is recalled and celebrated at length before it is found wanting. Preeminently, in *The First Anniversarie*, *proportion* was associated again and again with the loss of significant relationship, such as cosmic correspondence. At this point in *The Second*, it still is, as it will not be the next and last time it occurs, yet now it also participates in a strong, positive memory and avoids sinking into a wail of despair. The speaker, now more poet than dissector, more maker than deconstructor, has himself achieved a much better balance. Remembering that in Kepler's *Paralipomena* the principle (*ratio*) of balance is analogical—that is, proportional—lends another promising glimmer of renewed vision to the recurrence here of the intellectually charged word-concept *proportion*.

The geometrical analogies in this movement introduce a level of abstraction not seen before in *The Second Anniversarie* or in *The First.* There, mathematics applied to the astronomical measurement of the visible heavens by meridians and parallels whereas here mathematics involves pure geometry: lines, points, cubes, and circles per se. Donne's speaker employs this geometry to express the even greater purity, the perfectly proportioned balance, of Mistress Drury's constitution while living. His lines will have later repercussions:

And as, though all doe know, that quantities
Are made of lines, and lines from Points arise,
None can these lines or quantities vnioynt,
And say this is a line, or this a point,
So though the Elements and Humors were
In her, one could not say, this gouernes there.

(SA 131–36)

Her constitution might even have been compared favorably to the sun's or almost to a spirit's unity, and to her "proportions if we would compare / Cubes, th'are vnstable; Circles, Angulare" (SA 137–42). Donne's speaker seems almost to lack a Keplerian faith in geometric form, yet not quite. The most stable and most perfect of such forms, respectively the square and the circle, indeed pale in comparison with the even balance of Mistress Drury's elements and humors. Donne's speaker first summons pure geometric forms and, only by analogical extension, the physical bodies they could be applied to measure—here, the constitution of Mistress Drury's body—and thus in thought joins this pure abstraction to flesh, yet he does so while disqualifying geometrical analogy for comparison, unless by contrast. This is the negative way. Even so, his analogies cannot and do not attempt to deny that she, too, has sickened, which is evidence of a weakness, or to deny that she, too, has had to pass through death before claiming heaven. Directly acknowledging these inescapable human and religious conditions, he tactfully returns to his own soul to locate their reason in original sin and to try to understand its origin.

The speaker's explanation of inherited sin is a product of controverted thought, not of certain knowledge, the lack of which the speaker will later remind his soul. He will do so in no uncertain terms:

Poore soule in this thy flesh what dost thou know.
Thou know'st thy selfe so little, as thou know'st not,
How thou did'st die, nor how thou wast begot.
Thou neither knowst, how thou at first camest in,
Nor how thou took'st the poyson of mans sin.

(SA 254–58)

The emphasis on knowledge or, rather, on its acknowledged lack is evident in these lines, as is the fact that they back-reference both the speaker's earlier imagining his own death, as here, "in this thy flesh" and then his thinking about his soul's origin (SA 157–68). In contrast to certain knowledge, his earlier thoughts about conception and inherited sin were clearly speculative, offering one theory of origin after another.[58] Rather than contradicting the speaker's later acknowledgment of ignorance, they support it. Think, he tells his soul in the earlier passage, think on your making, the very instant of your conception in the filth of the body, the simultaneity of your origin with that of the infectious, fallen body; then, think about the competing theory, your origin directly from God, and the implication of your drawing into yourself the lower souls (vegetative and sensitive, that is, the organic soul) and thereby being infected by original sin (SA 157–68). After seven exhortations in roughly twenty lines to think thus disparagingly about his soul, the speaker shifts again to death and with it to positive thinking: "But thinke that Death hath now enfranchis'd thee, / Thou hast thy'expansion now and libertee" (SA 179–80).

What follows is generally acknowledged to be the affective high point of the poem, the imagined ascent of the speaker's soul to heaven. Since the soul's gender is traditionally feminine, a reader easily forgets that the ascending soul is not specifically the dead woman's, although hers is additionally implied by association. The expansive and soon ecstatic ascent takes off with an analogy of the soul to a bullet shot from a rusty gun, figuratively the body, which bursts with the shot; then comes an analogy of the soul to a newly hatched bird that breaks free from the shell of the body (SA 181–84). Free of the body's downward drag, this soul

> which late did cleaue,
> To'a body, and went but by the bodies leaue,
> Twenty, perchance, or thirty mile a day,
> Dispatches in a minute all the way,
> Twixt Heauen, and Earth.
>
> (185–89)

In the course of the next ten lines, she, the transcending soul, "carries no desire to know, nor sense" what might or might not be in the intervening "Ayre": meteors, the element of fire, inhabitants on the moon, the planets. In a word, she is unaffected by any of the concerns of astronomy and cosmology: "sweet Mercury / Works not on her" (SA 199–200). Here, Mercury is a notoriously erratic planet, the study of whose swift movement engaged Kepler, and simultaneously, Mercury is also a mythic figure, at once god of travel and "Hee that charm'd Argus eies" (SA 199). Mercury is also "the 'ingenious' god of the probing intellect, sacred to grammarians and metaphysicians, the patron of lettered inquiry and interpretation . . . the revealer of secret or 'Hermetic' knowledge, . . . the divine *mystagogue*."[59] Instead of being worked upon, or influenced, by Mercury, however, the soul herself "now is growen all Ey" (SA 200). Now all-seeing because no longer tied to data from the organs of sense, the soul has come into her own; punningly, she has become all "I," her unencumbered, true self, "Who, if shee meete the body of the Sunne, / Goes through, not staying till his course be runne" (SA 201–2). Faster than the speed of thought, his soul, "ere shee can consider how shee went, / At once is at, and through the Firmament" (SA 205–6). Climactically, after more than two dozen lines of ascent, the phrase "through the Firmament," followed by a full stop, is unexpected, quick, dramatic. "Firmament" indicates the limit of the visible world, beyond which is eternity or, in the old astronomy, the crystalline sphere(s), the Primum Mobile, and the Empyrean, the heavenly realm of God.[60]

Since the verb "consider" puns etymologically on stars, Latin *sidera*, which belong to the last sphere the soul leaves behind, it leads associatively into the lines that follow, which initiate the analogical sequel to the immediacy of transcendence. In some degree this sequel sustains yet also begins to explain, expand, and understand transcendence and in doing so begins to descend through decreasing intensity into reason: "as . . . as . . . so . . . for . . . (parenthesis) . . . as." (Analogy in Cajetan, as might be recalled from my fourth chapter, falls under semantics and logic.) Length alone suggests both the speaker's effort to sustain the vision for another dozen lines and its inevitable, gradual loss of thrust (SA 207–18). Reason becomes further evident with the causal conjunction "For" ("because") and conspicuous with the addition of an explanatory parenthesis:

> For when our soule enioyes this her third birth,
> (Creation gaue her one, a second, grace,)
> Heauen is as neare, and present to her face,
> As colours are, and obiects, in a roome
> Where darknesse was before, when Tapers come.
>
> (SA 214–18)

The final use of "as" in the penultimate line, along with the last line, returns the poem both to the earlier deathbed scene and specifically to its image of the groom with a taper, which began the speaker's meditation on the last things. Whereas the earlier taper offered "a little glimmering light," the present vision, though ending, through analogy remains focused elsewhere, on heaven. The point is that the vision has been accomplished, not that it has had to end, as do all earthly visions, which are subject to time and progression and must be temporary until and unless time itself stops.

The abiding value of such a vision, however, is necessarily also temporal. It is at once subject to and, more importantly, available in time. It is human.

The ascent of the soul is framed on either side by a wonderful, reflexive series of analogies, bullet and bird at the start, beads and bones prior to the concluding light of the taper:

> And as these stars were but so many beades
> Strunge on one string, speed vndistinguish'd leades
> Her through those spheares, as through the beades, a string,
> Whose quicke succession makes it still one thing:
> As doth the Pith, which, least our Bodies slacke,
> Strings fast the little bones of necke, and backe;
> So by the soule doth death string Heauen and Earth. . . .
>
> (SA 207–13)

In the first line of this passage, "as" is again counterfactual, signifying "as if," and "these stars" refer to the planets (also called stars) and the fixed stars in—not beyond—the firmament.[61] Using analogy, the speaker now reflects on the flight of the soul that he has already expressed excitedly in narrative. The verb "were" is a past subjunctive in accord with the opening condition "as if," and so the verb "leades," a historical present, occurs in a sentence that counterfactual fiction and verbal tenses qualify, if almost unnoticeably. The poem can no longer trace an ascent since the soul has passed beyond the firmament, the eighth sphere. The first analogy also moves from "these" to "those," from nearer to more distanced and from visible stars to conceptual spheres, the plurality of "spheres" including those beneath the firmament, not only the firmament itself. The vision is gradually winding down and losing the immediacy at the apex of its ascent.

The peculiarly striking phrase "speed vndistinguish'd" indicates an unearthly, unworldly speed because it is unmeasured and immeasurable by points of reference, in this instance the astral bodies and spheres. Like Keplerian light and the ascent of the disembodied soul, the speed it describes is instantaneous. It suggests Augustine's threefold comprehension of time, of past and future held together in the attentive present, and therefore offers an insight into timelessness, as well.[62] It also recalls Zeno's related paradox in which the linear trajectory of an arrow at any given moment consists of still points, in which case the arrow is motionless while moving, and it therefore combines movement and rest, speed and stillness. Most immediately, the soul's undistinguished and unmeasured speed recalls the speaker's earlier elevation of Mistress Drury's idealized yet embodied proportionality beyond that in geometry and, more exactly, beyond the conceptualized geometric forms that points and lines constitute. The speaker's puns, first "quicke,"—speedy, living—and then "still"—always, motionlessly, immutably—and their combination at once with "succession" and "one thing," the many and the one, movement and stasis, join together to straddle time and eternity, containing them both in a single line of verse, a continuity of words and rhythmic stresses within a single unit. Whereas in the narrative of the soul's ascent, the speaker earlier relied on bodies within the heavens to lend substance by negative illustration—what the soul did not see, do, or suffer—to the flight of the soul, now he employs these bodies figurally and constructively. The passage is metaphorical and more specifically analogical, not literal, and therefore not limited to physical bodies unless we choose to deny any working, transfiguring power to language—that is, any hypothetical, provisional, creative vision and, in short, any significant *poiēsis*.

The word "as" in the first line—"And as these stars were but so many beades"—does not mean just "in the same way"; it reaches poetically beyond this way to "as if" even while including the alternative meaning. Momentarily, "ere shee [the soul] can consider how," the word "as" thus lets us glimpse the nonfictive alternative before confidently excluding it upon notice of the subjunctive "were" a mere three monosyllabic words later. Psycholinguistics indicates that the seemingly instantaneous choice of words and meanings virtually always involves alternatives, although selection usually occurs prior to full consciousness.[63] The heightened awareness of composition, subject to revision and further heightened by artful *poēsis* in the present instance, surely crafts selection and variously can ignore, contain, or exploit the agency of the signifier. Moreover, because the opening analogy, like those that follow, participates in metaphor, it acknowledges difference, as well as similarity, not to mention the shaping imagination of Donne's analogizing speaker.

With each successive analogy, the speaker gradually settles down to earth. The first analogy of this series uses beads, which have been compared to prayer beads (not limited exclusively to Catholicism) but which are otherwise aesthetic objects, humanly crafted, made.[64] In the second, "bones" replace the beads. A "string," or thread, connects the first two analogies; in the first, the string is "speed vndistinguish'd," and in the second, it is the "Pith," or spinal cord. Here, "pith" is a carefully chosen word belonging to the natural world of plants and trees prior to its use in the English Renaissance also for the spinal cord. (*Cord* has a different, more artificial origin; the *OED* dates its specifically anatomical usage from the late eighteenth century.)[65] Perhaps surprisingly, even after earlier representations of death in the poem, it is death "by the soule," by means of the soul, which, analogized to both these strings, strings earth to heaven in the final, most abstract, least imagistic analogy. Again, death is the liberator and here also an agent, as earlier in line 179. Its agency is surprising since death is the epitome of passivity (being done to, suffering: Latin *pati/passus*). Yet it is only with death that the soul can take flight, truly coming into its own, and its flight is the third string, the complementary means-by-which, in the series. This means-by-which, like death, is active. Well and good, yet this last analogy seems to have something odd about or extra in it.

Here, flat-footed, formulaic representation is helpful: beginning with the translation of stars to beads and then to spheres, (A) these: (B) speed undistinguished :: vertebrae:pith :: death:soul :: earth:heaven, all of which are instances of stringing.[66] The third and final stage of the analogical series is structurally more complex than this alignment suggests, however: if the final line I have cited, "So by the soule doth death string Heaven and Earth," is reexamined, its syntax says two different things at once: the soul is the string between heaven and earth, and death is the string, or, death : soul :: soul:death. This last, inverted analogue should be added to those just above: these : speed undistinguished :: vertebrae : pith :: death : soul :: soul : death :: earth : heaven. Without the soul, death does nothing; without death, the soul goes nowhere. Alone, neither analogue has the power to string heaven and earth together. With such a heightened emphasis on death, it becomes more striking and significant that the exposed vertebrae in the second analogy constitute an image of death, the anatomized corpse, the backbone dissected and laid bare. Looking back at the image of beads that precedes this of bones, Shakespeare's line in *The Tempest*, "Those are pearls that were his eyes," comes to mind, illuminating the relation of poetic making to death and of death to poetic making that Donne's lines also engage.[67]

If establishing structural analogy formulaically can clarify meaning, by itself it also slights the dynamism of progress through the passage at hand. Structure, like other forms of analytic abstraction, is hardly identical to a reading experience: we quickly grasp the final analogy about death and the soul as a whole when we read it, although I suspect that we fully understand it only with thoughtful analysis. In the passage involved, the experience of reading responds especially to active verbs, "leades," "makes," and "strings"; to the quick movement through a succession of images, initially without end-stopped lines and then stopped only long enough to adjust the perspective earthward and further deathward; and to the range and emotive resonance of simple words like *speed* and *pith*, *beads* and *bones*, *heaven* and *earth*, together with the stimulus of intellectual conception, as in "speed vndistinguish'd" and its analogical sequel, the pithy string of the spine, as the passage descends from ecstatic vision, through the excitement of intellectual conception, to embodiment and more rational argument, only to return at the end to the light of tapers, with which the soul's paradoxically "long-short Progresse" earlier began and now begins anew (SA 219). The verbal tense at the end still lingers in the present: "colours are" and "Tapers come." Then the speaker, who has been in the background all along, intrudes openly, once more to address his soul.

As elsewhere in Donne's writing, perhaps most familiarly in *Deaths Duell*, his final sermon, which he delivered as a dying man, death is, through Christ's, the mother of redemption. Donne's and, in *The Second Anniversarie*, his speaker's engagement with death is conventional and Christian, but it is also in its emphasis, intensity, and purposiveness highly individual. So much is well known. What is less often stressed, however, is that the very *physicality* of death—ever granted the grace manifest in faith—is precisely what enables redemption. The mortal body is thus salvific, a necessary participant in the soul's immortality, its ultimate redemption. This is not simply to say that Christ suffered in the body but to grasp this knowledge as a truth felt along the very pith and bones of the spine. The difference is at once cognitive and affective, like faith itself and, in this way, also substantial. It is at once a remarkable reconception of the body that is achieved (and granted) in the progress of the poem. Martz noticed long since that the parenthesis in the lines "For when our soule enioyes this her third birth, / (Creation gaue her one, a second, grace,)" which adds a birth through the granting of grace, is a major difference between this *Anniversarie* and the end of the first one (SA 214–15). But what grace crucially enables is also the redemption *from*, *in*, and above all *through* physical death that Donne's speaker discovers in this second poem—Latin *à* (that is, *ab*), *in*, and *per* death, as again and again Donne the preacher will insistently and emphatically put this same point in his last sermon, indeed his last word, after a lifetime pondering death.[68]

The Second Anniversarie: Proportion in the Time That Remains

That the ascent of the soul occurs just prior to the middle of *The Second Anniversarie* leaves the speaker to make something of it in the time and space that remain. He is back in the body, though still addressing his soul, and the grieving parents are still there to be pleased and comforted. The first thing he does is to praise Mistress Drury's body at length, finding it so fair, so rich, and so perfectly unified that "one might almost say, her bodie thought" (SA 246). The narrative of the poem, its progress, from vision back to earth, is precisely what has enabled this perception "almost" of unity, which will return more elaborately near poem's end. The body, even the purest of postlapsarian bodies, was not and could not have been seen this way before—certainly not in *The First Annniversarie* and not even in the relatively more balanced *Fvneral Elegie*, in which, however, the body's substance is minimized, almost vaporized, in favor of the soul and its thinking:

One, whose cleare body was so pure, and thin,
Because it neede disguise no thought within.
T'was but a through-light scarfe, her mind t'enroule,
Or exhalation breath'd out from her soule.

(59–62)

In *The Second Anniversarie*, praise of Mistress Drury's near-unity next gives way to a still-longer reflection on the poor soul's lack of true knowledge not only of itself but also of the body or of anything else that is truly useful. Entomology, history, biology, and physiology are all brushed aside, as finally anything originating in sense data is. Instead, all knowledge and all libraries are now to be found in Mistress Drury's soul, now in heaven, which is "our best, and worthiest booke" (SA 320). Perhaps again surprisingly, when the speaker next addresses his soul imperatively, it is to urge her not to return "from this extasee, / And meditation of what thou shalt bee, / To earthly thoughts" (SA 321–23). I imagine that most modern readers might have thought that the ecstasy, this standing outside of one's usual place and self, had ended with the vision, but apparently not wholly. The condition of return to earthly concerns that the speaker imposes on his soul is a general reform of society: divines, aristocrats, courtiers, and everyone their corruption affects. Apparently not expecting this reform, he turns with reawakened energy to an unearthly state: "Vp, vp, my drowsie soule, Where thy new eare / Shall in the Angels songs no discord heare" (339–40). Now, the speaker, having earlier been uplifted by the imagined vision of his ascending soul, listens with an Augustinian's inner ear, and his thoughts are directed to higher things, including especially the exaltation of Mistress Drury's soul in heaven: he would hear "Angels songs" and the music of the spheres, which is inaudible to ears of flesh even in Kepler's harmonically proportioned representation of it in *Harmonice mundi* (340).[69]

I want to pause for a moment to emphasize that the occurrence of the ascent of the soul has made a real difference, that it is effectually a kind of sea change, and I do so because this fact, along with the importance of the narrative of the poem, is currently too little observed. Consider this example, complementary to the appearance of the imagined possibility of an inner ear: after the soul's ascent, the word *joy* (including cognates) appears no less than twenty times. In neither *Anniversarie* does it earlier appear, save the exception of one compromised, manic instance near the beginning of *The First Anniversarie* (20). The reverberation of joy in the latter half of *The Second Anniversarie* registers a transformation in tone and perspective. Words, which, as acoustic images, combine sound and space and which also convey concepts and afford their application, have an enhanced status in this *Anniversarie*. Their enhancement is a virtual signature of Donne's own faith, as it is later found in *Devotions* and the sermons.

Meditation on the inconstancy of earthly beauty and honor follows the reenergizing of the speaker's soul until he exhorts her, "to thy first pitch worke vp againe," that is, make an even greater effort to sustain higher thoughts and to recover true vision (SA 435). To observe that the speaker has to will this effort as an objection to its validity is both to overlook the traditional renunciation of worldly goods as part of the commitment to a religious calling and to forget the volition, the effort, faith asks on the part of the believer, ever allowing for prevenient and sustaining grace. Once again in the poem, a geometrical figure returns, whose abstraction in itself at this juncture is highly significant. The geometrical figure analogously relates diametrical lines within a circle to thoughts of heaven:

Know that all lines which circles doe containe,
For once that they the center touch, do touch
Twice the circumference; and be thou such.
Double on Heauen, thy thoughts on Earth emploied.

(SA 436–39)

"For once" means both "for a single time" and "for every time." It embraces the former ascent of the speaker's soul and ascents still to come. The recurrence of circular form in Donne's poems need hardly be stressed: "A Valediction: forbidding Mourning" affords a familiar example. A circle, the form of perfection and completion, is traditionally used to represent the movement of the soul, eternity, or God himself. Its center rests on a stable point equidistant from all points on its circumference; this is the center that was lacking in *The First Anniversarie*, where, since the center did not hold, coherence was gone. All diametrical lines cross this central point in the circle, and all radial lines spread out from and come back to it. It represents origin and return and the centrality of meaning and being, as in Kepler. The circumference, another figure of divine perfection since it has neither beginning nor end, is also the expansion, expression, and in some sense the completion of the central, stable point: as there is no circle without a center, so there is none without a circumference. As lines that touch the center touch the circumference twice as often and twice as much, thus evoking both a memory of time and number and a promissory Keplerian intimation of these in the very symbol of eternity, so the speaker urges his thoughts while on earth, which have already once ascended in soul-flight to the boundary of the firmament and affirmed a passage beyond it, to redouble their efforts to reach heavenward again and again.[70] It is hardly a coincidence that this speaker is heard next instructing his soul about the "essentiall ioye" of beatific vision, in which the sight of God is at last the unity of object and intellect (SA 440–42).

Returning to celebration of the deceased woman for a final time, the speaker focuses once more on her soul, which becomes an abiding *connection* between her joy, devotion, and commitment to heaven when she was on earth and its present fulfillment there. Explicit mention of her "precontract" and "Betroth[al]" to God while she lived and her present "marriage" in heaven recalls the meanings "dowry" and bodily "love" potentially available, along with the meaning "treasure," in the word *drury* and now translated into another register or, in short, transfigured (SA 460–62). The continuity of her earthly and heavenly life has also become exemplary, inspirational, and imitable on earth, not simply lost and gone forever or only a twilit memory, as in the earlier *Anniversarie*; instead, "to posteritee . . . [she will] for life, and death, a patterne bee" (SA 523–24). Of course the connection consists of another analogy. Of greater significance, however, is the fact that this time analogy is specifically acknowledged, its validity affirmed self-reflexively in the final movement of this poem, ever allowing for faith. For the last time in either *Anniversarie*, the word "proportion" recurs and does so uniquely as a positive synonym for "analogy" and thus, once more, as a positive connector between heaven and earth: "shee to Heauen is gone, / Who made this world in some *proportion* / A heauen" (SA 467–70; my emphasis).

In the earlier comparison of Mistress Drury's "proportions," as she "was," favorably to cubes and circles, her ideal balance was more simply a memory of her attainment on earth, as well as a plural and so relatively more material balance, at that (SA 141–46).[71] In the present passage, she is still gone, to be sure, and the remainder of the poem stresses differences between heavenly and earthly existence, but more than once it also suggests the possibility of a difference in degree and duration, not simply in kind, between them: "But could this low world ioyes essentiall touch, / Heauens accidentall ioyes would pass them much"; and again, "Onely in Heauen ioies strength is neuer spent; / And accidentall things are permanent" (SA 471–2, 487–88). In the first quotation, the subjunctive "could" opens the door a crack, as does the transformation of accidence to permanence in the second: a little earlier, the speaker has even allowed that "Shee . . . is gone before. / Shee, who had Here so much essentiall ioye" (SA 448–49). The possibility of a proportional relation between heaven and earth has evidently returned, although access to its realization or, indeed, to that of joy is neither easy nor universal: "All [that is, everyone] will not serue" (SA 440). The word "serue" suggests the multiple meanings of George Herbert's characteristic use of it in his poetry: "make do, suffice, satisfy" but also "do service, make an effort, work hard."[72] In the present context of Donne's poem, it dominantly signifies "work at it, have the capacity, find vision." The return of a proportional relationship is highly significant, but it is also thus qualified realistically.

Circling back to the general resurrection, that "last great Consummation," the speaker's attention shifts to glorified "earthly bodies," the bodies that will then ascend, and next to the admission that "Shee . . . left such a body, as euen shee, / Onely in Heauen could learne, how it can bee / Made better" (SA 491–94, 501–3). This is the immediate context in which another celebrated and controverted analogy occurs:

> for shee rather was two soules,
> Or like to full, on both sides written Rols,
> Where eies might read vpon the outward skin,
> As strong Records for God, as mindes within;
> Shee, who by making full perfection grow,
> Peeces a Circle, and still keepes it so. . . .
>
> (SA 503–8)

The word "rather" in the first line means both "more accurately," "more properly," and also "earlier, before the present time," and so when she was living. Succeeding lines extend and qualify the assertion of dual souls in this line: when living, her skin, figuratively the outer soul, is analogized to her intellect, figuratively the inner one, and the result is analogized to a parchment roll with writing on both its sides: outer soul:inner soul :: skin:intellect :: one written side:other written side.[73] Quite unlike the analogy of the woman's living body to a transparent scarf in Donne's initial

Fvneral Elegie, here the body has equal substance and, more broadly, has equality with the mind. Moreover, the juncture, the near-unity, of body and soul is as striking in Donne's image as in Saussure's modern analogy of concept and acoustic image, or signified and signifier, to the recto and verso of a single page. Yet the analogy in the passage at hand is also subject to reading, or interpretation, in line 3, and it involves writing, both as the surface of the present anniversary poem and as part of its content—namely, the "written Rols." The rolls, presumably made of parchment/vellum to endure, convey both an official and an antique association and most likely allude to a biblical, or sacred, context, perhaps to the double-sided scroll of Revelation 5:1.[74] The sanctity and the two-sidededness of Donne's image conceivably adapts a single-sided, parchment Torah scroll to the recto-verso relationship of the Old and New Testaments and thus also to the dominant emphases at the end of *The First* and *Second Anniversaries*, respectively. The "written Rols" further recollect the earlier imagery of words, writing, books, and libraries in *The Second Anniversarie* but do so with a notable difference: unlike paper, parchment is made from the skin of a dead animal.[75] Like the earlier image of the exposed vertebrae in the vision of the soul's ascent, it has a connection to life (Drury's) but only through death.

Yet "only" by itself is misleading here since the parchment in the image of the rolls has writing, words, on it, and these are specifically designated "Records for God"; that is, they figure forth evidence (expressions) of faith. "For God" can variously be taken as "for God to read"; "on behalf of God," "concerning God," and so effectually "of God"; and perhaps even as "for God's immanence" or providential presence.[76] Records are lexically also reminders of what is written on the heart, Latin *recordari*, "to remember," from *cor/cordis*, "heart." Memory and the heart, it need hardly be noted, are earlier much present in both *Anniversaries*. This image of written rolls is a recollection, a mnemonic gathering, and also, importantly and specifically—in my own play on *verbum*, "word"—a further reverberation.

The last two lines in the passage cited return again to a circle, this time connecting its geometric perfection to growth and thus to the dynamism of life itself: "Shee, who by making full perfection *grow*, / Peeces a Circle, and still keeps it so" (my emphasis). She enhances and completes a circle—piecing it—by bringing life to its abstract perfection; growth necessarily implies both living and change.[77] Whereas earlier Donne's speaker stressed the incommensurability of circles and cubes with her embodied proportions, here he recalls, corrects, and betters his former, negative comparison. With the verb "Peeces," followed and reinforced by the phrase "still keepes it so," his speaker also moves from the past, Mistress Drury as she was when alive, to the present. This is the time of growth and, indeed, of new life that is enduring. The line immediately succeeding "Shee . . . still keepes it so" is "Long'd for, and longing for'it [heavenly perfection, she], to heauen is gone" (SA 509). In the end, even desire, "longing," which was present only as a negative, a lack of interest in astronomy, during the ascent of the soul, is back as the erotic, Christian-Neoplatonic connector between heaven and earth, between the soul's longing for God and God's for the soul, and implicitly the mutual longing for reunion of the Drurys on earth and of their daughter Elizabeth, their treasure, their *drury*, in heaven.[78] Memory itself is redeemed by this eros.

CHAPTER 7

Milton's Twilight Zone

Analogy, Light, and Darkness in *Paradise Lost*

Culturally, associatively, and naturally, light brings with it reference to darkness, as day to night and life to death. The symbolic valences of night and light, together with their phonic likeness, occupied Milton throughout his writing, but especially after his blindness, as his blind protagonist eloquently testifies in *Samson Agonistes*. Various half-lights, such as twilight, dusk and dawn, shade and shadow, enter the picture, too. These belong neither to light nor to darkness, neither exclusively to day nor to night; in them, light and dark combine suggestively. In *Paradise Lost*, Milton's figure of Death is recurrently called a shade or shadow, but shades and shadows are often positive, enabling, and protective in the poem. After the Fall, for example, Milton's narrator rues Adam and Eve's disgrace in paradoxical terms: "thir Eyes how op'nd, and thir minds / How dark'nd; innocence, that as a veile / Had shadow'd them from knowing ill, was gon."[1] Underlying the ironic resonance of these lines is the traditional analogy between physical and mental vision, body and intellect, although its presence now conveys inner loss, not hypothetical access to supersensible knowledge. Because of this loss, the veil of innocence whose shade had at once protected and enabled the Adamic pair's vision of truth is gone.[2] Shadow and light evidently do not have fixed values in *Paradise Lost*. Nor does shadow's relative, darkness.

In Miltonic epic, darkness, light, and other related terms can radically be affected by situation, speaker, and perspective. In *Paradise Regained*, Milton's poetic narrator will borrow a term from astronomical optics, namely "Parallax," to refer to these modifying conditions of utterance; in *Paradise Lost*, his narrator will use an unstable series of perspectival analogies—bees, ancient giants, dwarfs, exotic pigmies, and moonlit elves—to sum, climax, and reflexively question the reality and location of hell as his own opening book portrays them.[3] Earlier in the same book, the Miltonic narrator had drawn a disorienting analogy between "the *Tuscan* Artist," Galileo, viewing the moon through an "Optic Glass," and his own and his readers' viewing "the broad circumference" of Satan's shield (286–89). All the analogies mentioned so far bear in some way on the satanic.[4] In the book that begins with the Invocation to Light and ends with Satan on the sun, the Miltonic narrator will similarly, if ironically, analogize this "Fiend" to a sunspot the like of which "Astronomer . . . Through his glaz'd Optic Tube yet never saw" (III.588–90). Such epic similes, to use the familiar term, are well known to function like multiple telescopic, microscopic, or otherwise affecting optical lenses, enlarging, diminishing, or in other ways both altering and enabling what we perceive in the text and the manner in which we respond to it.[5]

That epic similes are fundamentally analogical and that by being so they participate in a multidisciplinary cultural issue regarding knowledge is less familiar, perhaps, as is the fact that through optics this issue bears on light. To study light scientifically in the seventeenth century and prior to it was to study optics. While my present concern is not the influence of Kepler on Milton, although there is evidence of some, it is worth recalling again that a striking feature of Kepler's writings is his interest in the interface of the immaterial with the material that light effects, as well as with correspondences and other connections between the celestial and terrestrial realms.[6] Kepler's cosmos is interconnected and continuous despite his evaluative differentiation between material and immaterial phenomena. Not surprisingly, analogy pervades his work.

Books of Darkness

Because situation, speaker, and perspective, which in narrative are always developing, bear crucially on the values of light and darkness, there is every reason to approach Milton's book of light, the third, from the contrasting books of darkness, the first two. Beginning in medias res is an epic convention, but a poet intending to fly "Above th' *Aonian* Mount" knows both that convention must be shaped to his purpose and that any beginning has a defining effect (I.15). When reassessment occurs, moreover, it does so against the earlier background. A reader of Genesis 1:3–5 might even suppose that God had instituted as fundamental the principle of contrast between light and darkness, day and night, on the first day of creation. Shifting to the moral symbolism of light and darkness, which pervades the Bible, a reader of Milton's early prose will pertinently remember the famous passage in *Areopagitica* in which Milton endorses the principle of contrast for his own world: "As . . . the state of man now is; what wisdome can there be to choose, what continence to forbeare without the knowledge of evill?"[7] Momentarily, I also recall Kepler's "praises of shadow" because it is by contrast, Kepler explains, that we approach a scientific knowledge of light. Barring a few analogies, Milton's first two books hardly address astronomical discovery, but they offer instead an experiential knowledge of loss and evil.[8] It is only by traversing these books of darkness that we reach the books of light. In this sense, moreover, we initially accompany Satan. In fact, at the very start of Book III, "Hail holy Light," we might even suppose for a benighted moment that Satan—last seen beholding "This pendant world" and then, in the final line of Book II, hastening to it—is speaking (II.1052, 1055). This effect is surely deliberate. In *Paradise Lost*, we approach light from the darkness that the present section explores.

Well before Satan first raises his head from the burning lake in hell, what appears to be an alternative to pitch blackness comes from flames: indeed, hell itself "As one great Furnace flam'd" (I.62). Instead of firelight, however, the flames strangely provide "No light, but rather darkness visible . . . discover[ing]"—that is, uncovering, disclosing—only misery: "sights of woe, / Regions of sorrow, doleful shades" (63–65). Such darkness is preeminently inner, affecting "*The Passions of the minde*," yet it also indicates the sights, shades, and regions of this place that are seen—or rather, beheld, taken in.[9] It is pointedly available to vision understood as interior but also as exterior, or eyesight. Satan's prison is further depicted as a condition of "utter darkness . . . far remov'd from God and light of Heav'n" (71–73). Although "utter" means "outer," a reference to place, simultaneously it means "extreme" or "total" and thereby refers to an interior state as well. This extremity of darkness is defined by its distance at once from God and heavenly light, the two already linked in the narrator's lexical doublet "God and light of Heav'n." Notably, this extremity is not identified as *black*, a word reserved for the more specific uses subsequently to be discussed in this section.

Soon after these initial depictions of hell's landscape, Satan couches his first exclamation of loss solely in terms of light. He addresses Beelzebub, his closest ally and here a mirror of himself:

> But O how fall'n! how chang'd
> From him, who in the happy Realms of *Light*
> Cloth'd with transcendent *brightness* didst out-*shine*
> Myriads though *bright*.
>
> (I.84–87; my emphasis)

It is as if Satan, remembering this brightness, were already, almost instinctively, trying to replace God with light, in all its charged meanings, as if he could somehow hold onto the one without the other. Such thinking later takes a more open and perverse turn when, in the Great Consult of Book II, Mammon demands, "As he [God] our darkness, cannot we his Light / Imitate when we please?" and thus explicitly voices the dualist claim for a principle of light and another of darkness, which Satan will transport to the created world.[10] Mammon would not only embrace the darkness of hell but even deny the power of light to define it.

Extreme darkness in hell, "utter darkness" amid "the glimmering of . . . livid flames"—bluish flames (Latin *lividus*)—is still not pure blackness, the seemingly total absence of radiating light (I.182). Kepler's remarks about blackness, examined in Chapter 5, are suggestive here. Although Kepler is committed to the generative interface of the immaterial with the material that light effects, he can still write that "light practices deadly enmity with all matter," especially with blacks, "which are as it were materialized darkness."[11] The opacity of black, which blocks the penetration of light, provides the reason for this enmity, which Kepler hedges with the counterfactual expression "as it were." Kepler's evident Platonism causes some friction with Milton's Edenic monism, which further embraces materiality, but hell is not unfallen Eden, and in the course of *Paradise Lost* we witness in Satan not only an increasing loss of radiance but also a progressive descent into evil, mere sense, and lower forms of matter.[12] Satan's incarnation in a snake (twice) is a low point in this process, but his son/grandson Death is a sign of its ultimate end. Materialized darkness, blackness, evil, and death are terms and ideas—word-concepts—already in play with respect to Book I, and with Book II, Night will join them.

Plotinus, the founder of Neoplatonism, wavered regarding the identification of matter with evil, and Proclus, a major proponent of Neoplatonism, later rejected it since, in E. R. Dodds's words, at times the Plotinian view of matter "seemed too much like hypostatising the Devil." In order to avoid the equation of matter with evil, Dodds explains, Plotinus had to "subordinate ethical dualism to metaphysical monism, by denying negative values" such as pain, disfigurement, and moral evil. In this view, evil is simply privative and ontologically dependent on good. Dodds suggests two alternatives to this Plotinian solution, the one a subordination of "metaphysical monism to ethical dualism"—the dualism of Good and Evil, mind and matter—which he labels Manichaean, and the other a Spinozist-Hegelian reduction of "both good and evil to a relativity taken up and transcended in the Absolute."[13] This last possibility is particularly suggestive with respect to the shadings of light and darkness in *Paradise Lost*, but the second is evident in the dualist claims of the demons, and the first is pertinent as well, although Milton would have his metaphysical monism and still acknowledge pain and evil in the fallen world, including blindness and death. In *Paradise Lost*, Milton's subtlest, most complex, and most interesting thinking comes in figures of various sorts that can elude, exceed, and attempt to reconcile the binaries of traditional Western thought.

With respect to the first possibility, evil as the privation of good, I noted in Chapter 3 that for Milton sin is malevolent negation—active perversion, with privation as its necessary consequence—and that "all evil . . . is encompassed summarily in Scripture under the sign of *death*."[14] Although Milton derives matter from God's own substance in *Christian Doctrine* and insists that it originated in an "incorruptible," if also a "confused and disordered," state, he also holds that once it has "become the property of another," it can be corrupted, or "infected [*contagionem*: 'contagion'] and polluted," as he puts it. The same is true of forms, which for Milton "are themselves material."[15] The emphasis on individual agency and responsibility in Milton's Christianity enables his insistence that matter in itself remains incorruptible in its essence, while in practice and history, it has been and is profoundly tainted.[16] Post-Edenic matter might basically be the same thing as before, but it does not come with the same moral, ethical, or otherwise relative value. At the same time, it has been (or will have been) redeemed; it is not simply lost but still matters, as does time past, time present, and time still to come.

In *Paradise Lost*, blackness is consistently and specifically associated with hell, the only exceptions, if they are that, being the habits of Black Friars in the Paradise of Fools, the menacing "black . . . thunderous clouds," then the "black Air / Accompan[ying]" night after the Fall, and, in the Archangel Michael's closing narrative, the figuratively black-winged storm clouds gathering for the biblical Flood.[17] Otherwise, outside hell, but not really, Satan, seeking the snake, is "Like a black mist low creeping"; earlier, in the War in Heaven, the rebel angels reduce the "originals of Nature" to "blackest grain," or gunpowder, and the Godhead, creating this world, purges downward "The black tartareous cold Infernal dregs / Adverse to life."[18] Also nominally outside hell, "a black bituminous gurge / Boiles out from under ground, the mouth of Hell" in Michael's account of Nimrod's Tower of Babylon, and nominally inside hell, the Miltonic narrator refers to "black *Ghenna* . . . the Type of Hell" during the Catalogue of Devils.[19] These last two references to blackness, the first in Book I, the second in Book XII, connect hell to earthly history after the Fall, as, in fact, the preceding set of three does, too, since they variously figure the Fall, God's wrath, and earthly corruption.

The remaining occurrences of blackness cluster conspicuously in Book II, which is situated in hell, then Chaos, although one of them, which I treat in the next paragraph, comes in Book VII and effectually recalls a particularly striking use of "black" in Book II. Book II's cluster is at once the most numerous in the epic and the most important, and it is a sign that the demons' evil is already progressing. Two instances of blackness occur during the Great Consult, the first when Moloch, recommending renewed assault, pictures the "Black fire and horror" to be shot among the unfallen angels with a rage to equal the Godhead's (67); the second, when Belial rebuts Moloch's proposal by describing the inevitable failure that would follow even if, in support of such force, "all Hell should rise / With blackest Insurrection, to confound / Heav'ns purest Light" (135–37). Belial's superlative "blackest" tops Moloch's "Black fire," and Belial's term "Insurrection" implies a broader, more rational concept than fire (especially Moloch's oxymoronically black fire), one that acknowledges rebellion against an established authority. Chronologically, however, in Moloch's and Belial's words alike lies a memory of the "blackest grain" deployed in the rebels' cannonade during the War in Heaven.

Following the description of "Sad *Acheron* of sorrow, black and deep" during the trek of the diabolical explorers through their "dismal world" (572, 578), the word *black* occurs twice in the confrontation of Satan with Death. Most significantly, it first appears in the narrator's initial description of Death, where he questions whether Death might be called a shape or even a "substance . . . that shadow seem'd" and then adds, "black it stood as Night" (669–70). This is the striking connection of Death with blackness that Raphael's admonition to Adam and Eve in Book VII will later reinforce: "govern well thy appetite, least sin / Surprise thee, and her black attendant Death" (546–47). Subsequently in Book II, when battle is imminent between Satan and Death, the narrator compares their frowns to "two black" thunder clouds. Coming near the end of the narrator's inflated depiction of their mutual inflation, the blackness attributed to both of them effectually encapsulates the fact that they mirror each other, and it implicates the deliberately inflated art that at this moment depicts them, as well (714). In this instance, the art involves analogical epic similes. A similar implicating of rhetorical art conspicuously occurs when Milton employs allegory in depicting the unholy trinity of Satan, Sin, and Death together in Book II and, after this episode, when he recalls allegorical form in Satan's gradual degeneration into an abstraction of Evil and finally into a snake, this last a final allegory of the evil mind's own place.[20]

The care evident in Milton's deployment of blackness throughout *Paradise Lost* implies his deliberation. Blackness is hellish, but the only character directly called black is Death, described so by Raphael, but Night, to whose blackness the Miltonic narrator likens Death's, is a close and intimate competitor for the term, as is the progenitor of Death, Satan himself. To define the blackness of Death by Night's—"black it stood as Night"—is to signal the affinity of these two figures, attributing essential, incomprehensible blackness to Death while making Night its more accessible comparative. Apart from Death, Night is not called black in Milton's two books of darkness or in the rest of the epic, with the qualified exception of that "black Air / Accompan[ying]" night as a sign of the Fall, whose penalty and outstanding consequence, however, is death. The close connection of Death to Night is important, and it will return in later stages of my argument.[21]

Besides fire, other kinds of light-but-no-light in hell include the "Starry Lamps and blazing Cressets" in Pandemonium, which "yielded light / As from a sky," and two epic analogies involving sunlight, which intimate an excess of significance but do so elusively—teasing us out of thought, to borrow Keats's phrasing (I.728–30).[22] One of these analogies occurs when Satan is about to address his fallen comrades for the first time. The Miltonic narrator works into his analogy with a description of Satan's "form [that] had yet not lost / All her Original brightness, nor appear'd / Less than Arch Angel ruind, and th' excess / Of Glory obscur'd" (I.591–94). Notably, it is Satan's form, associated with essence in Milton's thinking, that accounts for his preeminence, as now for his "excess / Of Glory obscur'd."[23] Analogies for Satan's condition immediately follow:

As when the Sun new ris'n
Looks through the Horizontal misty Air

Shorn of his Beams, or from behind the Moon
In dim Eclips disastrous twilight sheds
On half the Nations, and with fear of change
Perplexes Monarchs.

(I.594–599)

The contemporaneity of the two astronomical analogies and their bearing on science, astrological predictions, and sociopolitical events are evident.[24] For a handy, contemporary example, Kepler, who served his imperial patron as astrologer, astronomer, and mathematician, spends whole chapters in his *Paralipomena*, or *Optics*, on lunar and solar eclipses and related parallaxes. He calls the "*darknesses*" of eclipses "the astronomers' *eyes*" and adds that such "*blemishes* illuminate the minds of mortals with most precious *pictures*," observations that reinforce the principle of contrast earlier invoked in this chapter (16).

Yet the force of Milton's analogy is clearly negative, not laudatory: the phrase "Shorn of his Beams" abruptly cuts short the promise of a newly risen sun, and in the second stage of analogy, the dim light of a solar eclipse portends disaster. The first analogy looks to the past, to Satan's former preeminence and recent fall, and the second looks to the future, to the threat he poses in his desire for revenge. In short, the paired analogies fittingly reflect his character and situation. Yet the presence of the sun in hell, if only in analogy, is still a little surprising, especially after all the depictions of hell's "darkness visible," "utter darkness," and "mournful gloom," not to mention my earlier chapter on Kepler, for whom the sun is a reflection, instrument, or vehicle of divinity.[25] In the first line of the inset, "The Sun new ris'n" is surely a resonant phrase, even before the epic reveals the extent to which Satan's vengeful mission will parody the Son's redemptive one. Further, if mention of the sun heightens the contrast between real sunlight and the nonlight of hell, it also connects the created world to hell simply by being where it is, not only by foreshadowing Satan's corrupting effect on nature. The paired analogies of the sun at dawn and behind the moon are hardly innocent (nor is my use of this word), and their presence, while easily explained away, exerts a sense of displacement when initially encountered. The presence of the sun, if only in analogy and with a hint of redemption, is, for the present and its context, compromising—shearing the sun of his beams.

A subsequent analogy involving sunlight intensifies the effect of displacement in the second book of darkness, right after Satan has volunteered to attempt the journey to Earth. This time, the analogy pertains to the rebels' "rejoycing in thir matchless Chief," whom they extol "as a God" (II.478, 487). They rejoice

> As when from the mountain tops the dusky clouds
> Ascending, while the North wind sleeps, o'spread
> Heav'ns chearful face, the lowring Element
> Scowls ore the dark'nd lantskip Snow, or showre;
> If chance the radiant Sun with farewell sweet
> Extend his ev'ning beam, the fields revive,
> The birds their notes renew, and bleating herds
> Attest thir joy, that hill and valley rings.
>
> (II.488–95)

This analogy—Satan is to the rejoicing demons as the sun is to the natural world—is even stranger, more dynamic, and more surprising than the earlier one involving sunlight. Like that earlier one, it first looks back to loss of light and then looks forward to the created world. It also recalls the stock example of compressed, analogical metaphor within the tradition of logical semantics, namely, "the field smiles" (blooms:field :: smile:face), which was discussed in Chapter 4; in this way it suggests that analogy itself is self-reflexively an issue, as it will be more openly when Raphael delivers his apology for recounting the War in Heaven.[26] What is most surprising about the present analogy, however, is the unabashedly positive feeling with which the narrator celebrates demonic rejoicing in its second half, despite his effort to back away from this breath of fresh air and sunlight in the moralizing lines that follow: these lines contrast diabolical concord to human strife, much to the latter's disadvantage (496–505). For a moment, a reader of the sunlit analogy could wonder whose side the narrator is on and whether he has adopted the demons' perspective. Before and despite his subsequent backpedaling, the analogy has achieved its disorienting emotional effect. If we were looking for passages to exemplify Milton's affection for the natural world, the passage at hand would be a candidate.[27]

Affection might be the problem, however. Attractive as is the sunlit world of the analogy, particularly amid the visible darkness of hell, the joy it brings is *only* affective and natural. These limitations invite reflection and, indeed, a double-take. Heightened awareness of a contrast between natural joy and demonic malice is one possible result since the demons have just voted for destruction and death over Moloch's revolt, Belial's resignation, and Mammon's construction. Yet the joy of reviving fields, chirping birds, and bleating herds undeniably also conveys—shares—their feeling of joy; like any other form of metaphor, analogy expresses connection as well as difference. As it stands, the present analogy extends such natural joy to the demons, bringing them close when they are at their worst, when they have, in effect, just chosen death over life and light. The analogy is further startling if numerous seventeenth-century lyrics have accustomed us to the homonymic play of *sun* with *son*, probably the most familiar pun of all in the period and always a potentiality, given Milton's subject. The combination "radiant Sun" hardly seems an accident in the present analogy, anticipating, as it does, the "radiant image" of God's "onely Son," not to mention the "radiant Shrine" of the Father himself, both in the next book (III.63–64, 379). The surety of such contrast comes only in retrospect, however; the nonpresence of the pun in the analogy at hand is on the level of the nonpresence of light in the flaming darkness: there yet not there.

But there is nothing wrong—positively wrong, as it were—with a joyous response to sunlight in the natural world, which includes the poet and his audience. Something of the same exhilaration occurs in the belated peasant's more mixed response of joy and fear to the moonlit reveling of faerie elves in the analogies that close Book I. There, as noted, perspectival analogies serve to comment on the preceding book of darkness and to displace our sense of it.[28] The imaginative displacement in these analogies, as in the sunlit analogies within both books of darkness, sharpens awareness of the competing values that these books introduce—metaphysical, ethical and moral, psychological, rhetorical, and political, to name familiar ones. But this displacement and the awareness it intensifies are also reminders that whatever the analogies drawn from outside the space and (relative) timelessness of these dark books, which lack the cycles of natural light, their dominant perspective belongs to hell and then to the "wild Abyss," the "vast vacuitie," the "crude consistence," and finally the pavilion of the reigning "Spirits of this nethermost Abyss, / *Chaos* and *ancient Night*."[29] If darkness is the overriding condition of utterance in Books I and II, nothing uttered within them is wholly unaffected or sacrosanct—certainly not sunlight.

Books of Light, Night, and Twilight

The Invocation to Light

At the opening of Book III, Milton's Invocation to Light introduces, or rather reintroduces, the relation between light and darkness evident in Books I and II. Before the Invocation is finished, this relation will be different, and it will develop further in the course of Book III, as well as extensively in subsequent books, to resituate Light and Night in relation to God.

In this and following sections, I intend to use Book III as a kind of home base, reaching back from it to the books of darkness and following its implications forward to passages that are crucially related to Light, Night, and their intermediate forms. Chaos, God's mount, and the creation will figure significantly in the following pages, which employ an unfolding argument to respond reflectively to the cumulative movement of narrative in *Paradise Lost*. In Milton's epic, as in Spenser's, forward progress and retrospective realization are an interwoven and interweaving process.

Milton's Invocation to Book III first surrounds the biblically warranted identification of God as light with alternative ways of approaching and expressing it. The very fact of alternatives bears initial emphasis:

> Hail holy Light, offspring of Heav'n first-born,
> Or of the' Eternal Coeternal beam
> May I express thee unblam'd? Since God is light
> And never but in unapproched light
> Dwelt from Eternitie, dwelt then in thee,
> Bright effluence of bright essence increate.
> Or hear'st thou rather pure Ethereal stream,
> Whose Fountain who shall tell?
>
> (III.1–8)

The poet frames the prayerful, hymnic opening of the Invocation with questions and in its brief course offers six ways to conceive of light: in the first sentence, as the first *holy* thing created (l. 1), as belonging to (that is, being "of") what has neither origin nor end and thus as forever existing in God (l. 2); in the second sentence, as expressly God himself in some way (l. 3), as that in which God dwells (ll. 4–5), as an outflowing ("effluence") of God's eternal essence, which is light (l. 6); and finally, at the beginning of the third sentence, which opens out to the creation of the cosmos, as a mysterious flow of rays whose source is inexpressible and, as such, finally beyond knowing except intuitively and affectively in faith (ll. 7–8).[30] The first line—"offspring of Heav'n first-born"—suggests God's son, "radiant image," visible appearance, substantial expression, and creative, freely redemptive Word, identified in Milton's *Christian Doctrine* as "first of created things."[31] This identification is remarkably close to Raphael's later description of "Light / Ethereal, first of things" in the creation of this world, phrasing to which subsequent discussion will return, comparing it with that of "*Night*, eldest of things" (II.962, VII.243–44).

The two opening sentences of the Invocation, procreative in line 1 and radiating outward in line 6, encircle God with his creation. The opening of the third sentence, "Or hear'st thou rather pure Ethereal stream, / Whose Fountain who shall tell?" echoes the same conjunction that begins line 2, "Or of th' Eternal Coeternal beam," and could be repunctuated with a colon or semicolon as a continuation of the preceding sentence. This same, third sentence is an open-ended question, to which the implicit response is the moment of creation in Genesis when light flowed from its source, or "Fountain":

> Or hear'st thou rather pure Ethereal stream,
> Whose Fountain who shall tell? before the Sun,
> Before the Heavens thou wert, and at the voice
> Of God, as with a mantle didst invest
> The rising world of waters dark and deep,
> Won from the void and formless infinite.
>
> (III.7–12)

This is the juncture of ethereal light with cosmic matter so important in Kepler's astronomy and optics, although for Kepler light is immaterial and, if somewhat mysteriously, for Milton material, as it generally was for the medieval perspectivists. What is striking in both the Keplerian and the Miltonic instances, however, is the *continuity* of light's dynamic occurrence, from God to radiating creation to the cosmos. This is what really matters.[32]

What follows next looks back to the preceding books of darkness as well as ahead to later descriptions of heaven, but this forward glance only becomes available retrospectively, and so I will briefly defer it as a premature presence, an abstraction from the progression of the poem that would be misleading *at this point*. Once again, as when the present Invocation began, we could almost suppose that Satan is addressing light. This time, however, the supposition is merely hypothetical—merely *as if*—rather than supported by the immediately preceding dramatic context:

> Thee I re-visit now with bolder wing,
> Escap't the *Stygian* Pool, though long detain'd
> In that obscure sojourn, while in my middle flight
> Through utter and through middle darkness borne. . . .
>
> (III.13–16)

The poet refers to the experience he has had of imagining and expressing—his word is "singing," that is, poeticizing—hell, "*Chaos*[,] and *Eternal Night*" for two books, and he does so as if he had actually been in their darkness, as, in a fictive, felt sense, he has been (18).[33] The experientially charged words "Escap't" and "detained," along with the first-person narrative, even in memory enforce the reality and immediacy of his participation. Only the eight lines hailing light separate this memory from the books of darkness. John Leonard has proposed that the reference to Night as "Eternal" in line 18 of the Invocation, the second of two times that Milton "in his own voice as poet" so describes Night, indicates that Night has neither been created nor given any other origin in the epic.[34] The earlier instance occurs when the narrative, rather than the present invocatory hymn, refers to the "Eternal *Anarchie*" of "eldest *Night* / And *Chaos*, Ancestors of Nature" (II.894–96). This instance happens suddenly, right "Before th[e] . . . eyes" of Sin and Satan when the gates of hell swing open, and it engages their vision—the eyes of these interlinked beholders—every bit as much and as significantly as do the "darkness visible" and "utter darkness" of hell itself, followed by the darkness of Night and Chaos.[35] Again, the second, later occurrence in the Invocation to Book III comes in the poet's dramatic memory of his participation in the perspective of darkness. That it occurs beyond "the *Stygian* pool," the void of Night, and the formlessness of Chaos does not mean that it is untouched by these. The perception of Night's eternity could be affected, a vestige of the extremity of that darkness, of which he "sung," now still so memorably, though nonetheless in the past tense. While this possibility does not make the reference to the eternity of Night vanish, it urges further consideration of context and affect with respect to it. Affect is hardly confined to feeling or mood swings alone. That it is perceptual and therefore bears on cognition was as familiar to Aristotelian psychology as to a modern theorist such as Paul Ricoeur.[36]

The poet who writes "What in me is dark / Illumine" in the invocation to Book I openly recognizes and acknowledges his credentials to imagine and express the things of darkness—whole books of them. His angelic narrator Raphael will later give us a potently creative God who "fill[s] / Infinitude" and unambiguously precludes a vacuum (VII.168–69); yet another of the Miltonic narrator's descriptions, often authoritative but not always unquestionably so, has already given us an evidently equivocal God, the creator of hell as "A Universe of death . . . for evil only good" and then, without reference to God, has presented us with the "vast vacuitie," effectually a vacuum, within the "Eternal Anarchie" of Night and Chaos (II.622–23, 894–96, 932). All these are instances of double voicing, to which the poet himself in the poem is not simply immune. More instances are to come, and I will return to Night's eternity in various contexts, eventually, I hope, shedding further light on Night's relation to God. At present, I want only to acknowledge that her eternity is indeed an issue.

Milton's "heav'nly Muse," the Invocation to III continues, has been his guide down into darkness and then safely up from it (19–21). The poet continues to address light, not the Muse, however, doing so in an ever more personal vein: "I . . . feel thy Sovran vital Lamp; but thou / Revisit'st not these eyes, that rowle in vain" (21–23). This lamp, clearly the sun, is the immediate source of life-giving sunlight, only whose warmth is felt by the blind poet, but its continuity with the preceding sources of light remains unbroken; that is, the shift to sunlight in the address to holy Light is smooth and unremarked—almost unspoken—yet it is considerable. In the privation of light that is blindness, warmth offers the comforting sensation of an illuminating presence, and a reaffirmation of purpose quickly follows: "Yet not the more / Cease I to wander where the Muses haunt . . . but chief / Thee *Sion*"—or *Zion*—the sacred mount of the Israelites as well as the traditional type of a sacred capitol, or city on a hill, with both its biblical and contemporary political and religious nuances (26–30).[37] When Milton adds "Nightly I visit" to his evocation of these sacred places, darkness returns, but Night is now creative, nourishing poetry: "Then feed on thoughts that voluntarie move / Harmonious numbers; as the wakeful Bird / Sings darkling, and in shadiest Covert hid / Tunes her nocturnal Note" (37–40). The final sentence of John Rumrich's essay on Chaos and Night, published together with Leonard's, is irresistible here: "Perhaps darkness is the price to be paid for Nightingales."

Transformed from the horrific emptiness of Book II and seen anew, Night now participates in constructive analogy: the wakeful, singing Nightingale is to "shadiest Covert," as Milton, the blind poet, to his poem. Before this stage of the Invocation, there have been few metaphors (for example, offspring, fountain) and only one that approaches proper analogy: light, "as with a Mantle," invests, or clothes, "The rising world of waters dark and deep" (10–11). Similarly, on the first Night in Milton's Eden, the moon will "unvail" her "light" to throw "o're the dark her Silver Mantle" (IV.608–9). The simultaneity of unveiling and mantling, revelation and cloaking, is noticeable in both instances; the mantling of darkness by light inheres in the creative act and in what is created; it is a form of enlightenment. The Invocation's dynamically rising "waters" refer to an element of the incipient Earth as well and, perhaps, to the more mysterious waters created above the firmament in Genesis 1:7. Prior to this Invocation, the cognates "water," "waters," or "watrie" appear in hell only to allude to Moloch as god of the Ammonites and to Tantalus and thus to a world temporally and spatially beyond this hell, and, if such waters are implicit in the embryonic, atomic potentiality of Chaos, they are absent as words or elemental forms there. Only much later do the Invocation's waters reappear in the "watrie calme" already responding to "the brooding wings [of] the Spirit of God" in the creation narrative of Book VII (234–35). What matters here, beyond another striking indication of Milton's amazingly deliberated care with his words, is the parallel in the Invocation to Light between its first metaphor of creation, which belongs to the mantle of "holy Light," and its second, the poet's Nightingale, which is more specifically analogical. In this respect, the present Invocation also circles back to the opening invocation of Book I, which similarly aligns God's creativity, to which Moses bore witness, with the poet's and does so analogically.

What follows next in the Invocation, however, is different. The poet returns to his loss of daylight and his envelopment by "ever-during dark." Lamenting both his loss of the ability to see the works of Nature and his loss of the "wisdome" they afford, he asks for the "Celestial light" that "Shine[s] inward" to irradiate his "mind through all her powers" and enable his access to "things invisible to mortal sight" (50–55). For my purpose, two issues stand out only now in the Invocation. The first is the clear distinction between celestial light and sunlight, divine and natural light, which blindness has enforced. Sunlight has gone dark for the poet, and his personal loss has disrupted the initial emphasis in his Invocation on unbroken continuity. Retrospectively, positioned at the outset of Book III, darkness again casts its shadow back over the first two books and particularly over their solar analogies. The second issue once more recalls the traditional analogy between physical and intellective vision with which this chapter began and which has played a role in earlier chapters. This analogy lies in the cultural background of the blind poet's inner light, but now, by contrast with the traditional content of the analogy, Milton's rewriting of it conveys both a breach and its compensation: outer loss and inner gain, physical deprivation but enhanced access to supersensible vision. Here, as elsewhere in *Paradise Lost*, reality has moved inward in the face of loss. Yet if eyesight is no longer the analogue for inner vision, the light that shines inward comes not only comes from the Spirit, but it also remains a figure derived from visible light, and the song of a blind bard is now analogous to a nightingale's. Analogy simply reappears in another register, resurrected and reborn. Nature's works still offer "wisdome." Contrast with Satan's self-enclosure is evident.

"Dark with excessive bright": God and God's Mount

In the middle of Book III, after the Father and Son have together resolved the future of fallen human beings, the angels hymn both of them, and their language recalls the Invocation to Light. In the Son, "th'Almighty Father shines" visibly and the "effulgence" of his "Glorie abides" (384–88). The Father is the

> Fountain of Light, thy self invisible
> Amidst the glorious brightness where thou sit'st
> Thron'd inaccessible, but when thou shad'st
> The full blaze of thy beams, and through a cloud
> Drawn round about thee like a radiant Shrine,
> Dark with excessive bright thy skirts appeer,
> Yet dazzle Heav'n, that brightest Seraphim
> Approach not, but with both wings veil thir eyes.
>
> (III.375–82)

This darkness, an "appeer[ance]" caused by a blinding excess of light, arises primarily from the visual incapacity of the angelic gazers, along with their visionary lyricist's, rather than from the divine essence. Like "darkness visible" in Book I, it is perceptual, but it is also natural, so to speak, as that earlier darkness is not: that is, even the ancients knew that gazing directly at the sun could be blinding, and implicitly this knowledge underwrites the powerful, protective image of dark light in the inset at hand. The earlier darkness in hell and Chaos also differs from the present, dazzling darkness because its cause is perversion and privation, not positive excess, and because it is an exterior, objectified place as well as a perceptual one.[38] It is both visible and all too accessible to demons, as God, visible only through his radiant image, the Son, cannot fully be to angels without destroying them. When all is said, however, dazzling darkness, or dark light, is still God as he actually is *perceived* directly.[39]

In Books V and VI, darkness and even Night are so close to God as also to seem part of him, and the same is true of a bright form of flame-light. In the exaltation of the Son, the Father speaks "as [as if] from a flaming Mount, whose top / Brightness had made invisible" (V.598–99). The volcanic potential of this counterfactual image is then actualized when the War in Heaven is imminent, and God speaks again: "Clouds began / To darken all the Hill, and smoak to rowl / In duskie wreathes, reluctant flames, the signe of wrauth awak't," and still again near War's end when the "Chariot of Paternal Deitie, / Flashing thick flames" rushes forth like the sound of a whirlwind (VI.56–59, 749–51). Such imagery, like the fire under heaven's floor, is a reminder not only of the Bible's angry God but again of the creator of hell, of that "Universe of death, which God by curse / Created evil, for evil only good."[40] This hell is a place "Where all life dies, death lives, and Nature breeds, / Perverse[ly]," where Nature is corrupt and where the "living Death" and "deathless Death" of Adam's fallen imaginings have their actual existence (II.622–25, X.788, 798).

A crucial passage involving God with Night appears in Book V, narrated by Raphael and covered by his apology for doing so in human terms. Nonetheless, this passage, which occurs just before Satan's envious response to the Son's exaltation, is profoundly suggestive:

> Now when ambrosial Night with Clouds exhal'd
> From that high mount of God, whence light & shade
> Spring both, the face of brightest Heav'n had changd
> To grateful Twilight (for Night comes not there
> In darker veile).
>
> (V.642–46)

Perhaps the adjective "ambrosial" neutralizes Night, contextualizes it as heavenly, or alone suffices to transform it into a period of delight and nourishment. Yet the subsequent pairing of light with shade, in this syntactic order, is a noticeable shift away from Night as such, and then the combination of light with shade as twilight—welcome, pleasurable, "grateful" twilight—is another. In view of the salience of light imagery in the epic, the combining in grateful twilight of its relative extremes as a cyclical recurrence in heaven is itself noteworthy. With twilight, however, the final, parenthetical step in Raphael's scene setting explicitly excludes Night altogether, at least as we have earlier encountered Night in the epic. Yet Night's "darker veile" still sounds somewhat ominous in this context, foreshadowing what is to come and recalling the books of darkness. Its parenthetical presence only ensures that Night's exclusion is noticed instead of suppressing this awareness. As a parenthetical remark, it subordinates Night but noticeably does not remove it.[41]

Before long in this book, under cover of Night, Satan will lead his followers north; in the next book, in a demonic parody of creation, they will fashion cannons and powder from "the materials dark and crude," "Th'originals of Nature," under heaven's floor, and "with shaddowing Squadrons Deep" to hide the product of their machinations, they will approach renewed battle with the loyal angels.[42] Satan's deleterious effect on matter is already evident, as is the similarity of the dark, crude, underlying materials, originals of Nature, to those of the "Ancestors of Nature," Chaos and Night, nethermost Spirits of the abyss (II.894–95, 969). Satan's treachery thus also affects and recontextualizes the nature and value of Night and its lighter relatives, such as shadow, in heaven. Yet in the passage immediately at issue, the narrative has clearly first located Night in the radiant mount of God and, as the parenthesis conspicuously signals at the end of the passage, it has done so deliberately, making sure that we notice.[43]

A later passage in Book VI, similarly describing the mount of God, reinforces the development of Night into shade or into a twilit combination of light and darkness, and it further urges the issue of Night's relation to Light and God:

> There is a Cave
> Within the Mount of God, fast by his Throne,
> Where light and darkness in perpetual round
> Lodge and dislodge by turns, which makes through Heav'n
> Grateful vicissitude, like Day and Night;
> Light issues forth, and at the other dore
> Obsequious darkness enters, till her houre
> To veile the Heav'n, though darkness there might well
> Seem twilight here.
>
> (VI.4–12)

Light and darkness—notably feminized, obsequious darkness—exist in an orderly rotation that is at least perpetual and, by attribution to God himself through his mount, both potentially eternal and likely so. Day and Night are present here only in comparison, again in this order, that of creation in Genesis 1:3–5, which has yet to occur in the poem. The ordering of light and darkness could also reflect the traditional primacy of something positive over its privative, light over dark. Or darkness could be simply a heavenly alternative to light—rest, a kind of Sabbath—as Raphael has earlier implied. But again, it is hard to ignore the facts that we have earlier witnessed two books of darkness and that this is not the first time that Raphael has associated darkness and Night with Light and God. Darkness looks here like a principle and potency of God.[44]

Momentarily, I want to move the difference between light and dark and the likely primacy of the one to the other into other contexts to illustrate more of its possible provocations. For one instance, Fredric Jameson has considered such material differences in an economic context, that is, in a culturally systemic one, which is the point here: he contends that we "cannot enumerate the differences between things except against the background of some more general identity." In other words, difference so conceived cannot be originary and, to return to Jameson, "at some level" difference must be merely the appearance of a fundamental unity.[45] In physical and historically scientific terms, however, original formlessness precedes form, disorder order, and difference concord. Repulsion precedes attraction: just think of the Big Bang Theory, or quite otherwise in Spenser's epic, think of the cosmic emblem of Concord, who stands between two youths, "brethren . . . Begotten by two fathers of one mother, the elder Hate and the younger Love.[46] In magnetism, however, whether William Gilbert's theory, Kepler's adaptation of Gilbert's theory, or Milton's allusion to Kepler's use of it, attraction and repulsion, love and hate, are simultaneous powers, as fusion and fission are in Lucretian theory. Contexts, like perspectives, change things. The context I have assumed for God's mount is cyclical, systemic, and differential: "light and darkness in perpetual round / Lodge and dislodge by turns." I have called it *natural*, hoping this term is acceptable for divinity. Its underlying unity is God himself, of whom dark light ("Dark with excessive bright") is an ecstatically intense, real perception.

In the Miltonic passage at hand, as similarly in Raphael's earlier description of heaven, the replacement of "Night" by "darkness" and "twilight" does not cancel the initial attribution of Night to God's mount or the comparison of it to the darkness that issues from that "Cave / Within the Mount of God, fast by his Throne." The mount, effectually God's capitol on a hill, is a metonymy for God, and Night, like Light, comes forth from it; Night, in fact, is "exhaled" in the earlier inset I cited from Raphael, much as is the breath of God in Milton's account of the creation of human life in *Paradise Lost* and *Christian Doctrine*.[47] Night is God's, if not quite within him or directly identified as him, and is either coeval with light or its privative sibling, at least as seen in Raphael's heaven, and perhaps this is once more the real point. Verbal registers shift with situations and contexts. When the loyal and rebel angels throw hills on one another, Raphael describes them as battling "under ground . . . in *dismal* shade , / Infernal noise; . . . horrid confusion heapt / upon confusion" (VI.666–69; my emphasis). Even as heaven thus nears the condition of Chaos and hell, shade takes on the condition of these spaces: *dismal* is a contraction of Latin *dies mali*, "day of evil." And when the wrathful Son attacks the rebels, the "Starrie wings" of his chariot make a "dreadful shade contiguous" underneath them, and he is himself "Gloomie as Night" (VI.828, 832). It is as if he were reclaiming all the shades of darkness, folding them back whence they issued.

At the same time, however, the nature and value of Night in the epic could be claimed as another instance of Milton's double voicing, something I have mentioned before in this chapter and the third. Such doubleness extends from puns to alternative explanatory traditions in science and theology; to characters' viewpoints, including the narrator's and Raphael's; and however briefly, even the poet's or narrator's as poet, and it also extends to the dominant perspectives of whole books. Milton's double voicing is thoughtful and exploratory, not just a game, a cynical trap, or an anxious expression of doubt.[48] Moreover, it is an invitation to interpretive choice insofar as "Reason is also choice" or informed by choice (III.108). If Milton did not value reason highly, his writings and his life would look very different. He is concerned at once to reach and to show the limits of reason, challenging his readers to recognize and try them and urging faith as the ultimate answer when reason "know[s] to know no more" (IV.775). Nonetheless, he considers reason as choice—enlightened, or right, reason—to be a bulwark of faith, and recurrently, he puts difficult choices before his audience.[49]

Satan on the Sun and Accounts of Creation

Returning again from Night to Light in Book III, I want to look at the final movement of this book, Satan's visit to the sun, "whose splendor likest Heaven / Allur'd his eye," and then at accounts of creation in and after Book III, first by Uriel and then by Raphael (572–73). These accounts raise further questions about the relation of Light, Night, and God and especially about the eternity of Night, as this eternity is voiced at the beginning of Book III and the end of Book II: respectively, the Invocation to Light ("*Chaos* and *Eternal Night*") and the characterization of "eldest *Night*," "eldest of things" (II.894, 962). These are the two relevant descriptions of Night that belong to Milton as narrator and poet rather than to one of the figures in his plot (for example, Belial's "uncreated night": II.150).

That Milton should send Satan to the sun is notable in itself; he did not have to. The visit presents another opportunity to introduce contemporary astronomy into the poem, complete with the ironic analogy of Satan to a sunspot. Among the functions of this analogy is once more the foreshadowing of Satan's corrupting effect on Nature: the sunspots Galileo's telescope revealed were by many considered disturbing signs of imperfection in heavens that were supposed to be changeless. The sun is Satan's first stop after escaping from hell and the darkness of Chaos and Night, again suggesting, as did his initial words in hell, the attraction light still holds for him, understandably given his former name, Lucifer, the light bringer, the star of morning and the promise of dawn. Another point of Satan's stop on the sun might be for the poet to represent in the narrative of Book III the progression from the radiant, godly light of heaven to the sunlit universe that was earlier evident in his Invocation to Light: this is effectually a movement from inspiration to unfolding event and as such a paradigm of creation. Satan's odyssey parodically foreshadows this paradigm in the narrative sequence but does so derivatively in history. Satan's solar stop also affords an opportunity for extensive attention to the properties of the sun itself and dramatizes the centrality of the sun in Milton's epic. Later, looking up from Earth at the sun in Book IV, Satan compares it to "the God / Of this new world," only next to declare that he hates its beams because its eminence among the stars reminds him of his lost glory (IV.32–39); he is unable to sustain an image of anything higher or other than himself. The later complement of Satan's renunciation of light in this moment is his further embrace of darkness in tempting Eve in a dream and in entering the snake, both under cover of night, whose creature he is fast and regressively becoming and further polluting.

Of additional note, Satan's visit to the sun occasions the first eyewitness account of creation, voiced by Uriel, and it includes another comment on the relation of darkness to Light:

> I saw when at his Word the formless Mass,
> This worlds material mould, came to a heap:
> *Confusion* heard his voice, and wilde uproar
> Stood rul'd, stood vast infinitude confin'd;
> Till at his bidding darkness fled,
> Light shon, and order from disorder sprung.
>
> (III.708–13)

Here, near the end of Book III, Uriel's voice recalls the poet's in the opening Invocation of this book, when he described creation as a mantle of light cast over "The rising world of waters dark and deep, / Won from the void and formless infinite." Uriel describes the shaping of cosmic matter from the embryonic atoms of Chaos, the purposeful gathering (heaping) of their random movement into a mass, and the enclosure of otherwise limitless space within the compass of cosmic form. The fleeing of darkness prior to the shining of light suggests the moving of the Spirit of God over the darkness of the deep in Genesis 1:2, right before God's command "Let there be light" in verse 3. It is nevertheless notable that in Uriel's account, darkness must first flee before light shines. This sequence gives power over the darkness to God's bidding directly rather than through created light. Uriel's account, whose purpose is to praise the creative power of God, also instances dramatic irony since it is unknowingly delivered to the adversary who seeks to appropriate or destroy God's creation.

The account of creation in Genesis evidently engaged Milton much as it had Augustine, who wrote and worried over five interpretations of it, including allegorical, literal, and figural versions, in his commentaries, *Confessions*, and *City of God*. For Augustine, as later for Milton, the question of the nature of earthly being and its relation to God was fundamental to understanding and belief. Augustine was especially concerned to interpret the account of creation in Genesis because belief in a material world that had originally been made good by God and rejection of a productive principle of Evil were fundamental differences between the biblical account and the Manichaeism to which he had formerly subscribed. For Manichaeans, as for Milton's Mammon, there is a kingdom of light and another of darkness. Milton's account of light recurrently contends with this possibility, too, as most immediately in Uriel's description of creation and, correlatively but contrastingly, in the imperial aspirations of hell's demons.

Raphael's narrative in Book VII, the main account of creation in the epic, begins where Uriel's did in Book III, but with notable differences. These start with the inclusion of the present reality of Books V and VI, Satanic rebellion and its aftermath, which precede the creation of this world in Raphael's account, not without lasting significance and a lasting effect. Raphael's mission, after all, is to warn Adam and Eve of Satan's approach and intention. Accordingly, having infused vital virtue and warmth throughout "the watrie calme," in his account the Spirit of God

> downward purg'd
> The black tartareous cold Infernal dregs
> Adverse to life: then founded, then conglob'd
> Like things to like, the rest to several place
> Disparted. . . .
>
> (VII.237–41)

Raphael's conglobing, or gathering and compacting into a round mass, is similar to the first part of Uriel's creation narrative, but it describes a more advanced state of development.[50] It specifies the appearance of ever-significant spherical form ("conglob'd"), the result of the imposition of likeness. In likeness it is even tempting to see an intimation of similitude, eventually including cosmological analogy, insofar as this whole verbal account bears witness to the creative w/Word, or Logos. Also new in the perspective of Raphael's narrative are "The black tartareous cold Infernal dregs / Adverse to life." Their connection with what Satan represents could hardly be clearer. Satan, as his fallen name testifies, is the Adversary—the "Adverse" embodied. In Paracelsus, tartar is the evil generated by the fall of

Adam and Eve, but, before Paracelsus, in gnostic-Manichaean doctrine, tartar "was created to bring about the 'de-Tartarization' of the world, and was destined to absorb everything devilish."[51] Neither view precisely fits the situation in Milton's poem at this point, but Satan's evil precedes and accounts for both; allusively and relatively, such historical views are available to readers, along with classical identifications of Tartarus with an infernal realm. Milton believes in the dynamism of God, which protects God's free will and his active potential for change, and the rebellious evil that the satanic Adversary freely embodies, like the generation and exaltation of the Son, looks to be part of this, all tending to a final end.[52]

Next in Raphael's narrative comes the climactic moment of creation from which the rest follows: "Let there be Light, said God, and forthwith Light / Ethereal, first of things, quintessence pure / Sprung from the Deep" (243–45). Among things, Light now has primacy, preceding the separation of the light from the darkness (250), and in its ethereal, quintessential essence, Light has a purity superior to things that are merely earthly. At the same time, it is from the Deep that Light has sprung, as if it were a potency liberated by God's command. With respect to Chaos, twice in Book II, Night has been called "eldest" or "eldest of things," phrasing that invites comparison to ethereal Light in the description by Raphael just cited (II.894, 962). Raphael's phrase "first of things" raises the question of whether primacy, which is here a systemic and metaphysical consideration, as well as a temporal one, differs from the significance of the word "eldest." This question returns my discussion to Night in Book II—"eldest *Night*," paired with Chaos in "Eternal *Anarchie*," and also called "eternal" in what could even be a transferred epithet, a Latinate practice, within the phrase "*Chaos* and *Eternal Night*" in the Invocation to Book III.

The first issue raised by the phrase "eldest of things," applied to Night in Book II, pertains to the Miltonic narrator's noun "things," which invites briefer comment than does the adjective "eldest." Leonard, finding Night, considered as void, "in some ways . . . more sinister" than Chaos, explains "things" in the phrase by invoking "the [ancient] atomists' theory that the void, though unreal, exists."[53] The void, since unreal, then, is nothing, no-thing—or better, perhaps, it represents existential nothingness—but then it still remains hard to square with the Miltonic narrator's "things." As existential nothingness, relation of the void to the "real nonentity" that is Death in Summers's much-cited designation of this figure is yet another question.[54] Death, thus seen, is real but not a thing that has being; it, too, is no-thing. In short, it resembles the nothingness of the atomists' void. Could Night be death in another guise? In Milton's epic, Death also exists as a formless figure that speaks, and Night is, or at least becomes, figuratively real as the Consort of Chaos (II.951–54). If only as paradoxes and apparent contradictions in the books of darkness, Night and Death appear related and conceivable, as before in their shared blackness.[55] Both have remarkable relations with God, creator of "A Universe of death" and "utter darkness" as well as of one of life and light. Such paradoxes in the books of light and darkness, however, are still near the beginning, not the end, of this question. Before further pursuit, however, the second issue the phrase "eldest of things" raises, which concerns the superlative descriptor "eldest," asks for attention.

Within a human context, wherein *eldest* usually occurs, it refers to age or the precedence that accrues to it and is therefore a reference to earthly time and matter; no one grows old in heaven or achieves status there by virtue of longevity, although change and some kind of time exist in Milton's heaven, and primacy of origin matters greatly there.[56] (Words such as "young," "youthful," and the like also occur with respect to angels when they are operating in the created universe.)[57] For *eldest*, whose etymology is Old English, the third definition the *OED* (2014) offers is "Earliest, first produced, most ancient," which is supported by relevant early modern examples, all of which refer to earthly time and culture, but the *OED* also offers a medieval example from *Ayenbite of Inwyt* (1340), a text in the Kentish dialect, in which God is "eldeste." This example occurs in a context in which God is unambiguously said to be everlasting, without end and without beginning.[58] Since one major depiction of God in *Paradise Lost* is that of a Father, "eldest" might figuratively apply to him as well as to Light either as another of his identities or as his first effulgence. The sense in which Night is "eldest" would similarly appear to depend on her figural—that is, metaphorized and more specifically personified—identity.

The initial appearance of the word "eldest" occurs when Sin and Satan—that significant coupling—gaze into the "deep, . . . dark / Illimitable Ocean" of Night and Chaos, where even the word "Ocean" functions as a metaphor, and time and place are lost, again making "eldest" a curious term at best. Here "eldest" is likely a prefiguration of the personified form Night will more clearly assume in the "dark Pavilion" of Chaos (960). In fact, "eldest Night" might already be said to participate in some degree of figuration from the moment that Satan and Sin peer into the deep, where Night and Chaos are already *holding* "*Anarchie*" figuratively, as if holding court, as this phrasing signifies (II.895: "hold . . . *Anarchie*").[59] Moreover, if the embryonic atoms belong to Chaos, and if Night represents the Lucretian void, then Chaos exists within Night. Together, they also figure "the dissolution and death" of the natural world, which is as yet uncreated in the epic narrative, a figuration based on the end of Lucretius's second book.[60] That Milton's figures of Chaos and Night cohabit in a pavilion simply makes sense of the fact that they are combined in his narrator's depiction of them within the "deep," as they are in Lucretius. Separated from the embryonic atomic matter of Chaos, Night as Void might exist theoretically but otherwise would be merely an abstraction from real existence and still not, as we say, the real thing.[61] But the very name of Night, if Night represents the void, also becomes a displacement or concealment—a veil or a mantle, metaphorically speaking. Why the name "Night" at all rather than simply "Void," a monosyllable like Night? What does Night offer that Void does not? One answer lies in the relations of Night already examined, which crucially include its pairing with Light, various shades, degrees, and kinds of darkness, and associations ranging from inner to outer and from natural to spiritual, physical to metaphysical. Night is even linked by sound to Light. In comparison to void, Night also has a far richer past as a mythological figure, most often sinister.[62]

The Miltonic narrator's second use of the word "eldest" comes with the more pronounced figuration of the "dark pavilion" itself, where with the Anarch Chaos, "Enthron'd / Sat sable-vested Night, eldest of things, / The Consort of his Reign" (II.901–3). With this fuller figuration and clearly wedded to Chaos, Night's thingness is less problematical because more openly metaphorical, or "as if"—a real illusion, if you will, and as such a foretaste of the reptilian vignette featuring an illusory reality in Book X. Another relevant occurrence of "eldest," this time in Eden, appears in the hymnic aubade of Adam and Eve after she recounts the story of her dream, which is actually a nightmare. The human pair call on the heavenly bodies that move in "mystic Dance" to resound "His praise, who out of Darkness call'd up Light," and then on the "elements the eldest birth of Natures Womb" to join their song (V.178–81). In the innocent poetry of Eden, within worldly time, the personified figure of Nature, herself traditionally the eldest of things, bears as her eldest offspring the created elements collectively, which, as the narratives of Uriel and Raphael suggest, the random, embryonic atoms of Chaos underlie more darkly in Night. But Adam and Eve as yet know nothing of Chaos and his consort Night, with the unsettling exception of Eve's nightmare, which situates within her something dark, perhaps a principle or potency, like that in God's mount. Adam's immediate, intuitive recognition of the evil origin of Eve's dream suggests that he shares this principle—as does their God, into whose mind "Evil . . . May come and go"—like Night and Day, perhaps (V.98, 117–18).

In the Edenic pair's aubade of praise to God, the word "eldest" specifies only a particular family relationship within the created world they experience, intuitively here, it appears. Their use of "eldest," too, suggests that this word, applied to Night and thereby binding her to Chaos in Book II, likewise obtains within the wedded relationship of these two personified figures: theoretically, empty space may precede the restless atomic matter operating in it, but physically, they are simultaneous. Like context, the *figuration* of Night in Book II really makes a *significant* difference.

God's Creative Secrets

Returning from Night, "eldest of things" in Book II, to Raphael's creation narrative in Book VII, in which Light is "first of things" in the created universe, I want to conclude by glancing at two other passages in this book, both bearing on Night, and once more at the relation of this figure to Milton's God. The first passage pertains to the origin of Day and Night in Raphael's account of creation and ties back to his description of "ambrosial Night with Clouds *exhal'd* / From that high mount of God," following the exaltation of the Son (V.642–43; my emphasis). This time, the celestial choirs celebrate "when Orient Light / *Exhaling* first from Darkness they beheld; / Birth-day of Heav'n and Earth" (VII.254–56; my emphasis). Whereas in the earlier instance God effectually exhales darkness, in the latter, darkness exhales light. The parallel retrospectively suggests God's darker side in the earlier exhalation, but now, in the latter angelic perception, it suggests his redemptive paradigm, the bringing (exhaling—breathing?) of good out of evil, light out of darkness. Once again, in the latter, the emergent creativity of Light proves dominant, but not without an unsettling hint of cyclicality. A natural cycle of light and dark is orderly and pleasing, but when, as here, these alternating phenomena take on moral, ethical, or otherwise symbolic values, they become significantly different. In the chronology of the epic they have become so ever since Satan's revolt in heaven, and in the order of its narration, ever since the poet's invocation at the beginning of Book I.

Raphael's second passage bearing on Night's relation to God leads to another combination of light with dark, as twilight was earlier, but this time to a bisexualized one. This sexualized combination offers the fullest account of God's cave in the epic and goes far to resolve the relation of Light and Night. Raphael, shortly before his creation narrative, warns Adam and Eve to stay within the lawful bounds of knowledge, framing a narrative of being with a reminder about limits. He cautions the human pair against speculation about "things not reveal'd, which th'invisible King, / Onely Omniscient, hath suppress in Night" (VII.122–23). His words hold yet another striking revelation about the relation of God and Night. When together Sin and Satan initially peer into Chaos, a moment whose importance only increases as the epic unfolds and recurrently triggers memories of it, "Before thir eyes in sudden view appear / The secrets of the hoarie deep" (II.890–91). These are the secrets of eldest Night and Chaos, which Satan, although he has already witnessed them, later assures these two figures he has no intention to explore or disturb (II.970–72). Leonard, citing Edward Le Comte, notes and accepts in "secrets" a pun on "secret parts, genitals" and in "hoarie" perhaps another pun on "whore."[63] Granted these puns and the fact that the unrevealed things "suppress [by God] in Night" are not called secrets by Raphael in his warning, his own word for them, "things," is elsewhere a common pun on the genitals of either sex in the period—one throughout Shakespeare's writings, for example; moreover, in a related passage, Raphael does use the word "secrets" to designate what God in Book VII has "suppress in Night," the deity in Book VIII thus acting "wisely to conceal, and not divulge / His secrets to be scann'd" (73–74).[64] Given such puns, the several, intimate associations of God with Night, together with Light, in the epic support the possibility that in the secrets of Night *and* Chaos, Milton updates and explicitly and tactfully sexualizes a common medieval reference to "goddes privete(s)," God's privacy, mystery(ies), secret(s), into which one spies or pries at one's peril.[65] Meanings of *privete(s)* range from "privacy, secrecy, concealment, and private place, hiding place" to both "male and female genitals."[66] In connection with the violation of *privete(s)*, the biblical story of Ham's viewing his father Noah's nakedness also comes to mind; for it, Ham's son, Canaan, who implicitly participates in the violation, is cursed (Genesis 9:20–27). The Bible, in short, authorizes a precedent, the sort of thing on which Milton brooded—like that dove "on the vast Abyss" (I.21).

Only Sin and Satan are explicitly said to peer into such secrets in the poem, and even Satan, subsequently battered by his flight through "utter and through middle darkness," distances himself from any intention to probe further "The secrets of your Realm [Night's and Chaos']" (II.970–72, III.16). For the poet who celebrates the "two great Sexes," whose "Male and Female Light, / . . . animate[s] the World," ensouling (*anima*) and vitalizing it, the sexuality of God, secreted within the deep darkness of Chaos and Night, the one figured male, the other female, is hardly inconceivable (VIII.150–51). As many readers have noticed, accounts of creation in *Paradise Lost* depict the bisexual Spirit of God, like a female dove "brooding on the vast Abyss" and, like a male one, impregnating it.[67] *Spirit* is itself a common word for *semen*, "seed," in this period as well as for the Spirit of the Godhead, and in this latter sense, traditionally an expression for the bond of love between God and his offspring, the Son.[68] Moreover, "seed" is present in the reproductive organs of both males and females, according to the best medical authorities in the period.[69] Respecting Milton's unified vision in *Paradise Lost* and particularly his conception of prime matter, Harinder Singh Marjara connects the "materials dark and crude / Of spirituous and firie spume" under heaven's floor, notably including "Sulphurous and Nitrous Foame," both with the stuff of hell and Chaos and with generation.[70] Milton's descriptions themselves thus suggest the connection with semen.

Returning to that "Cave / Within the Mount of God, fast by his Throne," whence light and darkness, "Like Day and Night," alternate "in perpetual round," I want to take into account its systemic participation in a radiating pattern of bisexuality.[71] My explanation begins by recollecting and developing familiar passages, most of which I have invoked before. In God's mount, Light and Dark are complementary principles, potencies, attributes. Their alternation suggests creation and rest, life and privation, for which another term, introduced by Kepler, might be inertia, a version of stasis.[72] Privation in extremis brings death, but if not in extremis and instead combined with light, it becomes provident shade or grateful twilight. It seems fitting, given principles of life and privation, movement and its lack, that God should create not only a world of life on earth but also in hell one of death, if he chooses to do so. Death has no place in Raphael's narrative of creation, however, aside from the purging of those black infernal dregs adverse to life, which certainly glance at it. In the narrative sequence of the poem and likely also in its chronology, such dregs are the aftereffects of rebellion and hell. To Raphael's God of Infinitude, "*because*" of whom "the Deep" is "Boundless," also belongs the deep, dark "illimitable Ocean without bound" of Night and Chaos.[73] This same "wilde Abyss" of atoms warring in otherwise empty space the Miltonic narrator earlier described as "The Womb of nature and perhaps her Grave"; doing so, he merely interpreted the abyss from a later, postlapsarian, earthly perspective in which death is physically real and not from one too simply opposed to Raphael's sense of the Deep in his account of creation (II.911).[74] Instead, the narrator's description of Chaos and Night provides the darker underside, at once more primitive and privative, of God's cave of light and dark within the mount, or metonymy, of God in Raphael's later narration.

From the vantage point of both their narrations, the boundless deep of Night and Chaos is itself within the eternity and potentiality—the potency—of Milton's dynamic God.

The cave and the deep of Milton's God, the cave above and the deep below (and more generally outside) both heaven and the created universe, together participate in what might be the most common, deeply rooted pairing in Western myth, that of the womb with the tomb. Each, cave and deep, is both or potentially both. Milton's enactment of this pairing also has a more specific and highly relevant early modern ancestor in Spenser's mythic Garden of Adonis, whose existence illuminates and further supports its likelihood. Leonard notes that the precedent for Milton's "wilde Abyss" even after the Creation—thus, I would add, as a basic component of being as such—is Spenser's "huge eternal *Chaos*," which lies "in the wide wombe of the world . . . In hatefull darknes and in deepe horrore" and continues to supply "the substances of Natures fruitfull progenyes."[75] What Leonard does not note is that within Spenser's account of this same mythic Garden of Adonis, a place of eternally cycling life and death, is a bisexual mount of perpetual generation, underneath which is a cave of sex and death containing the wild, empassioned, deadly boar (III.vi.48).[76] Milton could hardly have failed to notice the obvious symbolism of these components.

Analogies of Being, Temptation, and Redemption

What remains to consider is fundamental both to Milton's plot and to the role of analogy in it: specifically, God's analogy of redemption in Book III; the scale of Edenic being in Book V, which is ontological and analogical; Satan's explicit and effective parody of this analogical scale in Book IX; and in books XI to XII, starting with the Angel Michael's reassertion of the Father's redemptive analogy, the analogical reading of history at once within the poem and in the world before and beyond it, an extension that the analogical series of occasions implies. This final reading, like the earlier uses of analogy in the epic, participates in the debate about the valid use of analogy that is evident in medieval and early modern culture as well as in ancient and modern times. The analogical reading of history in Milton's final books is also more narrowly figural, as Erich Auerbach has employed this term but without his effort to defend the historicity of biblical writing by unrealistically walling it off from extended metaphor, whether in its allegorical or analogical form or in their combination.[77] The relation of history more generally to rhetoric, story, and fictive narrative has been revised since Auerbach theorized figuration while not simplistically rejecting his brilliant, seminal work.[78] My own reading of Milton's epic continues to argue and to show that analogy, like metaphor, is typically creative and cognitive, not merely illustrative, and that it is also a way of seeing and understanding the world that is deeply rooted in the history of intellectual culture. Above all, it is the constructive instrument by which different registers of meaning and being are related. Its relational function, which inheres in its structure, is neutral with respect to content, as the greater materiality of twilight, for example, is not, yet both serve in their own ways to mediate extremes and opposites.[79]

In Book III, the leading book of light, God explains his plan of redemption for human beings in analogical terms whose structure is at once conspicuous, familiar, and basically Aristotelian: "As [as if] from a second root," God's Son, who is to be the New Adam and new "Head of all mankind," though born of woman and therefore also "*Adams* Son," is to restore what fallen Adam lost (286, 288). God combines the analogy of a root and its branches, at once organic and traditionally genealogical, with procreative descent and rhetorical transference, with relationship and renewal, and with similarity and difference: the New Adam is to a flourishing root as the Old Adam is to a dying one; the New is to life as the Old to death. The next stage of analogy continues the organic image and develops a social analogy whose terms are financial and legal: the merit earned by the Son's sacrificial love will be "imputed"—attributed, again transferred—to all Adam's guilty sons (290–91). The verb *imputed* derives from Latin *in*, "into," and *putare*, "to cleanse, purify; to settle an account, to reckon." Reaffirming the original, organic image of a root but implicitly developing it into a graft, which more clearly involves artful re-creation, at once botanical and metaphorical, God then promises his own Son that Adam's sons will "live in thee transplanted, and from thee / Receive new life" (293–94). He continues, combining the social with the organic imagery:

> So Man, as is most just,
> Shall satisfie for Man, be judg'd and die,
> And dying rise, and rising with him raise
> His Brethren, ransom'd with his own dear life.
>
> (III.294–97)

The analogies are layered and intertwined, but the logic of analogy is obvious. Talionic justice (an eye for an eye) will at once be satisfied and transmuted to redemption. It will be transplanted, translated, carried across to another reality through the potency of self-sacrificing charity—love. All this through analogical metaphor, has been, now is, and will be actualized in history by God through his Son.[80]

God's use of analogy is not an isolated instance of his method, indeed of his thinking, insofar as his method represents it. Analogy pervades his creation. In the analogical scale of being, which Raphael voices in Eden, and later in Satan's abusive parody of it in his temptation of Eve, a form of the word *proportion* even occurs, explicitly marking the presence of its traditional synonym, *analogy*. The same word, *proportion*, also recurs in Adam's self-abusive replay of Eve's reasoning just before he, too, falls. Satan, by exploiting analogy in his temptation of Eve, is intent on fully upsetting God's applecart, on stealing, possessing, and polluting not only his fruit but also his method, the promise inhering in it, and finally his creative and re-creative Word. The questions become, in short, to whom does analogy belong, and how is it valid? The same questions have risen recurrently in this volume in diverse historico-cultural contexts.

The introduction of Raphael's full-scale analogy of being is prefaced by an explanation of his capacity as an angel to taste, digest, and assimilate the "earthly fruits" of Eden, turning fleshly body to spirit, corporeal to incorporeal—lower to higher, heavier to lighter, denser to rarer, one degree or form of substance to another (V.412, 464): this is the familiar Aristotelian formula for analogy, A:B :: C:D :: E:F, and so on, which offers the possibility of exchanging certain of its terms, such as B with D. We have seen it in Cajetan, Descartes, Kepler, Donne, and numerous others. The Miltonic narrator's theologically loaded verb for this assimilation is *transubstantiate*—carry across, transfer, or translate one substance to another, as literally and objectively in the Catholic Eucharist but figuratively and subjectively in Reformed Protestant ones (V.438). This term, as Milton's narrator punningly uses it, refers at once to a physiological process and to the actualizing of a structural metaphor. Both involve basic, material continuity and formal difference: on the one hand, form is material for Milton, but on the other, fleshly body is not spirit absolutely, or incorporeal corporeal absolutely, any more than denser is rarer or a stone is a gas. Both involve substantial change, if we take "substantial" to indicate a change in form. The change is now organic and natural rather than supernatural and mystical; it is also formal rather than *simply* material.[81] (Note again both that the word *substance* itself is controverted and notoriously unstable in this period, not simply identical to *matter*, and that form is material for Milton.)[82]

The same—that is, basic continuity and formal difference—is true of the Miltonic narrator's succeeding analogy for this substantive change, the alchemical turning of "Metals of drossiest Ore to Perfect Gold" (V.438, 442). Yet the narrator's use of the alchemical analogy is qualified by skepticism ("or holds it possible to turn": 441), a small dissonance but worth noting, given tensions in Raphael's later development of his own analogical argument. At base, the process the Miltonic narrator describes in his alchemical analogy remains material, but through it, the form (material in Milton but not the same as the base) is different. Form is what shapes, distinguishes, and actualizes a thing. Associated with essence in Milton's thinking, it is real. In a striking perversion of the upward transubstantiation that Raphael and the Miltonic narrator together describe, Satan's form degenerates from higher to lower, and he sinks increasingly into base matter, perhaps, from an ontological perspective, returning to Chaos and Night in the end. Morally, he also sinks ever deeper into sin, and the result is increasing privation, deeper and deeper darkness, whose last extremity is loss of being. From this perspective, Night as void would suitably at last mean annihilation for him, if anything created by God could be reduced to nothing.[83] At the same time, since void in actuality exists with and for atomic matter, the annihilation of satanic matter is no more than a theorized, hypothetical speculation whereas Satan's reduction to his chaotic, material base is an increasingly real likelihood. Kepler's suggestive phrase "materialized darkness" for blackness comes again to mind.

Raphael's analogy of being soon follows his analogical explanation of angelic digestion, doing so in response to Adam's request for further comparison—analogical food for thought (467). Raphael provides this but not before warning Adam that all things return to "one Almightie . . . If not deprav'd from good"—that is, if not morally and physically corrupted: *de*, here as intensifier, "completely," and *parvus*, "crooked"—and therefore, if not bent against (a Lucretian atomist might say "if not swerving away" from) its true nature (V.469, 471).[84] Raphael's explanation begins, "All things" have been created, all

Such to perfection, one first matter all,
Indu'd with various forms, various degrees
Of substance, and in things that live, of life;
But more refin'd, more spiritous, and pure,
As neerer to him plac't or neerer tending
Each in thir several active Sphears assignd,
Till body up to spirit work, in bounds
Proportiond to each kind. So from the root
Springs lighter the green stalk, from thence the leaves

More aerie, last the bright consummate floure
Spirits odorous breathes: flours and thir fruit
Mans nourishment, by gradual scale sublim'd
To vital Spirits aspire, to animal,
To intellectual.

(V.471–85)

At first, Raphael's explanation might seem to repeat what he has earlier said about angelic digestion. But its first sentence includes an emphasis on boundaries and a tension between this and aspiring growth or between circular and upwardly spiraling movement not evident in his disquisition on digestion, although present elsewhere in the epic. Moreover, until the word "Proportiond" enters near the end of this sentence, to be followed by the proper analogy of a growing plant, a reader could overlook the fact that the passage is specifically analogical. To begin with, all are more or less basically, or materially, the same in Raphael's explanation (471–74), although, in an apparent reflection of the Miltonic narrator's alchemical metaphor, which noticeably intervened between Raphael's digestive analogy and the present one, each has or attains greater refinement and purity, greater spirituality, the closer to God it gets or, alternatively, already is (475–76). Some tension now exists between "is" ("plac't") and "gets" ("tending"), respectively alluding to fixed position and aspiring growth and to stasis and dynamic, upward movement. It is as if the intrusion of alchemy (441), a human effort to go nature one better, had introduced another consideration, a complication in the assimilative yet still upward movement. Sequence has again invited attention, suggesting a process of thought.

The process of transubstantiation then returns after additional emphasis on the formal differences among substances, each active within the purer or less pure sphere assigned it and thus within a hierarchy. At the same time, however, the phrase "Till body up to spirit work" right after a reference to spheres, whether in the precise sense of three-dimensional circles or more loosely as a domain of activity, implies motion that spirals upward and suggests its imaginative origin in Neoplatonism (477–78). This motion is bounded, however—"Proportiond to each kind" or form of substance and still implicitly hierarchical with respect to purity (479). Tension, in short, continues.

Proportionality might be said to reach full bloom in the analogy of root, stalk, leaves, and flower exhaling odorous spirits, which follows and is next applied to similar aspiring growth within the physiology of human being.

This analogy at once suggests the articulation of each part and the cooperation, continuity, and wholeness attained in the completion of perfect growth. Combining rhetorical technique with botanical content, the analogy of a growing plant materializes Raphael's initially more abstract explanation, and doing so resolves the apparent tension between established order and aspiring growth. It is functional and cognitive, not merely illustrative. It resembles the concluding *ergo* of a syllogism, but with the vitally important difference that it is dynamic and physical, not merely logical: the stalk "springs" and the flower "breathes," as do the vital spirits of human kind.[85] Analogy itself has thus been trans-shifted from abstract logic to more holistic poetic invention in the analogy of root-to-floral spirits with the gradual, laddered spirits within human beings. Time and process, barely implied before, have specifically entered the equation. Aspiring growth has first been contained within the plant, then this level of being has been absorbed into human being through the consumption of "flours and thir fruit" (482), thereby becoming the most basic kind of spirit within human kind and through stages rising to animal and intellectual forms. Aspiring growth and bounded kinds have both been respected. This is neither logic nor metaphysics; it is analogical poetry, and it involves substantial change.

Raphael exercises the traditional cognitive function of analogy in moving from the visible to the invisible, from the sensible plant to the spirits hidden within the human body, spirits real in theory but, as can too readily be forgotten, not accessible and convincingly demonstrable to sense. As in Kepler's mental tribunal, the actual process of change is hidden and has to be taken on (scientific) faith. With all the pluses of Raphael's analogy, it is still limited by its confinement within the botanical and physiological, a limitation that he only transcends ambiguously or after the fact: ambiguously—indeed, in a logical term, *equivocably*—as Stephen Fallon and N. K. Sugimura alike have observed, by seeming to equate the intellectual spirits with the soul, from which they are otherwise distinct in Raphael's phrasing.[86] Further ambiguity arguably results from the addition not only of discursive but also of intuitive reason, or intuitive intellection (*nous*, traditionally distinguished from *dianoia*), as a possible endpoint of the process of aspiration (sublimation) and still later the addition of conscious choice. This last addition and arguably the one before it come after the angel's analogy.

Raphael holds that discursive and intuitive reason differ "but in degree, of kind the same," relating this fact to his evident consumption of Edenic fruit (V.490–92). As Joad Raymond observes, however, Raphael's claim at this point that the fruit is converted into the angel's "proper substance" is also ambiguous, meaning either "'suitable for angels' or 'superior in substance to paradisal vegetation'" or, perhaps, I would add, poised between both (287). In other words, further analogy, doubleness, or difference, rather than pure identity, between angelic and human digestion is evident, as the explanation moves between extremes of flesh and spirit, corporeality and incorporeality. The ultimate form of incorporeality would presumably be God, a Spirit, yet one in whom Milton's *Christian Doctrine* nonetheless posits "some bodily force" or some principle of corporeality: the word "force" (in Latin *vis*) is another multipurpose word in this period; commonly words, too, have force, as do fictions and figures in Milton's *Logic*.[87] Like *transubstantiate* or *transform*, Raphael's word *convert* (< *cum*, "together," and *vertere*, "turn") carries with it the intimation of metaphor—the archtrope, or archturn (< *tropos*, "turn"). Etymological metaphor is further notable in Raphael's explanation; for a familiar example, the word *fruit* itself traces back to Latin *frui*, "enjoy," which, in the conversation of an angel about eventual beatitude, invites memory of Augustine's association of this Latin verb with spiritual and ultimately beatific enjoyment, an association that is commonplace knowledge throughout the early modern period (and firmly distinguished by Augustine from the things of this world). If the tie of metaphor itself to the material earth is highlighted in such ways, a familiar function of analogy has always been to bridge the visible and invisible; there is nothing extraordinary about its doing so here. Even the word *degree*, which the *OED* derives from Latin *de*, "down," and *gradus*, "step," has this tie; *degradation*, "a step down," is its most obvious cognate. In the beatitude that Raphael promises, what some might see as a realistic shadow, or fallen possibility, is thus also present.

To maintain, as Raymond does, that Raphael's mode of representation, or accommodation, "is not a form of metaphor . . . but a means of representing truths in figurative manner" is un-Miltonically to deny metaphor in its very presence.[88] Effectually, it is also to return this issue to the discussions of light and metaphor by Aquinas and Cajetan, examined in Chapter 4, which try, through the semantics of logic, to separate "true" light from its figurative uses. To do so on the basis of faith is not to remove the presence of structural metaphor any more than right reason, or reason enlightened by faith, simply cancels the use of reason. When, treating "the Similar," or what "is called proportion, in Greek usually *analogia*," Milton held in his *Art of Logic* that "A fictitious similarity (*similitudo*: similitude) has the same force (*parem vim*)" as the nonfictitious ones he has considered; the most immediate example of the latter is an analogy between a helmsman, a storm, a wise man, and fortune, and the forceful fiction he specifies is a fable by Aesop. "Force," a proper, historical translation of *vim*, indicates not truth as such but the "Power to convince or persuade the reason or judgement: convincing or appealing power."[89] Milton, a logician and grammarian as well as a poet, knew better than most when he was using metaphor and metaphorical analogy, with or without the enlightenment of faith.

Raphael, now clearly outside his botanical analogy, adds obedience to his explanation of fruition. This is the moral dimension of conscious choice, which, like intuition, is informed by faith.[90] Stopping with the transubstantiation of vegetable into fully animated human being, Raphael would have finessed the more complex issues that occur at the higher levels of conscious being and come to a head first in Satan's rebellion, then in Eve's, then Adam's. These involve reason but finally and crucially faith, which is affective and intuitive, and in Milton's epic finally inseparable from right reason: etymologically, obedience is the capacity to hear and heed (< Latin *oboedire*, "to listen to"), or in biblical terms I have cited before, to have the enlightened ears that hear and eyes that see with faith (for example, Matthew 13:16–17). Again, however, obedience is notably *outside* Raphael's analogy, beyond it, an exteriority, or otherness, to prove consequential in the temptation of Eve, then Adam. The limits of analogy, not only its cognitive potencies, are part of the Miltonic story.

Continuing, the angel speculates that someday, if the human pair remain obedient, their "bodies may at last turn all to Spirit . . . and wingd ascend / Ethereal, as wee" (497–99). His speculation is innocent enough yet also potentially misleading, an opportunity to overlook his earlier emphasis on hierarchical order and his subsequent reminder of the obedience that this order assumes: transubstantiation, as Raphael represents it, normally operates within the appointed spheres of being and only extraordinarily beyond them, yet flowers and their fruits turn dynamically into human being. Raphael's message is double-sided. Possibly the extraordinary, which is outside established order, would in time have become the ordinary for human beings had Adam and Eve not fallen, and this is evidently Raphael's meaning, but it remains subject to misunderstanding. Again a tension—perhaps more charitably and holistically a finely calibrated balance—between hierarchy and aspiration is once more evident, as in Raphael's initial exposition of being in the passage.[91]

Raphael's speculation that human bodies might turn to spirit also seems a poor fit for Milton's stress in *Christian Doctrine* on the inseparability of body and spirit: "The idea that the spirit of man is separate from his body, so that it may exist somewhere in isolation, complete and intelligent, is nowhere to be found in scripture and is plainly at odds with nature and reason."[92] Such separation and Raphael's vaporizing the body into spirit are alike in denying that such bodily existence is *definitively* human. Even a Neoplatonist like Ficino considered a body essential to ultimate human happiness, as noted in Chapter 3. With respect both to the issue of Raphael's bounded sphere and to that of the human body, one might apply the angel's explanation that every superior faculty or form contains within it the inferior ones (V.409–10), but is this more than a theoretical containment? Can a spiritual being be aware in practical terms of the bodily appetites and members it includes or has transcended without serious schizophrenia, or, in Raphael's instance, without ever having been aware of them to begin with? Raphael may eat with "real hunger" (V.437), but is hunger here a substitute for "relish" or "enjoyment" (*fruitio*) rather than a felt need, lack, or craving? How is his hunger real rather than translative in some sense, that is, figurative? When in *Christian Doctrine* Milton explains that "spirit, being the more excellent substance, *virtually*, as they say, and *eminently* contains within itself what is clearly the inferior substance," he adds by *analogy*, "in the same way as the spiritual and rational faculty contains the corporeal, that is the sentient and vegetative faculty."[93] In *Paradise Lost*, as in *The Faerie Queene*, phrases like "they say" indicate myth or speculative fiction, and there is surely something circular about Milton's illustrating the terminology (*virtually*, *eminently*) he derives from Scholasticism with an analogy based on the medical-philosophical *theory* of spirits or faculties, which is not far distant ontologically from the traditional analogy between physical and intellectual vision recurrently met in this volume.[94] The Scholastic terms *virtually* and *eminently* mean "in a more perfect or higher manner," not "essentially," as Marjara, for one, assumes.[95] Virtual or eminent inclusion is also to be distinguished from formal inclusion, which is more *specifically* defining. *Virtually* is another term that played a conspicuous role in Eucharistic controversy involving figurative and literal meaning. Nicholas Ridley, the learned bishop credited by Archbishop Thomas Cranmer with his conversion to Protestantism, crucially used it so.

Another misunderstanding a listener (or reader) could take from Raphael's speculation is that Adam and Eve would necessarily turn into angels if they were to turn all to spirit. Presumably, the human pair would still differ from angels, formally as a different kind of spirit, perhaps different, for instance, as hydrogen differs from helium.[96] In this sense, they would still be distinct while fully spiritual. As later, near the end of Book VIII, when Raphael evokes from Adam a spirited defense of human love by seeming to reduce it to sexual appetite, mere sense, we are reminded again that this sociable angel has his own bounded sphere assigned. Even in the spiritual realm there are different degrees of spirit, although, in the Son, God unites (has united, will unite) all "as one individual."[97] This is the paradox that I pursued in Chapter 3.

No more than Night and Light does analogy have a single, uniform value in the epic. In Book IX, a crucial argument in Satan's temptation of Eve pointedly recalls the analogical order of creation that Raphael has earlier laid out. Satan perverts the cognitive use of analogy and by doing so parodically demonizes the order it creates: again, the two are intimately connected in his hands. If unwittingly or intuitively, Satan also exploits Raphael's speculation about Adam and Eve's ascent to pure spirit, which he has not overheard. Satanically possessed, the snake, already having presented himself as living proof, at once visual and biological, that eating the forbidden fruit does not bring death, proposes to Eve that by tasting the apple she and Adam "should be as Gods, since I as Man, / Internal Man, is but *proportion* meet, / I of brute human, yee of human Gods" (IX.710–12; my emphasis). Satan's ironic confusion of the conditional "should" with "ought"—obligation or duty—is almost funny: the devil is a skilled analogist, aware of the qualified, speculative, partial aspect of analogical method. Of course, he also omits the framing condition of faithful obedience, basing his analogy on scientifically demonstrable, physical evidence alone and inadvertently exposing the limitation in this context of doing so. Annabel Patterson notices that the continuation of Satan's analogy of godhood, "So ye shall die perhaps, by putting off / Human, to put on Gods, death to be wisht," suggests that "death is merely a metaphor," but she does not remark that the more specific historical issue concerns analogy and, here, a parody of Raphael's analogical promise of beatitude (105). Raphael's speculation, as earlier observed, could, but need not, be misunderstood as transformation from human to angel or indeed to demigod (cf. IX.937, III.341).

When Adam, like Eve, succumbs to temptation, he attempts to rationalize his doing so in much the same terms the snake used. He remarks the tangible evidence that the snake's tasting has not proved deadly to him; instead, he supposes, or pretends to suppose, that the snake

gaines to live as Man
Higher degree of Life, inducement strong
To us, as likely tasting to attaine
Proportional ascent, which cannot be
But to be Gods, or Angels Demi-gods.

(IX.933–37; my emphasis)

If analogy, or proportion, is the way up, it is also the way down, as Satan has already proved by his debasement—his degradation—in the form of a reptile and as the human pair will prove by their reduction of love to bestial appetite in the immediate aftermath of their fall. Analogy in itself is relative in value, and it always includes difference, as well as likeness.

In a still later, more promising aftermath, Adam and Eve penitently utter prayers that fly upward aspiringly, but they arrive

> Dimensionless through Heav'nly dores; then clad
> With incense, where the Golden Altar fum'd
> By thir great Intercessor, [come] . . . in sight
> Before the Father's throne.
>
> (XI.17–20)

The Son, intercessor for the human pair, offers the prayers to his Father in an argument that recalls the charitable logic (and merely rational illogic) of biblical parables, such as that of the prodigal son who finally comes home to a welcoming father and of the single lost sheep, for whose recovery there was more rejoicing than for ninety-nine faithful ones (Luke 15:1–7, 11–32). It also looks forward to the familiar paradox of the Fortunate Fall, the extraordinary, redemptive union of God and man in the Christ as cause for rejoicing despite its seeming dependence within earthly time on Adam and Eve's original sin. In particular, the parables allusively evoked, which are analogical in structure, begin the process of reclaiming analogy from satanic abuse, but the preceding excerpt describing the reception of the sinners' prayers may be even more significant for the Miltonic narrator's purpose.

Sound is considered material both anciently and in early modernity. But noticeably, the arriving prayers are "Dimensionless," as is the "Illimitable Ocean without bound, / Without dimension" that Satan and Sin first view in Book II.[98] They are identified conspicuously as immaterial here, at least in any but the basest, embryon-atomic sense in which the realm of Chaos and Night is so. Alternatively, they are *pneuma*, "breath," and therefore pure spirit—material but fully rarefied—and distinguished from this basest sense, indeed, in moral terms the binary opposite of it. Having clad the prayers, as he earlier clothed the fallen human pair, the filial intercessor dresses them with incense, making them available to vision and smell and, in short, to sense (17–18). He is transubstantiating them, his actions strongly suggesting a religious ceremony, whether Old Testament or New, a suggestion of ritual that "the Golden Altar" and his "Golden Censer" intensify (18, 24). Words like "savour" and "the smell of peace" that follow and invite God to taste and inhale the prayers further suggest not only the material sensibility of God but also a repossessing of physical corporeality in heaven after the fall of Adam and Eve. (Notably, even death sniffs the smell of "mortal change" with a single nostril once the Fall occurs, something heaven noticeably now counters: X.272–73, 280.) Parallels in the War in Heaven include the filial Word's restoration of the "uprooted Hills" to their places, the flowers to "Hill and Valley," and his repossession of darkness.[99] Perhaps both the senses of "dimensionless" that I have suggested are therefore pertinent, and the Son, the filial Redeemer, is returning them from opposition to harmonious continuity. What is clear is that the dimensionless prayers of sinners need transubstantial help to find acceptance in heaven.

The Son's intercession for Adam and Eve's penitent utterances includes another highly resonant word: "ingraft" (XI.35). The Son refers to the ingrafting of all their works, good or bad, on himself as their representative and the redeemer of their debt, and his botanical image pointedly recalls the transplantation and renewal in God's earlier analogy of redemption. Grafting signals the need of art, godly art, to remedy and improve fallen nature. Once again, this is an analogical art. The reception and acceptance of Adam and Eve's prayers at the opening of Book XI then eventuates in Michael's mission to reveal the course of biblical history to the human pair before expelling them from the Garden. A conspicuous thread in this revelation concerns the recurrence of instances in which a single, just individual stands out. Earlier in the narrative, this thread started with the Son himself in Book III, then the loyal angel Abdiel in Book V; in the chronology of the poem, it started with the Son's providential exaltation. On the fallen earth, the same thread includes Enoch, Noah, Abraham, Moses, and Jesus—all in analogical order and serving both to correct and balance the catalogue of demonic figures, gods of the fallen world, in Book I and to reconnect earth to heaven: so the Son is to Abdiel as Enoch to Noah, as Abraham to Moses, as Moses to Jesus. The poet, "In darkness, and with dangers compast round, / And solitude," who had begun his epic by paralleling it with Genesis and himself with Moses, participates in this thread as well. The thread is figural, but it is even more significantly analogical, involving structural contrast and difference, not only similarity, and embracing connections that are not only or simply biblical but also extend well beyond biblical times. Both figure and analogy are fundamentally and perceptually metaphorical as well, even where they are also historically real or real by divine fiat or both. God writes in history and, in this poem, in any other form of existence he chooses, even when he declines to choose or chooses negatively. Analogy connects cognition with rhetoric, God's creation with the poet's, and both of these with darkness and light.

In Book XII, near the end of Michael's revelation, the angel describes the Resurrection of Christ in terms that recall both the financial aspects of God's analogy of redemption in Book III ("Thy ransom paid": XII.424) and the analogy of Satan as a sun newly risen but shorn of its beams in Book I. Significantly, the Son of God rises while it is still night, although starlit: "*ere* the third dawning light / Returne, the Starres of Morn shall see him rise / Out of his grave, fresh as the dawning light" (XII.421–23; my emphasis). The phrase "Starres of Morn" glances as well at Satan's former name, Lucifer, also the name of the morning star, and now grants this name to all the faithful angels. As so often elsewhere in this epic, redemption and repossession extend to the coordinates of light and darkness and to their analogous applications in poetic memory. After Michael's account of redemption and resurrection, the last third of Book XII holds a final occurrence of the word "light," which comes in Adam's enthusiastic, if still naïve, response to the producing of good out of evil and "Light out of darkness!" (469–71, 473). Yet for all the comfort of promised redemption and the poignant memories and intimations of hope in the final verse paragraph of *Paradise Lost*, the Fall has greatly darkened the poem by its end. Here the only light comes with the "bright array" of descending cherubim, while from below dark mist gathers. Like a comet, the "Sword of God" blazes before the cherubim, but threateningly and destructively, scorching the Garden (633–36). The sword flames with the wrath last seen in the War in Heaven and first seen in the flaming darkness of hell. Like Providence, a promise is still there in the final lines, but otherwise they are at best open-ended.

The Introduction to this volume, which began with death, darkness, and evil, concluded with a précis of the present chapter, which began the same way. In ending, it noted that the issue of analogical figuration is still with us—still an issue debated and still one whose outcome, or issue, is productively and creatively open. In my own play on the word *issue* and its Latin root (*exire*), my introduction also remarked an extension of Donne's play in *Deaths Duell*, his own funeral sermon, which puns hauntingly and provocatively on this word.[100] An analogical series of the sort evident in Kepler's harmonic law, the forerunner of Newton's theory of gravity, suggests the possibility of an infinite extension, although Kepler himself did not embrace this. For logical and speculative thought, for the extension of possibilities—whether constructive or parodically deconstructive—analogy arguably vies in importance with the introduction into Western culture of the mathematical sign zero, a relative latecomer in the thirteenth century.[101] Essentially creative, imaginative, fictive, hypothetical, and metaphorical, analogy remains a fundamental resource for poetic brooding across cultures—historical, geographical, and intellectual.

Acknowledgments

Over a decade, this project has incurred numerous debts to those who have kindly read it in whole or in part, including the anonymous readers for Fordham University Press and for its first three chapters, which were published in earlier forms, namely, "Body of Death: The Pauline Inheritance in Donne's Sermons, Spenser's Maleger, and Milton's Sin and Death," in *Rhetorics of Bodily Disease and Health in Medieval and Early Modern England*, ed. Jennifer C. Vaught (Farnham: Ashgate, 2010), 171–91; "Mutability and Mortality: Reading Spenser's Poetry," in *Celebrating Mutabilitie: Essays on Edmund Spenser's Mutabilitie Cantos*, ed. Jane Grogan (Manchester: Manchester University Press, 2010), 246–74; and "Satanic Ethos and Envy: The Origin of Evil and Death in *Paradise Lost*," *Milton Studies* 51 (2010): 137–64. I am grateful to the publishers of each of these volumes and to their editors, in the instance of *Milton Studies*, to Duquesne University Press and to Laura L. Knoppers. In addition, my introduction was read, with helpful comments, by Sarah Massey-Warren and later by my Fordham editor, Tom Lay. Two colleagues at Indiana University, Joan Pong Linton and William R. Newman, the first in early modern English and the second in the History and Philosophy of Science, offered comments and helpful suggestions for Chapters 4 and 5, respectively. My greatest debt is to Tamara Goeglein, who read the whole manuscript over the course of some years and responded critically to it. I am grateful to all my readers, each of whom contributed measurably to an improved book. I also thank those who heard talks derived from parts of the manuscript at various conferences and responded to them: Spenser at Kalamazoo, the Newberry Library Milton Seminar, the International John Donne Society Conference, and the Milton Conference in Murfreesboro. Julian Lethbridge stands out among these responders, insofar as his question in Kalamazoo led to my chapter on Spenser's Mutability. In addition to these collegial contributions, a grant from the Office of the Vice-Provost of Research at Indiana University partially funded the final stage of my project.

My deepest expression of gratitude is for the life and work of the late Helen Tartar, my former editor at Stanford and Fordham. Helen's intelligence, judgment, and dedication meant a world of difference to me. Her friendship, celebrated with our annual MLA dinner, was a delight, mutually expressed in quirky jewelry and, when possible, colorful garb. Now I read her posthumously published poems with a renewal of memories, and I dedicate this volume to her.

Notes

INTRODUCTION. ISSUES OF DEATH, LIGHT, AND ANALOGY

1. *Online OED*, s.v. *Issue n.* and *v.* (accessed 8/11/15); also Charlton T. Lewis and Charles Short, *A Latin Dictionary* (Oxford: Clarendon, 1966), s.v. *Exeo*. The suggested meanings extend the English and Latin definitions, as does Donne in the sermon mentioned next in this paragraph and in note 2, below.

2. John Donne, *The Sermons of John Donne*, ed. George R. Potter and Evelyn M. Simpson, 10 vols. (1959; rpt. Berkeley: University of California Press, 1984), 10:229–48. Donne delivered the sermon less than a month before his death and in anticipation of it.

3. *The Faerie Queene*, ed. A. C. Hamilton, 2nd ed., with text by Hiroshi Yamashita and Toshiyuku Suzuki (Harlow: Pearson, 2001), VII.vii.45–46. On Spenser's figure of Death, see also David Lee Miller, "Death's Afterword," in *Imagining Death in Spenser and Milton*, ed. Elizabeth Jane Bellamy, Patrick Cheney, and Michael Schoenfeldt (Houndmills: Palgrave Macmillan, 2003), 185–99, here 188–89.

4. *Online OED*, s.v. *Visage*, 1a, 3a, 7, 8a, b (accessed 1/31/15).

5. The original title was "Satanic Ethos and Envy: The Origin of Evil and Death in *Paradise Lost*," *Milton Studies* 51 (2010): 137–64. Originally, it was invited for circulation by the Newberry Milton Seminar, organized by David Loewenstein and Regina Schwartz.

6. *De Doctrina Christiana*, ed. James Holly Hanford and Waldo Hillary Dunn, trans. Charles R. Sumner, 4 vols. (New York: Columbia University Press, 1933), 15:202–3 (Bk. I.xii), in *The Works of John Milton*, ed. Frank Allen Patterson et al.: "Mala . . . omnia [the singular main verb implies the collective 'evil'] et quicquid ducere ad interitum videtur, sub mortis nomine [literally, 'name'; modern 'word'] summatim Scriptura complectitur"; translation mine. Cf. *Christian Doctrine*, ed. Maurice Kelley, trans. John Carey, in *Complete Prose Works of John Milton* (New Haven, Conn.: Yale University Press, 1973), 6:393.

7. On prosopopeia and personification, see Richard A. Lanham, *A Handlist of Rhetorical Terms* (Berkeley: University of California Press,

1969); Aristotle, *Poetics*, trans. Richard Janko (Indianapolis, Ind.: Hackett, 1987), 207, s.v. *ēthos, ethos*; and James J. Paxson's admirable study *The Poetics of Personification* (Cambridge: Cambridge University Press, 1994), 8–34.

8. These books ranged from Philippe Ariès, *The Hour of Our Death*, trans. Helen Weaver (1981; rpt. Oxford: Oxford University Press, 1991); Jonathan Dollimore, *Death, Desire, and Loss in Western Culture* (London: Penguin, 1998); Robert N. Watson, *The Rest Is Silence: Death as Annihilation in the English Renaissance* (Berkeley: University of California Press, 1994); *Imagining Death in Spenser and Milton*, ed. Elizabeth Jane Bellamy, Patrick Cheney, and Michael Schoenfeldt (London: Palgrave Macmillan, 2003); Louis Schwartz, *Milton and Maternal Mortality* (Cambridge: Cambridge University Press, 2009); Susan Sontag, *"Illness as Metaphor" and "AIDS and Its Metaphors"* (New York: Doubleday, 1990); Elaine Scarry, *The Body in Pain: The Making and Unmaking of the World* (New York: Oxford University Press, 1985); Dale B. Martin, *The Corinthian Body* (New Haven, Conn.: Yale University Press, 1995); Ovey N. Mohammed, *Averroës' Doctrine of Immortality: A Matter of Controversy* (Waterloo, Ont.: Wilfred Laurier University Press, 1984); Norman T. Burns, *Christian Mortalism from Tyndale to Milton* (Cambridge, Mass.: Harvard University Press, 1972); Brian Cummings, *Mortal Thoughts: Religion, Secularity, and Identity in Shakespeare and Early Modern Culture* (Oxford: Oxford University Press, 2013). On psychological and conflicting medical definitions of death, or rather on their effectual lack, cf. Annabel Patterson, *Milton's Words* (Oxford: Oxford University Press, 2009), 101–11, here 99–100. This list is supplemented in my notes to the various chapters.

9. C. P. Snow, *The Two Cultures* (1959; rpt. Cambridge: Cambridge University Press, 1993).

10. William Shakespeare, *King Lear*, ed. R. A. Foakes (London: Thomson Learning, 2000), IV.vi.198–99. For the ancient commonplace that whatever lives, moves, see *The Faerie Queene*, VII.vii.55: "all that moueth, doth mutation loue"; cf. vii.17–19, 47, 58: Mutability's claim is that everything that moves, that is everything that lives, changes.

11. Aristotle, *On the Soul* (*De Anima*), trans. J. A. Smith, in *The Complete Works of Aristotle*, ed. Jonathan Barnes, rev. ed., 2 vols. (Princeton, N.J.: Princeton University Press, 1984), 1:682 (429a3–4).

12. Hans Blumenberg, "Light as a Metaphor for Truth at the Preliminary Stage of Philosophical Concept Formation," trans. Joel Anderson, in *Modernity and the Hegemony of Vision*, ed. David Michael Levin (Berkeley: University of California Press, 1993), 30–62, here 31; cf. "Light remains what it is while letting the infinite participate in it" (31).

13. Jacques Derrida, "White Mythology: Metaphor in the Text of Philosophy," trans. F. C. T. Moore, *New Literary History* 6 (1975): 5–74, esp.

46–53; "La mythologie blanche (la métaphore dans le texte philosophique)," *Poétique* 5 (1971): 1–52, esp. 31–36. For background, see David C. Lindberg, *Theories of Vision from Al-Kindi to Kepler* (Chicago: University of Chicago Press, 1976), 89–97: Lindberg recounts how the study of physical light was thought to enable the comprehension of "the origin and structure of the material universe" (97).

14. Fernand Hallyn, *The Poetic Structure of the World*, trans. Donald M. Leslie (New York: Zone, 1993), 13–14, 20; also Paul Ricoeur's *Time and Narrative*, trans. Kathleen McLaughlin and David Pellauer, 3 vols. (Chicago: University of Chicago Press, 1984), 1:31. Cf. Karsten Harries, *Infinity and Perspective* (Cambridge, Mass.: MIT Press, 2001), e.g., 198: "there is a poetic component to all human knowledge."

15. Johannes Kepler, *Mysterium Cosmographicum: The Secret of the Universe*, trans. A. M. Duncan, introduction and commentary by E. J. Aiton (New York: Abaris, 1981), 149: text of the second edition, published in 1621, which appends notes to the text of the 1596 edition.

16. Charles Sanders Peirce, *Peirce on Signs: Writings on Semiotic by Charles Sanders Peirce*, ed. James Hooper (Chapel Hill: University of North Carolina Press, 1991), 251–52. De Man, "The Epistemology of Metaphor," in *On Metaphor*, ed. Sheldon Sacks (Chicago: University of Chicago Press, 1979), 11–28, here 24–25 (my emphasis).

17. Sir Philip Sidney, *An Apology for Poetry or the Defence of Poesy*, ed. Geoffrey Shepherd (1965; rpt. Manchester: Manchester University Press, 1973), 124.

18. Aristotle, *Poetics*, trans. I. Bywater, in *The Complete Works of Aristotle*, 2:1457b16–23.

19. The medium of poetry distinctively *heightens* linguistic tropicality, figuration, wordplay, and the like, but it remains on a continuum with ordinary language, differing in degree, not kind: e.g., cf. Graham Allen, *Intertextuality* (London: Routledge, 2000), 44. On metaphor (i.e., *translatio*, "translation") as the umbrella term for all the tropes, see Judith H. Anderson, *Translating Investments: Metaphor and the Dynamic of Cultural Change in Tudor-Stuart England* (New York: Fordham University Press, 2005), chap. 7.

20. See especially Chapter 4 in this volume.

21. John Donne, *Sermons of John Donne*, 4:87. Cf. Owen Gingerich on the difference between "ontological laws of nature" and "epistemological laws of nature": "Kepler and the Laws of Nature," *Perspectives on Science and Christian Faith* 63, no. 1 (2011): 17–23.

22. E.g., see Mary Poovey, *A History of the Modern Fact: Problems of Knowledge in the Sciences of Wealth and Society* (Chicago: University of Chicago Press, 1998).

23. For the connection with Newton, see W. Pauli, "The Influence of Archetypal Ideas on the Scientific Theories of Kepler," trans. Priscilla Silz, in C. G. Jung and W. Pauli, *The Interpretation of Nature and the Psyche* (New York: Pantheon, 1955), 155–56. The names and explanations of Kepler's laws vary somewhat from writer to writer: e.g., Thomas S. Kuhn, *The Copernican Revolution: Planetary Astronomy in the Development of Western Thought* (Cambridge, Mass.: Harvard University Press, 1957), 212–17 (with helpful diagrams); Owen Gingerich, *The Eye of Heaven: Ptolemy, Copernicus, Kepler* (New York: American Institute of Physics, 1993), 43, 47; John North, *The Norton History of Astronomy and Cosmology* (New York: W. W. Norton, 1994), 321, 323; Albert van der Shoot, "Kepler's Search for Form and Proportion," *Renaissance Studies* 15, no. 1 (2001): 59–78, here 65.

24. Translation by William H. Donahue (Santa Fe, N.M.: Green Lion, 2000); Donahue's title page precedes Kepler's title (*Paralipomena* . . . Astronomy) with the word *Optics*. Donahue discusses the meaning of *Paralipomena* and his decision not to translate it on pages xiv–xv. Olivier Darrigol offers the translation "omitted things" for *Paralipomena* in his *History of Optics: From Greek Antiquity to the Nineteenth Century* (Oxford: Oxford University Press, 2012), 31.

25. In *Mortal Thoughts*, Cummings gradually and progressively offers an explanation of the focus and organization of his study that resonates with the views I express in this Introduction (e.g., vii, 15, 17, 41, 43).

26. Charles Monroe Coffin argues for Donne's awareness of Kepler's *Paralipomena* and *Dioptrice*: *John Donne and the New Philosophy* (New York: Humanities Press, 1958), 166–72.

27. John Donne, *Ignatius His Conclave*, ed. T. S. Healy (Oxford: Clarendon, 1969), 7; *Biathanatos* (1646[?]; rpt. New York: Facsimile Text Society, 1930), 146 (Donne's marginalia, Pt. 2, Dist. 6, Sect. 8); and *Paradoxes and Problems*, ed. Helen Peters (Oxford: Clarendon, 1980), 33 (Problem X, variously numbered in other editions) and 105n: the reference is to Kepler's *De stella tertii honoris in cygno* (1606), i.e., *De stella nova*. Cf. Howard Marchitello, *The Machine in the Text: Science and Literature in the Age of Shakespeare and Galileo* (Oxford: Oxford University Press, 2011), 116–17. In *Ignatius His Conclave*, Donne might also conflate reference to Galileo's descriptions of the moon in *Sidereus Nuncius* (1610) with Kepler's *Astronomia nova* (1609). Compare *Ignatius*, 81 ("he may draw the Moone, like a boate floating upon the water, as neere the earth as he will"), with Kepler's *New Astronomy* [*Astronomia nova*], trans. William H. Donahue (Cambridge: Cambridge University Press, 1992), 405: the sun's motive power provides a current in which the planets steer as do ferrymen in skiffs. On the broad concerns of *De stella nova*, see Coffin, *John Donne and the New Philosophy*, 82–83, 124. For Galileo, see

Sidereus Nuncius or the Sidereal Messenger, trans. Albert Van Helden (Chicago: University of Chicago Press, 1989).

28. See Hallyn, *The Poetic Structure of the World*, who reports that Kepler thought Donne, when writing *Ignatius His Conclave*, also to have been acquainted with Kepler's *Somnium* (253). Although the *Somnium* (*Dream*) was only published posthumously in 1634, drafts of it circulated in Germany from 1610 (253). Marjorie Hope Nicolson, in her *Science and Imagination* (Ithaca, N.Y.: Cornell University Press, 1956), 63, 67–74, prints the relevant note by Kepler in the *Somnium* and proposes that Henry Wotton, Thomas Harriot, or Henry Percy might have acquainted Donne with the *Somnium*. See also the introduction by John Haffenden, ed., in William Empson, *Essays on Renaissance Literature*, vol. 1: *Donne and the New Philosophy*, 2 vols. (Cambridge: Cambridge University Press, 1993), 1:1–61, esp. 25–44, on Donne's awareness of the new astronomical science. John North reports that in 1606 Harriot sent Kepler a table of angles of refraction but not the sine formula that he had discovered: *Norton History of Astronomy and Cosmology*, 323.

29. See Wilbur Applebaum, "Donne's Meeting with Kepler: A Previously Unknown Episode," *Philological Quarterly* 50 (1971): 132–34.

30. Applebaum, "Donne's Meeting," 234; also Max Caspar, *Kepler*, ed. and trans. C. Doris Hellman (1959; rpt. New York: Dover, 1993), 288. Within the scientific community, communication between England and the Continent appears to have been good: according to Van Helden, a translator of *Sidereus Nuncius*, news of Galileo's book reached England about a month after its publication (87). Cf. Frédérique Aït-Touati, *Fictions of the Cosmos: Science and Literature in the Seventeenth Century*, trans. Susan Emanuel (Chicago: University of Chicago Press, 2011), 8: Aït-Touati's view that science was "a European enterprise that transcended national boundaries" applies to Kepler.

31. Gerald Curzon, *Wotton and His Worlds: Spying, Science, and Venetian Intrigues* (n.p.: Xlibris, 2003), 241. On Wotton's report to Bacon, see Stephen Straker, "The Eye Made 'Other': Dürer, Kepler, and the Mechanization of Light and Vision," *Science, Technology, and Culture in Historical Perspective* 101 (1960): 7–25, here 21–22. Curzon notes that in 1584, while at Oxford, Wotton wrote *De oculo*, concerning the structure of the eye (25). In another coincidence, in 1638, Wotton provided Milton with a letter of introduction for use on his grand tour (269); cf. Barbara K. Lewalski, *The Life of John Milton* (Oxford: Blackwell, 2000), 77, 88.

32. On the Fludd-Kepler controversy, see Caspar, *Kepler*, 290–93; Hallyn, *The Poetic Structure of the World*, 246–52; and esp. Pauli, "The Influence of Archetypal Ideas," 151–240, here 157, 190–201, 237, and appendices 1 and 2 (213–35); also, on Kepler and Fludd, see J. V. Field, *Kepler's Geometrical Cosmology* (Chicago: University of Chicago Press, 1988), 179–87.

33. On simile and similitude, see Anderson, *Translating Investments*, 131, 135, 143–44.

34. On visualization in early modern science, see Catherine Wilson, *The Invisible World: Early Modern Philosophy and the Invention of the Microscope* (Princeton, N.J.: Princeton University Press, 1995), 113. In tempting Eve, the reptilian Satan both presents himself as visible evidence that any penalty for eating the fruit is void and offers the visibly fertile Earth as evidence that "The Gods" have not created it: John Milton, *Paradise Lost*, ed. Barbara K. Lewalski (Oxford: Blackwell, 2007), IX.687–89, 718–22.

1. "THE BODY OF THIS DEATH": DONNE'S SERMONS, SPENSER'S MALEGER, MILTON'S SIN AND DEATH

1. In the Vulgate, Romans 7:24 reads, "Infelix ego homo, quis me liberabit de corpore mortis huius?" The transliterated Greek text of Romans 7:24 reads, "talaipōros egō anthrōpos tis me rusetai ek tou sōmatos tou *thanatou toutou*" (my emphasis).

2. *Eerdmans Dictionary of the Bible*, ed. David Noel Freedman, Allen C. Myers, and Astrid B. Beck (Grand Rapids, Mich.: William B. Eerdmans, 2000), 194, s.v. *Body*; 464: s.v. *Flesh*. I have retained such aids as transliterative stress and length marks. See also Dale B. Martin, *The Corinthian Body* (New Haven, Conn.: Yale University Press, 1995), esp. 6–25, 132 (identity through participation); and John A. T. Robinson, *The Body: A Study in Pauline Theology* (1952; rpt. London: SCM, 1966), 9–11, 15–18.

3. Liddell and Scott, s.v. *ΣΑ΄ΡΞ*. Robinson, *The Body*, 22–25, explains *sárx* not specifically as fleshly passions but more holistically as "living for the world" rather than for God.

4. Lancelot Andrewes, *The Works of Lancelot Andrewes* (1854; rpt. New York: AMS, 1967), V:451 (except Latin, my emphasis).

5. I am grateful to Peter Zervos for help with the New Testament Greek. Cf. Dale B. Martin, *The Corinthian Body*, 135: "Death and sin are not abstract states [in Paul's epistles] but demonized beings." Much of the social and cultural history that so richly informs Martin's study was likely unavailable to the writers I treat, but they knew the primary languages of the Bible that orient my argument.

6. Erich Auerbach, *Figura*, trans. Ralph Mannheim, in *Scenes from the Drama of European Literature* (New York: Meridian, 1959), 11–76, here 51; *Neue Dantestudien, Instanbuler Schriften* 5 (Istanbul: Robert Anhegger, Walter Ruben, Andreas Tietze, 1944), 11–71, here 45. The translator's phrase "creative poetic faith" is more exactly "poetic, formative, belief-force." In *The Body in Pain: The Making and Unmaking of the World* (Oxford: Oxford University Press, 1985), 182, Elaine Scarry excludes "allegory" from the Old Testament events she treats.

7. Mannheim translates Auerbach's "konkret Innergeschichtlichen" as "concrete historical reality" (42; cf. 37 for the German), on which see my *Reading the Allegorical Intertext: Chaucer, Spenser, Shakespeare, Milton* (New York: Fordham University Press, 2008), 9–15.

8. Paul Ricoeur, "The Metaphorical Process as Cognition, Imagination, and Feeling," in *On Metaphor*, ed. Sheldon Sacks (Chicago: University of Chicago Press,1979), 141–57, here 142, 147.

9. Jean-François Lyotard, "The Dream-Work Does Not Think," trans. Mary Lydon, in *The Lyotard Reader*, ed. Andrew Benjamin (Oxford: Blackwell, 1989), 19–55, here 29–30.

10. William Shakespeare, *Troilus and Cressida*, ed. David Bevington (Walton-on-Thames: Thomas Nelson and Sons, 1998), II.iii.174–75.

11. William Shakespeare, *King Lear*, ed. R. A. Foakes (1997; rpt. London: Thomson Learning, 2000), IV.vi.126–27: Lear to Gloucester at Dover, just before wiping his hand because "it smells of mortality."

12. Ben Jonson, *Works*, ed. C. H. Herford and Percy and Evelyn Simpson, 11 vols. (1925–52; rpt. Oxford: Clarendon, 1963–75), 8:122; I have substituted italics for small capitals in *Paul*, also changing *v* to *u*.

13. For Jonson, see my *Words That Matter: Linguistic Perception in Renaissance English* (Stanford, Calif.: Stanford University Press, 1996), 121–23.

14. John Donne, *The Sermons of John Donne*, ed. Evelyn M. Simpson and George R. Potter, 10 vols. (1962; rpt. Berkeley: University of California Press, 1984), 10:232–33.

15. R. C. Bald notes that the title of Donne's last sermon "(beginning *Deaths Duell*)" was not entered in the Stationers' Register and that at least its first two words were given to it by the printer: *John Donne: A Life* (Oxford: Oxford University Press, 1970), 529n1. Be that as it may, the subtitle aptly describes the sermon.

16. Izaak Walton, *The Lives of John Donne, Sir Henry Wotton, Richard Hooker, George Herbert, and Robert Sanderson* (1927; rpt. London: Oxford University Press, 1966), 78.

17. For examples, see Anderson, *Words That Matter*, 189–231, e.g., 205–7, 211–13.

18. In Stephen Pender's word, Donne's knowledge of medicine was "profound": "Essaying the Body: Donne, Affliction, and Medicine," in *John Donne's Professional Lives*, ed. David Colclough (Cambridge: D. S. Brewer, 2003), 215–48, here 217–18. Pender examines Donne's use of "embodied experience as one key to knowledge of the soul" (220).

19. Edmund Spenser, *The Faerie Queene*, ed. A. C. Hamilton, 2nd ed., with text by Hiroshi Yamashita and Toshiyuki Suzuki (Harlow: Pearson Education, 2001), I.iv.27–29. Unless specified otherwise, citations of Spenser are to this edition and citations to "Hamilton, ed.," are to its notes and other apparatus.

20. In order, quoted from William Shakespeare, *Julius Caesar*, in *The Riverside Shakespeare*, gen. ed. G. Blakemore Evans, 2nd ed. (Boston, Mass.: Houghton Mifflin, 1997), I.ii.194 (Caesar describing Cassius); *The Faerie Queene*, I.iv.30 (Envy).

21. For theories about Maleger's name, see Philip Rollinson, "Arthur, Maleger, and the Interpretation of *The Faerie Queene*," *Spenser Studies* 7 (1986): 103–21, here 106. On evil, cf. John Milton's discussion of the Fall and sin in *Complete Prose Works of John Milton*, vol. 6: *Christian Doctrine*, ed. Maurice Kelley, trans. John Carey (New Haven, Conn.: Yale University Press, 1973), 388–91: "it is only its [the action's] misdirection or deviation from the set course of law which can properly be called evil" (391, I.xi); for the Latin, see *The Works of John Milton*, 14–17, ed. James Holly Hanford and Waldo Hilary Dunn, trans. Charles R. Sumner (New York: Columbia University Press, 1933), 15:198.

22. In "Renaissance Psychology and the Defense of Alma's Castle," *Spenser Studies* 19 (2004): 135–57, James W. Broaddus describes Maleger's troops as "vices resulting from habitual exercise of passions rather than the passions themselves [since] . . . the attack is from the outside, but passions are inward forces" (144). I follow Hamilton, ed., 262nn6.1, 8–13 (introduction), in associating these troops not only with the senses, which *affect* us by causing *affections* in us, but also with the Deadly Sins, insofar as I find the number seven (plus five senses) an irresistible evocation of them in this context. *Affection(s)* does not signify only passions in the popular modern sense, namely powerful emotions; it indicates something(s) or action(s) that acts upon or influences us psychically or bodily or both. Spenser refers to *passion* in introducing the House of Alma (II.ix.1), but this is his sole use of the word in II.ix–xi, excepting the behavior of Shamefastness *within* the House of Alma (ix.41, 43).

23. On this topos, see Georgia Ronan Crampton, *The Condition of Creatures: Suffering and Action in Chaucer and Spenser* (New Haven, Conn.: Yale University Press, 1974), chap. 1; also Sean Kane, *Spenser's Moral Allegory* (Toronto: University of Toronto Press, 1989), 73.

24. Harold L. Weatherby, "Two Images of Mortalitie: Spenser and Original Sin," *Studies in Philology* 85 (1988): 321–52, here 327, 330–32: That "the wages of sin is death" is a familiar biblical commonplace (Romans 6:23); Weatherby reverses the entailment: the wages of death are sin. Cf. Harold L. Weatherby, *Mirrors of Celestial Grace: Patristic Theology in Spenser's Allegory* (Toronto: University of Toronto Press, 1994), 158–61.

25. Weatherby, "Two Images of Mortalitie," 345–46 (my emphasis); for further discussion of Maleger, see 345–52, cf. 336, 338. Weatherby reprises and redistributes the argument of his essay in *Mirrors of Celestial Grace*, 155–87, esp.

179–87 (Maleger); he also cites Hamilton as representative misreader more than two dozen times, noting that, whereas Hamilton is interested primarily in imagery, he is interested in ideas (13–14). I concentrate on Weatherby's essay because it is more focused, more specifically textual, less elaborately polemical, and, for my purposes, more contained than the book: unless otherwise specified, subsequent citations of Weatherby are to the essay, with additional annotations, as appropriate, specifically to the book. Broaddus follows Weatherby in associating Maleger primarily with death, not sin. Broaddus's notes afford a judicious review of current and older scholarship on Maleger; Weatherby's notes afford another review.

26. In "Flesh, Spirit, and the Glorified Body: Spenser's Anthropomorphic Houses of Pride, Holiness, and Temperance," *Spenser Studies* 15 (2001): 17–52, Kenneth Borris considers Maleger theologically to signify the flesh (32). Rollinson glosses Maleger as Misrule, his hags as the concupiscible and irascible supports thereof, and his troops as passions (107–13).

27. In *Mirrors of Celestial Grace*, Weatherby is more circumspect about identifying Maleger as passion per se: "passibility and mortality are *virtually* synonymous. . . . To be fallen is to be mortal is to be passionate" (181; my emphasis).

28. In Borris's view, Alma "anagogically prefigures attainment of the glorified or spiritual body" (39) despite description of her as besieged, "dismayd," "affright[ed]," and implicitly vulnerable to Maleger and his vicious troops (II. xi.16). In psychological theories of Neoplatonic coloring within Christianity, the intellective soul answers before God for the entire soul, passible and vegetative soul(s) included, as further noted in Chapter 2 of this volume.

29. On resemblances, see Spenser, *The Faerie Queene*, Hamilton, ed., 244–45nn50–52; Kane, *Spenser's Moral Allegory*, 72; Dorothy Stephens, *The Limits of Eroticism in Post-Petrarchan Narrative: Conditional Pleasure from Spenser to Marvell* (Cambridge: Cambridge University Press, 1998), 55–56; Lewis H. Miller Jr., "Arthur, Maleger, and History in the Allegorical Context," *University of Toronto Quarterly* 35 (1966): 176–87.

30. In *Mirrors of Celestial Grace*, Weatherby describes Maleger as "a personification of death who is passionately alive" (181). For a positive view of Alma's relation to the world of passion ("yet"), see William Junker, "Spenser's Unarmed Cupid and the Experience of the 1590 *Faerie Queene*," *ELH* 79 (2012): 59–83, esp. 70–75.

31. Broaddus is persuasive regarding the physiological connection of brain (or mind) with body.

32. Weatherby, "Two Images of Mortalitie," 349, notes that the Eastern Patristic tradition "confers Christ's own impassibility upon both the soul and the body." The emphasis on Arthur's infirmity is a difficulty for this view.

Christ's impassibility is itself another complicated theological issue of long standing.

33. James Carscallen, "The Goodly Frame of Temperance: The Metaphor of Cosmos in *The Faerie Queene*," in *Essential Articles for the Study of Edmund Spenser*, ed. A. C. Hamilton (Hamden, Conn.: Archon, 1972), 347–365, here 360–61.

34. John Calvin, *Institutes of the Christian Religion*, ed. John T. McNeill, trans. Ford Lewis Battles, 2 vols. (London: SCM Press, 1960), 2:1312 (IV.xv.11); cf. 1:512 (II.xvi.7); 1307 (IV.xv.5). Quotation in the following sentence is from *Institutes*, 2:1313 (IV.xv.12).

35. How stagnant water becomes an image of vivification or rebirth rather than specifically of mortification (implied by baptism but not temporally restricted to it) is beyond me: see Weatherby, "Two Images of Mortalitie," 345, 348; in *Mirrors of Celestial Grace*, 186, Weatherby refers to the lake as "passive" and "static" but objects to Anthea Hume's description of Arthur's victory as a memory of baptism rather than as baptism "in any unqualified sense" since she also considers the lake "static" and "lack[ing] . . . vitality,": *Edmund Spenser: Protestant Poet* (Cambridge: Cambridge University Press, 1984), 125–26; by 186–87 of *Mirrors*, Weatherby effectually sees the lake as death, and so his difference from an Augustinian position looks moot. Borris, "Flesh, Spirit, and the Glorified Body," 34, interprets *standing* as "still, not ebbing," hence constant, and the lake as baptismal. For A. C. Hamilton's discussion of connections between I.xi and II.xi, see his *Structure of Allegory in "The Faerie Queene"* (Oxford: Clarendon, 1961), 102–4, 114–15.

36. E.g., Weatherby, "Two Images of Mortalitie" and *Mirrors of Celestial Grace*. Cf. Carol V. Kaske on the occurrence in biblical allegoresis of radical incongruity between image and meaning: *Spenser and Biblical Poetics* (Ithaca, N.Y.: Cornell University Press, 1999), 174.

37. Mortification is an important Protestant concept. Even aside from the mortifications of Redcrosse in the House of Holiness, consider Calvin's pronouncements on the subject in *Institutes*, 1:512 (II.xvi.7: "Christ's death and burial . . . [affords] liberation from death . . . and mortification of our flesh"); 1:595 (III.iii.3: "*Mortification and vivification*"); 2:1307 (IV.xv.5: we are "sharers in his [Christ's] death" and feel its "working in the mortification of . . . [our] flesh"); 2:1312 (IV.xv.11): "we are baptized into the mortification of our flesh, which begins with our baptism and which we pursue day by day and which will, moreover, be accomplished when we pass from this life to the Lord."

38. The suffix of condition or quality *-hed*, often spelled *head*, in time is crossed with and eventually (with a very few exceptions) replaced by the suffix *-hood*, which was originally a substantive meaning "person, quality, condition": *Online OED*, s.v. *-hood*, suffix; *-head*, suffix (accessed 1/2/15).

39. *Online OED* s.v. *Fairly*, 1, 7; cf. I.1a, 5; II.7a (accessed 2/17/15). The tonal possibilities that the *OED* uncharacteristically notes seem timeless—like irony and its soberer cousins implicit qualification and indirect reservation. Consider, too, the psycholinguist Jean Aitchison's experiments regarding verbal selection and recognition: subjects activate all plausible meanings and, for an ambiguous word, "simultaneously activate more than one meaning . . . even when the context strongly biases them in one direction." They "briefly activate both meanings of a homonym, even in cases where one of them is inappropriate": *Words in the Mind*, 2nd ed. (Oxford: Blackwell, 1994), 214–21; chaps. 17–18 are pertinent.

40. John Milton, *Paradise Lost*, in *The Riverside Milton*, ed. Roy Flannagan (Boston: Houghton Mifflin, 1998), I.84–87. References in this chapter to Milton's writings are to this edition. Alastair Fowler, ed., *Paradise Lost*, rev. 2nd ed. (Harlow: Pearson, 2007), 65n84: Satan's opening response also recalls Isaiah 14:12.

41. *Substance*, together with its cognates, is an overdetermined term in the sixteenth and seventeenth centuries: e.g., see Stephen M. Fallon, *Milton Among the Philosophers: Poetry and Materialism in Seventeenth-Century England* (Ithaca, N.Y.: Cornell University Press, 1991), 3; also my *Translating Investments: Metaphor and the Dynamic of Cultural Change in Tudor-Stuart England* (New York: Fordham University Press, 2005), 49–51; and my *Words That Matter*, indexical entries for *Substance*, *Thing*, and *Res*.

42. See Louis Schwartz, *Milton and Maternal Mortality* (Cambridge: Cambridge University Press, 2009), 211–32, for an excellent discussion of Sin and Death.

43. *Online OED*, s.v. *Phantasm*, 2: "An apparition, a spirit, or supposed incorporeal being appearing to the eyes, a ghost." *OED* cites Milton's line as an example of this meaning. Cf. *OED*, 2a (accessed 2/15/15).

44. Barbara K. Lewalski, *The Life of John Milton: A Critical Biography* (Oxford: Blackwell, 2000), 508; see also 344.

45. Flannagan, ed., *The Riverside Milton*, 675n161, conjectures that Milton enlarged his list of diseases because "he had encountered more [ailments] himself."

46. In *villify*, I find a play on the verb *vilify* (from Latin *vilificare*), alternatively with two letter *l*'s, and the noun cognates *villain/villein* (used interchangeably by Spenser). *Villain* also exists as a verb, according to the *Online OED* (accessed 2/15/15).

47. E.g., *Paradise Lost*, XI.517, 531, 805, 807; XII.583. On Miltonic temperance, see Michael Schoenfeldt, *Bodies and Selves in Early Modern England: Physiology and Inwardness in Spenser, Shakespeare, Herbert, and Milton* (Cambridge: Cambridge University Press, 1999), 131–68.

48. *Paradise Regained*, IV.285–364; "To Mr. Cyriac Skinner Upon his Blindness": "What supports me, dost thou ask? / The Conscience, Friend, to have lost them [his eyes, hence sight] over ply'd / In Liberties Defence" (Flannagan, ed., *The Riverside Milton*, 29). Schoenfeldt cites a letter from Milton to Leonard Philaris suggesting that his blindness may have been caused by his poor digestion (152, cf. 159–61); see also William Kerrigan, *The Sacred Complex: On the Psychogenesis of Paradise Lost* (Cambridge, Mass.: Harvard University Press, 1983), 202. Lewalski, *The Life of John Milton*, 304–5, reports that when Milton wrote to Philaris in 1654 he wished not to "'seem to refuse aid whensoever offered, perhaps divinely.'" Milton was afflicted by gout as well as by blindness and gastritis while engaged with *Paradise Lost*.

49. In modernity, a similar analogy informs Susan Sontag's exposure of the cultural tropes applied to cancer and AIDS, which interpret disease as carelessness or depravity rather than as sin. *"Illness as Metaphor" and "AIDS and Its Metaphors"* (1978, 1989; rpt. New York: Doubleday, 1990).

2. MUTABILITY AND MORTALITY IN *THE FAERIE QUEENE*

1. Edmund Spenser, *The Faerie Queene*, ed. A. C. Hamilton, 2nd ed., with text by Hiroshi Yamashita and Toshiyuku Suzuki (Harlow: Pearson, 2001). Unless specified otherwise, citations of *The Faerie Queene* are to this edition, and citations to "Hamilton, ed." are to its notes and other apparatus. Spenser's shorter poems are cited from *The Yale Edition of the Shorter Poems of Edmund Spenser*, ed. William A. Oram, Einar Bjorvand, Ronald Bond, Thomas H. Cain, Alexander Dunlop, and Richard Schell (New Haven, Conn.: Yale University Press, 1989).

2. In "Two Images of Mortalitie: Spenser and Original Sin," *Studies in Philology* 85 (1988): 321–52, Harold L. Weatherby asserts as his grounding position that Spenser's Titaness in the Mutabilitie Cantos is not only "the cause of death" but "as change and decay she [really] is death" (321). Weatherby's argument regarding sin and death in his book *Mirrors of Celestial Grace: Patristic Theology in Spenser's Allegory* (Toronto: University of Toronto Press, 1994) is cited at least five times in Hamilton's edition, whose citations are influential: *The Faerie Queene*, ed. Hamilton, 154n37.4–5, 171nn3–4 (headnote), 267n40, 692n6.5, 702n7.4–9 (twice). I concentrate on Weatherby's original essay "Two Images," with additional annotations, as appropriate, specifically to *Mirrors*. Carol V. Kaske has a review of Weatherby's *Mirrors* in *The Spenser Review* (1995), accessible via *http://www.english.cam.ac.uk/spenser online/*.

3. Richard A. McCabe notes that "the wheel of fortune was often identified with the wheel of life": *The Pillars of Eternity: Time and Providence in "The Faerie Queene"* (Black Rock: Irish Academic Press, 1989), 200. Edwin

Greenlaw, "Spenser and Lucretius," *Studies in Philology* 17 (1920): 439–64: "Ovid deals with change; Spenser and Lucretius with Mutability" (458). Sean Kane sees the Mutabilitie Cantos "enact[ing] the dilemma of being and becoming": *Spenser's Moral Allegory* (Toronto: University of Toronto Press, 1989), 211–12; cf. 222, 224.

4. Cf. Angus Fletcher, "Complexity and the Spenserian Myth of Mutability," *Literary Imagination: The Review of the Association of Literary Scholars and Critics* 6 (2004): 1–22, esp. 18–22.

5. The name of Cronus gets conflated with Chronos, or Time, and in this way Cronus/Saturn becomes the god of time: Oskar Seyffert, *Dictionary of Classical Antiquities*, rev. and ed. Henry Nettleship and J. E. Sandys (1956; rpt. Cleveland, Ohio: World, 1963), 167–68. Alastair Fowler considers Mutability herself "actually an embodiment of time" seen "as alteration or inconstant change": *Spenser and the Numbers of Time* (London: Routledge and Kegan Paul, 1964), 227.

6. See Harry Berger Jr., "The *Mutabilitie Cantos*: Archaism and Evolution in Retrospect," in *Spenser: A Collection of Critical Essays*, ed. Harry Berger Jr. (Englewood Cliffs, N.J.: Prentice-Hall, 1968), 146–76, esp. 155–57. Similarly, Louise Gilbert Freeman, "Vision, Metamorphosis, and the Poetics of Allegory in the *Mutabilitie Cantos*," *Studies in English Literature* 45 (2005): 65–93, here 82. Berger's essay is a classic to which I owe much.

7. The accuracy of Weatherby's readings of the Fathers is beyond the scope of my discussion. Like Augustine, the earlier and Eastern Fathers are subject to variant interpretations.

8. John Milton, *Christian Doctrine*, in *Complete Prose Works*, vol. 6, ed. Maurice Kelley, trans. John Carey (New Haven, Conn.: Yale University Press, 1973), 388–91, here 391 (I.xi): "it is only its [the action's] misdirection or deviation from the set course of law which can properly be called evil." For the Latin, see *The Works of John Milton*, 14–17, ed. James Holly Hanford and Waldo Hilary Dunn, trans. Charles R. Sumner (New York: Columbia University Press, 1933), 15:198.

9. Kaske, *Spenser Review*, 8, cites two lines in Spenser's third book as an additional reference to original sin: Belphoebe is "Pure and vnspotted from all loathly crime / That is ingenerate in fleshly slime" (III.vi.3). See *Mirrors*, 167–70, for Weatherby's reprise of the references to sin in Spenser's corpus.

10. Cf. Ephesians 1:1–14, 4:4–5. On baptism, see, for example, John Calvin, *Institutes of the Christian Religion*, ed. John T. McNeill, trans. Ford Lewis Battles, 2 vols. (London: S.C.M Press, 1960), 2:1317 (IV.xv.17): "for a long time [we] did not grasp the promise that had been given us in baptism; yet that promise . . . ever remained fixed and firm and trustworthy." Patrick Grant suggests that the red cross might allude to the function of the cross in

the Anglican rite of baptism as well as the mark of St. George: *Images and Ideas in Literature of the English Renaissance* (Amherst: University of Massachusetts Press, 1979), 36, 42–46.

11. For an extensive list of sources and analogues for the serpent woman, see *The Faerie Queene*, Hamilton, ed., 35n14.7–9; and John Milton, *Paradise Lost*, ed. Alastair Fowler, rev. 2nd ed. (Harlow: Pearson, 2007), 142–43n653–59.

12. Redcrosse abandons Una after a full night of eating "his stout heart" out; he has time to think about what he has seen before assenting to, and acting on, it (I.ii.6). His is not *merely* a crime of blinding passion, "The eie of reason . . . with rage yblent" (ii.5).

13. Debora Shuger, *Censorship and Cultural Sensibility: The Regulation of Language in Tudor-Stuart England* (Philadelphia: University of Pennsylvania Press, 2006), e.g., chap. 1, esp. 43–44, 47, bears on Error's emission of books and papers.

14. Erich Auerbach, "*Figura*," trans. Ralph Mannheim, in *Scenes from the Drama of European Literature* (New York: Meridian, 1959), 11–71, here 51; *Neue Dantestudien, Instanbuler Schriften* 5 (Istanbul: Robert Anhegger, Walter Ruben, Andreas Tietze, 1944), 11–71, here 45.

15. E.g., "lyke a lysard with a ladyes visage / Thefly ["thiefly," like a thief] thou me robbest": William Langland, *The Vision of Pierce Plowman* (London: Robert Crowley, 1550), sig. C.c.ii^{r}. On Ruddymane's stain, see *OED*, rev. 1933, s.v. *Red hand, red-hand, a.* and *sb.* A. *adj.1.* Sc. (Orig. *Law*), = "RED-HANDED 1. (Common in 16th c.)." A 1535 instance reads, "That samin carle . . . Come the thrid nycht, . . . To steill the irnes ["irons," likely tools, implements], and was tane reid hand." Also B.*sb.*1: 1577–78: "The said Alexander Winsister wes not takin with reidhand," i.e., with clear evidence of guilt; cf. *Online OED*, 3rd ed. (accessed 1/15/15). The ties between Scotch and Irish history are numerous and deep, and Spenser used Scotch sources. He also had exposure to law in Ireland.

16. Analyzing the Mordant/Amavia/Ruddymane episode in *Spenser and Biblical Poetics* (Ithaca, N.Y.: Cornell University Press, 1999), Carol V. Kaske interprets Ruddymane's stain as original sin (159–79), as earlier does A. C. Hamilton, *The Structure of Allegory in "The Faerie Queene"* (Oxford: Clarendon, 1961), 108–9.

17. Hamilton, ed., glosses *feculent* as "covered with faeces" (223n61.4–9); this meaning of the word, while not wrong, is too limited. That the *OED* ascribes this meaning, as well as simply "filthy," to Spenser's line is not decisive; we often forget that *OED* ascriptions are interpretive. The *OED* also offers the broader sense "Containing or of the nature of faeces or dregs; abounding with sediment or impurities; thick; turbid" (s.v. *Feculent*, 1); the online version is

unchanged in substance (accessed 1/5/15). The two examples in the *OED* most relevant chronologically to Spenser's line are further revealing: in 1578, "The grosse and faeculent part of blood"; in 1607/8, "Any feculent or dreggy refuse." Pilate's hands are bloody.

18. Tantalus, in other versions, steals nectar and ambrosia from the gods or fails to keep secret Zeus's decrees: Seyffert, *Dictionary of Classical Antiquities*, 612. Weatherby, *Mirrors*, 144–45, interprets Tantalus as boundless desire. Tantalus's blood-guilt for Pelops, who was served up to the gods to test their omniscience, is most to the contextual point.

19. Spenser often uses narrative sequence ironically: e.g., the appearance of Sans Foy and Duessa when Redcrosse abandons Una or the appearance of the Giant of materialism in Book V.ii, right after the nailing of Lady Munera's ambiguously human/metal hands and feet on high. Here the narrator's line of vision, along with Artegall's and our own, moves from the brutally dismembered tokens of Munera hung high to the monstrously elevated Giant. It thus moves from the visual display of severed members, a gruesome materializing of justice, to a gigantic embodiment of materialism: see my discussion in *Words That Matter: Linguistic Perception in Renaissance English* (Stanford, Calif.: Stanford University Press, 1996), 167–89.

20. Weatherby, "Two Images," 339–41; *Mirrors*, 174–77.

21. Geoffrey Chaucer, *The Works 1532, with Supplementary Material from the Editions of 1542, 1561, 1598, and 1602* (1969; rpt. London: Scolar, 1976), f. Bvi[v].

22. In "Dame Nature and the Nymph," *English Literary Renaissance* 26 (1996): 243–58, here 256, Weatherby attributes interpretation of lines 6–9 as Eastern to Carol Kaske, insofar as the Eastern Fathers used the imagery of debilitating disease to characterize fallen Adam's legacy; but see his "Two Images," 130, *Mirrors*, 175. Such imagery is hardly limited to Eastern Patristic characterizations of fallen nature: consider the diseases of Langland's Deadly Sins, Job's afflictions, or Calvin, *Institutes of the Christian Religion*, 1:246 (II.i.5): "filthy plagues, blindness, impotence, impurity"; 249 (II.i.6: my emphasis): Adam "*infected* all his posterity"; 250 (II.i.250: my emphasis): "*contagion . . . a corrupted nature*." Also Carol V. Kaske, *Spenser and Biblical Poetics* (Ithaca, N.Y.: Cornell University Press, 1999), 167–68.

23. Cf. Weatherby's exceptional suggestion (*Mirrors*, 205) that in Book III.vi (Chrysogone's birthing and the Garden of Adonis), Spenser may have been "testing the relative claims of the two theologies [Western and Eastern] which he recognized to be at issue . . ." In "Two Images" and *Mirrors*, Weatherby is unwilling to consider such a possibility in Books I and II.

24. On the anciently established interpretation of Ovidian metamorphoses as an indication of the psychological state of a character, see S. K. Heninger Jr.,

Touches of Sweet Harmony: Pythagorean Cosmology and Renaissance Poetics (San Marino, Calif.: Huntington Library, 1974), 268.

25. Hamilton, ed., 704n18.5, offers the gloss "sin of mortality; i.e., death and corruption; or, from Lat. *crimen*, judgment and sentence (of death)." S. P. Zitner, in Edmund Spenser, *The Mutabilitie Cantos*, ed. S. P. Zitner (London: Nelson, 1968), likewise invokes *crimen* and suggests "'corruption,'" with perhaps overtones of 'judgment'" (136n18.5).

26. Spenser, *The Faerie Queene*, VII.vii.18, I.i.21, II.ix.21, II.x.50, cf. III.iv.35; on the use of *slime* (*limus terrae*) rather than dust in the Latin Vulgate and by Spenser, see John Erskine Hankins, *Source and Meaning in Spenser's Allegory: A Study of "The Faerie Queene"* (Oxford: Clarendon, 1971), 137, 278. Weatherby (*Mirrors*, 167) finds references to "mutable, mortal flesh as slime" among his Fathers. He does not mention Hankins's evidence. On "slime," see also Phillipa Berry, "Renewing the Concept of Renaissance: The Cultural Influence of Paganism Reconsidered," in *Textures of Renaissance Knowledge*, ed. Phillipa Berry and Margaret Tudeau-Clayton (Manchester: Manchester University Press, 2003), 17–34, here 30–31.

27. Gordon Campbell, Thomas N. Corns, John K. Hale, and Fiona J. Tweedie, *Milton and the Manuscript of* De Doctrina Christiana (Oxford: Oxford University Press, 2007), 117–18, distinguish three varieties of the ancient heresy of mortalism within Protestantism: namely, the beliefs that the soul dies with the body until the time of the general resurrection, that the soul sleeps from death to general resurrection, and that the soul dies with the body—period. Luther and Tyndale defended the notion of soul sleep; Calvin rejected any form of mortalism. The view of Campbell et al. is that Milton thought the soul died with the body but was resurrected at the last judgment (118). On Averroës, see Ovey N. Mohammed: *Averroes' Doctrine of Immortality: A Matter of Controversy* (Waterloo, Ont.: Wilfrid Laurier University Press, 1984), 131–32; also chap. 3n45 in the present volume (on reception).

28. Paul Alpers, *The Poetry of "The Faerie Queene"* (Princeton, N.J.: Princeton University Press, 1967), 209–14, 231–32.

29. Lodowyck Bryskett, *A Discourse of Civill Life*, ed. Thomas E. Wright (Northridge, Calif.: San Fernando Valley State College Renaissance Editions, 1970). Jerry Leath Mills discusses the relation of Spenser's interest in mortalism to cruxes in the famously difficult geometrical stanza describing the House of Alma: "Spenser, Lodowick Bryskett, and the Mortalist Controversy: *The Faerie Queene*, II.ix.22," *Philological Quarterly* 52 (1973): 173–86; I have benefitted greatly from his explanation. In the context of the *Discourse*, Bryskett's geometrical explanation of the structure and functioning of the soul is a rejection of mortalism, as it appears to be in Spenser's geometrical stanza. That it is so hardly indicates that mortalism is no longer Spenser's (or

Mutability's) concern in the Mutabilitie Cantos. For further context, see Andrew King, "Lines of Authority: The Genealogical Theme in *The Faerie Queene*," *Spenser Studies* 18 (2003): 59–77, esp. 60, 75.

30. Bryskett's *Discourse* is mainly a translation of Giovanni Battista Giraldi Cintio's [i.e., Giambattista Giraldi Cinthio's] "Tre dialoghi della vita civile," in his *De gli hecatommithi* (1565); Bryskett also inserts material drawn from the work of Alessandro Piccolomini and Stefano Guazzo: see Bryskett, *A Discourse of Civill Life*, Wright, ed., x–xvii, xx–xxinn15, 18–19. Bryskett and his Spenser persona refer at length in the *Discourse* to *The Faerie Queene* (21–23), and in *Amoretti* XXXIII Spenser addresses Bryskett directly.

31. Another relevant context for Spenser's questions is the work of Pietro Pomponazzi, an Italian Aristotelian who rejected the influential Averroism (variant of mortalism) of his time (1462–1525) to argue for the individual soul but also rejected the Thomist argument for individual immortality. For Pomponazzi and other Paduans, the atomism of Lucretius was unscientific. Pomponazzi nonetheless argues that the soul is mortal insofar as it cannot function or exist without the body. But he still thinks that the intellective soul understands universals and truth. Only in faith, not reason, he embraces individual immortality: *The Renaissance Philosophy of Man*, ed. Ernst Cassirer, Paul Oskar Kristeller, and John Herman Randall Jr. (1948; rpt. Chicago: Chicago University Press, 1956), 266–73, 281–381, e.g., 305, 315–16, 319. On Averroism and Spenser, cf. McCabe, *The Pillars of Eternity*, 133–36, 208.

32. Mills ("Spenser, Lodowick Bryskett, and the Mortalist Controversy," 177–178) observes that Bryskett's outline of the soul is "not unusual in Elizabethan psychology," especially among those of Neoplatonic inclination.

33. Mills, "Spenser, Lodowick Bryskett, and the Mortalist Controversy," 181–83.

34. Mills, "Spenser, Lodowick Bryskett, and the Mortalist Controversy," 177–78.

35. The Giant's threat has been linked to crop failures, inflation, enclosures, poverty, and vagrants in England, to worries about the importation of Anabaptist communism from the Continent, and to recurrent Irish uprisings closer to home.

3. SATANIC ETHOS: EVIL, DEATH, AND INDIVIDUALITY IN *PARADISE LOST*

1. David Quint's study of Milton's allusions, *Inside "Paradise Lost": Reading the Designs of Milton's Epic* (Princeton, N.J.: Princeton University Press, 2014), 124–32, resembles my argument about envy in this chapter, especially my prioritizing of Satan's envy and negation and my emphasis on satanic circularity and sight. Quint was clearly unaware of my argument, published as

"Satanic Ethos and Envy: The Origin of Evil and Death in *Paradise Lost*," *Milton Studies* 51 (2010): 137–64. Although the manuscript of the present book was complete when Quint's appeared, I have incorporated some reference to it in notes to Chapter 7.

2. On the positive potential of pride, cf. Johan Huizinga's claim that the "formalized pride [of the chivalrous idea] gives rise to a conception of honour." Characterizing honor as a "strange mixture of conscience and egotism," Huizinga quotes Jacob Burckhardt on its compatibility "with many vices and [its being] susceptible to extravagant delusions" yet its undeniable support of much that has historically been considered "pure and noble": *The Waning of the Middle Ages: A Study of the Forms of Life, Thought, and Art in France and the Netherlands in the Dawn of the Renaissance* (New York: Doubleday, 1924), 69. Controversy about Burckhardt aside, there is human truth in Huizinga's observations. See also the meaning of *pride* as "prime, flowering, splendor" early in Spenser's epic: Judith H. Anderson, *The Growth of a Personal Voice: "Piers Plowman" and "The Faerie Queene"* (New Haven, Conn.: Yale University Press, 1976), 24.

3. See Stella Purce Revard's study of theological (and literary) background in *The War in Heaven: "Paradise Lost" and the Tradition of Satan's Rebellion* (Ithaca, N.Y.: Cornell University Press, 1980). Subsequent reference to Revard in the body of the text is to this book. For a thoughtful treatment of Satan's rebellious will, free or not free, see Andrew Escobedo, "Allegorical Agency and the Sins of Angels," *ELH* 75 (2008): 787–818.

4. John Milton, *Paradise Lost*, ed. Barbara K. Lewalski (Oxford: Blackwell, 2007): the bracketed pronoun is mine. To the anonymous reader for *Milton Studies* who contributed five pages of helpful suggestions for an earlier version of this chapter, I remain indebted and grateful.

5. John Milton, *De Doctrina Christiana*, ed. James Holly Hanford and Waldo Hillary Dunn, trans. Charles R. Sumner, 14–17 (New York: Columbia University Press, 1933), 15:202–3 (Bk.I.xii), in *The Works of John Milton*, ed. Frank Allen Patterson et al: "Mala . . . omnia [the singular main verb implies the collective "evil"] et quicquid ducere ad interitum videtur, sub mortis nomine [literally, 'name'; modern 'word'] summatim Scriptura complectitur"; translation mine. Cf. John Milton, *Christian Doctrine*, ed. Maurice Kelley, trans. John Carey, in *Complete Prose Works of John Milton* (New Haven, Conn.: Yale University Press, 1973), 6:393. Mendele Anne Treip notes Milton's designation of sin (evil) in *Christian Doctrine*, I.xi, as *active* obliquity, deviation, perversion from right and good, and privation as its consequence: *Allegorical Poetics and the Epic: The Renaissance Tradition to "Paradise Lost"* (Lexington: University Press of Kentucky, 1994), 315n3; see *Christian Doctrine*, ed. Kelley, 6:388–91, here 391. Separate in theory, the act and its consequence

coalesce in practice. Dennis Richard Danielson similarly recognizes that the theoretical abstractions of theology become "blurred" in practice: *Milton's Good God: A Study in Literary Theodicy* (Cambridge: Cambridge University Press, 1982), 5, 39.

6. Brian Cummings, *The Literary Culture of the Reformation: Grammar and Grace* (Oxford: Oxford University Press, 2002), 174, 301. Cf. Samuel Fallon, "Milton's Strange God: Theology and Narrative Form in *Paradise Lost*," *ELH* 79 (2012): 33–57, here 35–39.

7. The *Rhetoric to Alexander*, long associated with Aristotle, is now thought to be by a pseudo-Aristotle, likely Anaximenes of Lampsacus. I refer to it as Aristotle's since this would likely have been the early modern view.

8. Aristotle, *The Complete Works of Aristotle*, ed. Jonathan Barnes, 2 vols. (Princeton, N.J.: Princeton University Press, 1984), 2:1444b37–1445a29.

9. Cicero, *On the Ideal Orator (De Oratore)*, trans. James M. May and Jakob Wisse (Oxford: Oxford University Press, 2001), 179 (Bk. II.208–10).

10. Francis Bacon, *The Works of Francis Bacon*, ed. James Spedding, Robert Leslie Ellis, and Douglas Denon Heath, 15 vols. (Boston: Brown Taggard [imprint varies; vols. 6–10: Taggard and Thompson], 1860–64), 11:311.

11. *OED*, s.v. *Envy sb.*, 1, 1c, 3; renumbered in the online version: 1a, 1c, 3a (accessed 12/12/14). My later discussion uses the earlier numbering, readily available online as well.

12. By "equivocal monism," I want simultaneously to invoke the following: Neil D. Graves, "'The whole fulness of the Godhead dwells in him bodily': The Materiality of Milton's God," *Christianity and Literature* 52 (2003): 1–20, here esp. 1, 4, 9–12, 14–15, http://galenet.galegroup.com; also Danielson, *Milton's Good God*, 38; cf. N. K. Sugimura's rejection of monism, particularly a material one, in Milton's writing: *"Matter of Glorious Trial": Spiritual and Material Substance in "Paradise Lost"* (New Haven, Conn.: Yale University Press, 2009), 113–14, 119, 129, 156; also William B. Hunter's earlier observations regarding Milton's equivocal use of *spirit*: *Visitation Unimplor'd: Milton and the Authorship of "De Doctrina Christiana"* (Pittsburgh, Penn.: Duquesne University Press, 1998), chap. 8, esp. 130–32. Noting that even a vital materialism is dualist (186), Sugimura particularly challenges the monism argued by Stephen Fallon in *Milton Among the Philosophers: Poetry and Materialism in Seventeenth-Century England* (Ithaca, N.Y.: Cornell University Press, 1991). Harinder Singh Marjara, *Contemplation of Created Things: Science in "Paradise Lost"* (Toronto: University of Toronto Press, 1992), e.g., 68–76, 210–19, 222; throughout, Marjara embraces a Milton who is monist and materialist but without the complications Sugimura and, to a lesser extent, Stephen Fallon locate in the text. In general, much disagreement centers on definitions/conceptions of *spirit*, *substance*, *matter*, and other basic terms. My

use of *monism* indicates that I have *not* renounced it in Milton, though I find unqualified generalizations about his dualism or monism dubious. For a finely balanced essay bearing on Milton's "rather . . . anxious and fragile monism," see Joe Moshenska, "'Transported Touch': The Sense of Feeling in Milton's Eden," *ELH* 79 (2012): 1–31, here 3; also Rachel Trubowitz's argument for Milton's turn away from his earlier Hebraic monism: "Sublime/Pauline: Denying Death in *Paradise Lost*," in *Imagining Death in Spenser and Milton*, ed. Elizabeth Jane Bellamy, Patrick Cheney, and Michael Schoenfeldt (Houndmills: Palgrave Macmillan, 2003), 131–50, here 144.

13. William Langland, *Will's Visions of Piers Plowman, Do-Well, Do-Better, and Do-Best*, ed. George Kane and E. Talbot Donaldson (London: Athlone, 1975), V.75–130, here V.77, 81: I have changed thorns to *th* and removed editorial brackets; brackets remaining are my glosses. The Kane-Donaldson edition includes variants from Robert Crowley's two editions and subsequent reprint of 1550. Crowley's last edition was reprinted by Owen Rogers in 1561, along with *Pierce the Ploughman's Creed*. Milton's wording in his *Apology Against a Pamphlet* . . ., etc. (i.e., *Apology for Smectymnuus*) suggests that this is the edition of *Piers Plowman* that he encountered: *The Vision of William Concerning Piers Plowman*, ed. Walter W. Skeat, EETS 81 (London: N. Trübner, 1884), 4:869*n*43; and *Complete Prose Works of John Milton*, ed. Don M. Wolfe et al.; vol. 1, ed. Don M. Wolfe (New Haven, Conn.: Yale University Press, 1953), 868–953, here 915–16.

14. Edmund Spenser, *The Faerie Queene*, ed. A. C. Hamilton, 2nd ed., with text by Hiroshi Yamashita and Toshiyuku Suzuki (Harlow: Pearson, 2001), I.iv.30.

15. Morton W. Bloomfield, *The Seven Deadly Sins: An Introduction to the History of a Religious Concept, with Special Reference to Medieval English Literature* (n.p.: Michigan State College Press, 1952), 223: Bloomfield cites *Jacob's Well* [ca. 1440], ed. Arthur Brandeis, Part 1, EETS 115 (London, 1900), 81ff. Cf. Revard, *The War in Heaven*, 72: Martin Luther, analyzing Satan's resentment of the Son's superiority, "remarks that implicit in . . . [his] initial conception of evil was a strong antipathy for the good."

16. Joseph Epstein, *Envy: The Seven Deadly Sins* (Oxford: Oxford University Press, 2003), 7.

17. *Online OED*, s.v. *Phantasm*, 2: "An apparition, a spirit, or supposed incorporeal being appearing to the eyes, a ghost"; updated online, 2a: "An apparition, spirit, or ghost; a visible but incorporeal being" (accessed 1/2/15). Joseph H. Summers considers Sin and Death "real nonentities" in *The Muse's Method: An Introduction to "Paradise Lost"* (Cambridge, Mass.: Harvard University Press, 1962), 39. When Yaakov Mascetti observes that "Satan's encounter with Sin [and Death] . . . takes place in his own psyche," he is and

is not right: "Satan and the 'Incomposed' Visage of Chaos: Milton's Hermeneutic of Indeterminacy," *Milton Studies* 50 (2009): 35–63, here 56. Sin is a product of Satan's brain, but once she exists, she has objectivity as a figure with ties to a real past and future, one of which is Death; she is consequential. Satan does not fully control Sin and Death, although he cooperates with them. In Milton's tenth book, they are revealed to be working for God—"My Hell-hounds," as he calls them (629–32). The mind is and is not "its own place" in Milton's epic (I.254).

18. On dialogism in *Paradise Lost*, see James Dougal Fleming, *Milton's Secrecy and Philosophical Hermeneutics* (Aldershot: Ashgate, 2008), e.g., 156–58.

19. The *Online OED* does not recognize the abstracted sense of *fraught* as "distressed" until 1966: s.v. *Fraught, pa.pple., adj., v.* (accessed 1/2/15).

20. With John Leonard, I find hints in the text that Satan's response to the Son's manifestation has deeper roots than those Raphael addresses: *Naming in Paradise: Milton and the Language of Adam and Eve* (Oxford: Clarendon, 1990), 147–62. With respect to angelic prescience, consider *Paradise Lost*, II.346–47 and I.650–54. For additional views, cf. Diane Kelsey McColley, *Poetry and Ecology in the Age of Milton and Marvell* (Aldershot: Ashgate, 2007), 204–5; William Poole, *Milton and the Idea of the Fall* (Cambridge: Cambridge University Press, 2005), 159.

21. I would modify the gloss that Lewalski, ed., offers on *Paradise Lost*, III.108, with the alternative reading that reason itself is informed by choice, as it is by faith. On sin, see *Christian Doctrine*, ed. Kelley, 6:388 (citation), 389–91 (I.xi). The Latin text reads, "concupiscentia mala seu male faciendi libido, et malefactum ipsum": *De Doctrina Christiana*, 15:192, in *Works*, ed. Patterson. Milton has already explained that the subdivisions of sin might be called degrees, parts, or modes or considered among themselves as cause and effect. He thus invites various readings in this regard. See relevant definitions of desire/will in the *Oxford Latin Dictionary*, ed. P. G. W. Glare (Oxford: Clarendon, 1996), s.v. *libido*: 2: "One's will or pleasure (as an overmastering force in determining one's conduct). b a passion"; s.v. *concupisco*: "To conceive a strong desire for, desire ardently, long for, covet." Milton uses neither *voluntas* nor *liberum arbitrium* in this passage, either of which would signal a more rational form of desire, and he modifies both irrational desires, *concupiscentia* and *libido*, with a cognate of *malus*, "bad, evil."

22. For *OED*, s.v. *Envy*, see note 11 in this chapter. Cf. Escobedo's assumption that envy is involuntary, to which one might object that ill will, which is volitional, is its most distinctive form ("Allegorical Agency," 791–92, 797). At the least, the nature of envy is more complex than simply being willed or unwilled.

23. E.g., Alastair Fowler, ed., *Paradise Lost*, rev. 2nd ed. (Harlow: Pearson, 2007), 325n669.

24. Neil Forsyth finds a pun in the word "impaird": *The Satanic Epic* (Princeton, N.J.: Princeton University Press, 2003), 173. Cf. R. A. Shoaf's focal puns in *Milton: Poet of Duality: A Study of Semiosis in the Poetry and Prose* (New Haven, Conn.: Yale University Press, 1985), ix. Treating the same passage and pun in *How Milton Works* (Cambridge, Mass.: Harvard University Press, 2001), Stanley Fish skips over initiating envy to focus on pride and subsequently discusses envy as an offshoot of pride (chap. 15, esp. 511–513, 528).

25. Cf. the Archangel Uriel's failure to penetrate Satan's disguise or his hypocrisy, the thoughts of his heart (III.681–91); also Feisal G. Mohamed, *In the Anteroom of Divinity: The Reformation of the Angels from Colet to Milton* (Toronto: University of Toronto Press, 2008), 110; Robert H. West, *Milton and the Angels* (Athens: University of Georgia Press, 1955), 133–36; Sugimura, "*Matter of Glorious Trial*," 173–74.

26. On Milton's employment of *reduce* in *Paradise Lost*, see Judith H. Anderson, *Reading the Allegorical Intertext: Chaucer, Spenser, Shakespeare, Milton* (New York: Fordham University Press, 2008), 295–98, 412nn52–53, 307–8: Michael's *reduction* of his army under the leadership of the Son, exalted on the chariot of God, is paradigmatic (VI.775–79): cf. John T. Shawcross, in "Forum: Milton's *Christian Doctrine*," *Studies in English Literature* 32 (1992): 155–62, here 157–58; David Norbrook, *Writing the English Republic: Poetry, Rhetoric, and Politics, 1627–60* (Cambridge: Cambridge University Press, 1999), 475–77. Milton's God has earlier explained that "under thee as Head Supream / Thrones, Princedoms, Powers, Dominions I reduce" and that ultimately "God shall be All in All" (III.319–20, 341). Victoria Silver perceives that Satan performs an ironic reduction on himself when he confronts Abdiel: *Imperfect Sense: The Predicament of Milton's Irony* (Princeton, N.J.: Princeton University Press, 2001), 233–37, here 234.

27. The citation of "All in All" (1 Cor. 15:28) in Milton's Book III pertains directly to humankind. As in the Bible, its larger context also addresses the fallen angels and the Son. It bears on the related passage in Book V, God's proclamation of the Son's role. See also *Paradise Lost*, VI.731–34, Ephesians 1:9–10, and Danielson, *Milton's Good God*, 222–23.

28. See Stella Revard, "Milton's Critique of Heroic Warfare in *Paradise Lost* V and VI," *Studies in English Literature* 7 (1967): 119–39. Also John Rogers, *The Matter of Revolution: Science, Poetry, and Politics in the Age of Milton* (Ithaca, N.Y.: Cornell University Press, 1996), 122–29, on Satan's claim to self-authorization.

29. Satan's mention of "The debt immense of endless gratitude" (IV.52) anticipates Melanie Klein's essay "Envy and Gratitude": *Envy and Gratitude*

and Other Works, 1946–1963 (New York: Delacorte Press/Seymour Lawrence, 1975), 176–235; for an assessment of Klein's and Freud's views, see Melvin R. Lansky, "Jealousy and Envy in *Othello*: Psychoanalytic Reflections on the Rivalrous Emotions," in *Jealousy and Envy: New Views About Two Powerful Emotions*, ed. Léon Wurmser and Heidrun Jarass (New York: Analytic, 2008), 25–47, here 26–28, 44.

30. Fowler, ed., notes that the passions ire, envy, and despair all cause pallor (221n115–17).

31. See Cheryl H. Fresch, "'Aside the Devil Turned for Envy': The Evil Eye in *Paradise Lost*, Book IV," in *Living Texts: Interpreting Milton*, ed. Kristin A. Pruitt and Charles W. Durham (Selinsgrove, Penn.: Susquehanna University Press, 2000), 118–30. Fresch overlooks New Testament occurrences that associate the evil eye with perception. Cf. Benjamin Kilborne, "The Evil Eye, Envy, Shame: On Emotions and Explanation," in *Jealousy and Envy: New Views About Two Powerful Emotions*, ed. Léon Wurmser and Heidrun Jarass (New York: Analytic, 2008), 129–48.

32. Satan's opening response also recalls Isaiah 14:12 (Fowler, ed., 65n84).

33. Revard, *The War in Heaven*, 85.

34. See my *Reading the Allegorical Intertext*, 292–98, for additional development of the view that "Satan's sin changes everything" (292); likewise, John Donne's *First Anniversarie*, lines 191–95: *The Variorum Edition of the Poetry of John Donne*, ed. Gary A. Stringer et al. (Bloomington: Indiana University Press, 1995), 6:11; and chapter 7 in this volume.

35. See Jonathan Goldberg, *Endlesse Work: Spenser and the Structures of Discourse* (Baltimore, Md.: Johns Hopkins University Press, 1981), 91–94n7; also *OED*, s.v. *Self pron.*, *a.*, and *n.* C. *sb.* I. From the pronoun: 3. "Chiefly *Philos*. That which in a person is really and intrinsically he (in contradistinction to what is adventitious); the ego (often identified with the soul or mind as opposed to the body); a permanent subject of successive and varying states of consciousness." The first example cited is "1674 TRAHERNE Poet. Wks. (1903) 49 'A secret self I had enclos'd within, That was not bounded with my clothes or skin.'" The relevant *OED* entry online is unchanged in substance or relevant numbering since 1911 (accessed 12/13/14).

36. *Works*, Patterson, ed., 15:40–41(Bk. I.vii); *Christian Doctrine*, ed. Kelley, trans. Carey, 6:318. Sumner's translation "compound" for *duplex* could be misleading, and so, with Carey, I have indicated the more literal translation "double." Cf. Moshenska, "'Transported Touch,'" 9–11.

37. John Donne, *The Sermons of John Donne*, ed. Evelyn M. Simpson and George R. Potter (1962; rpt. Berkeley: University of California Press, 1984), 3:109.

38. *OED*, 2nd ed. (1989), s.v. *Individual a.* and *sb.* A.1, 3a–c., 4; B.b, in all instances including examples. The second edition is still available online, as is the 3rd ed., updated June 2014: in the 3rd ed., cf. A.1, 4a–b, 5; B.1a, 2a (accessed 12/13/14). Individuation in the early modern period has been much discussed: e.g., in relation to inner being, see Anne Ferry: *The "Inward" Language: Sonnets of Wyatt, Sidney, Shakespeare, Donne* (Chicago: University of Chicago Press, 1983), Introduction and 31–70; Katharine Eisaman Maus, *Inwardness and Theater in the English Renaissance* (Chicago: University of Chicago Press, 1995), 1–34. More recently, see Ramie Targoff, *John Donne: Body and Soul* (Chicago: University of Chicago Press, 2008), esp. 16–24; Meredith Anne Skura, *Tudor Autobiography: Listening for Inwardness* (Chicago: University of Chicago Press, 2008), esp. 1–18; Brian Cummings, *Mortal Thoughts: Religion, Secularity, and Identity in Shakespeare and Early Modern Culture* (Oxford: Oxford University Press, 2013), 46–50.

39. See esp. chap. 4 of Cassirer, *The Individual and the Cosmos in Renaissance Philosophy*, trans. Mario Domandi (New York: Harper and Row, 1963). Cf. Joel B. Fineman, *Shakespeare's Perjured Eye: The Invention of Poetic Subjectivity in the Sonnets* (Berkeley: University of California Press, 1986), esp. 1–48; and for responses to Fineman, see *Shakespeare's Sonnets: Critical Essays*, ed. James Schiffer (New York: Garland, 2000), 3–71 (Schiffer's Introduction). A sustained effort to counter Fineman's view comes from David Schalkwyk, *Speech and Performance in Shakespeare's Sonnets and Plays* (Cambridge: Cambridge University Press, 2002).

40. Claire Colebrook, *Milton, Evil, and Literary History* (London: Continuum, 2008), 85, updates Satan as "the first new historicist" (self-fashioning himself) before situating him in the Cartesian mold.

41. In *"All in All": Unity, Diversity, and the Miltonic Perspective*, ed. Charles W. Durham and Kristin A. Pruitt (Selinsgrove, Penn.: Susquehanna University Press, 1999), 48–66, here 50. See also Kenneth Borris, *Allegory and Epic in English Renaissance Literature: Heroic Form in Sidney, Spenser, Milton* (Cambridge: Cambridge University Press, 2000), esp. parts 1 and 4. Borris locates Milton's Idea of unity in the corporate body (a version of the traditional *corpus mysticum*) of the Son. Also Fish, *How Milton Works*, 287; cf. 253–56, 262, 278; but additionally on the Pauline body, note 57 in this chapter.

42. *Online OED*, s.v. *Person* II.2a: "An individual human being" (accessed 12/15/14).

43. *Eerdmans Dictionary of the Bible*, ed. David Noel Freedman, Allen C. Myers, and Astrid B. Beck (Grand Rapids, Mich.: William B. Eerdmans, 2000), s.v. *Life, Soul.* Also *Christian Doctrine*, in *Complete Prose Works*, 6:411: "in biblical idiom, the word *soul* is regularly used to mean the whole animate body" (Bk. I.xiii); and Caroline Walker Bynum, *The Resurrection of the Body in*

Western Christianity, 200–1336 (New York: Columbia University Press, 1995), 4–6, including notes 9–10.

44. See *A Milton Encyclopedia*, ed. William B. Hunter Jr. et al. (Cranbury, N.J.: Associated University Presses, 1978–83), 5:194–99, s.v. *Neoplatonism* (by Jack Ashley). Irene Samuel is persuaded of Milton's interest in Florentine Neoplatonism but offers few specifics: *Plato and Milton* (Ithaca, N.Y.: Cornell University Press, 1947), 35, 36, 41–43; cf. 103, 151. Sugimura extensively demonstrates Milton's knowledge of early Neoplatonic commentaries on Aristotle: *"Matter of Glorious Trial,"* 113, 115, and throughout.

45. On Pico and Padua, see Edgar Wind, *Pagan Mysteries in the Renaissance*, rev. ed. (Harmondsworth: Penguin, 1967), 65. On Milton's mortalism, see Gordon Campbell, Thomas N. Corns, John K. Hale, and Fiona J. Tweedie, *Milton and the Manuscript of "De Doctrina Christiana"* (Oxford: Oxford University Press, 2007), 117–18: Milton thought the soul died with the body but both were resurrected at the last judgment; cf. Sugimura's skepticism regarding Milton's belief in mortalism (*"Matter of Glorious Trial,"* 142, 145–47, 153, 156); also the widespread interest in mortalism documented by Norman T. Burns, *Christian Mortalism from Tyndale to Milton* (Cambridge, Mass.: Harvard University Press, 1972), e.g., 14–18, 73–75, 169–83. Milton's view accords with that of Averroës, as Ovey N. Mohammed describes the latter: *Averroes' Doctrine of Immortality: A Matter of Controversy* (Waterloo, Ont.: Wilfrid Laurier University Press, 1984), 131–32; the historical reception of Averroës is not the same as the thinking of Averroës himself for a reader of Arabic (e.g., 11, 138–39). In Stoicism, for a related example, the human soul is part of the world-soul only temporarily dwelling in the physical body. S. K. Heninger Jr., *Touches of Sweet Harmony: Pythagorean Cosmology and Renaissance Poetics* (San Marino, Calif.: Huntington Library, 1974), 265.

46. Burns, *Christian Mortalism*, 57, 68, 75.

47. Wind, *Pagan Mysteries*, 63, 65n46.

48. Pico della Mirandola, *On the Dignity of Man*, in *On the Dignity of Man, On Being and the One, Heptaplus*, trans., respectively, Charles Glenn Wallis, Paul J. W. Miller, and Douglas Carmichael (Indianapolis, Ind.: Bobbs-Merrill, 1965), 11–12.

49. Wind, *Pagan Mysteries*, 63.

50. Marsilio Ficino, *Platonic Theology*, trans. Michael J. B. Allen, Latin text by James Hankins, with William Bowen, 6 vols. (Cambridge, Mass.: Harvard University Press, 2004), 4:291 (Bk. XIV.viii.8). Further citation of Ficino is to this edition, unless otherwise noted. With respect to Pico's denial of the possibility of a mutual friendship with God, cf. Thomas H. Luxon, *Single Imperfection: Milton, Marriage, and Friendship* (Pittsburgh, Penn.: Duquesne University Press, 2005), 113.

51. Wind, *Pagan Mysteries*, 63–64. See note 26 in this chapter and my earlier discussion of *reduce* in this chapter. Cf. Pseudo-Dionysius, the Areopagite, *The Complete Works*, trans. Colm Luibheid (New York, Paulist, 1987), 77: from the One "derives the existence of everything as beings, what they have in common and what differentiates them. . . . Hence, the harmony and the love which are formed between them but which do not obliterate identity."

52. Marsilio Ficino, "Five Questions Concerning the Mind," trans. Josephine L. Burroughs, in *The Renaissance Philosophy of Man*, ed. Ernst Cassirer, Paul Oskar Kristeller, and John Herman Randall Jr. (Chicago: University of Chicago Press, 1948), 185–212, here 204–5.

53. Wind, *Pagan Mysteries*, 64–65.

54. *John Milton: Complete Poems and Major Prose*, ed. Merritt Y. Hughes (New York: Odyssey, 1957), 680.

55. Marsilio Ficino, "Five Questions Concerning the Mind," 210–12. Cf. Valery Rees, "Ficinian Ideas in the Poetry of Edmund Spenser," *Spenser Studies* 24 (2009): 73–134: the final step of contemplative ascent in Ficino is descent to the soul and the body (80), but "'no one ascends unless God has in some measure first descended to him'" (84); eros in Ficino is vitally dynamic. Rees treats the available works of Ficino and Plato and their place in the university curriculum of early modern Cambridge (98–124).

56. Cf. Diane Kelsey McColley, *Poetry and Music in Seventeenth-Century England* (Cambridge: Cambridge University Press, 1997), 216–17; but cf. Marshall Grossman, *"Authors to Themselves": Milton and the Revelation of History* (Cambridge: Cambridge University Press, 1987), 64–66.

57. A gloss of 1 Cor. 15:28. *The Geneva Bible* (1560; rpt. Madison: University of Wisconsin Press, 1969) offers, "We shal be perfectly fulfilled with his glory and felicitie." In 1622, Lancelot Andrewes explains the end of time as "the fulness indeed, when God shall be not, as now He is, somewhat in every one, but 'all in all' . . . the fulness of eternity and in it the fulness of all joy"; in 1609, however, he further specifies that "the measure shall be so full as it cannot enter into us, we cannot hold it. We must enter into it; *Intra in gaudium Domini tui*": *The Works of Lancelot Andrewes*, ed. John Henry Parker (1854; rpt. New York: AMS, 1967), i.63, 283. In Milton's *Christian Doctrine*, in *Complete Prose Works*, 6:626–27, 1 Cor. 15:28 is taken to mean that after all things have been made subject to the Son, he will be made subject to God, "*so that God may be all in all*" (626). The end of the Son's kingdom "will not be one of dissolution but of perfection and consummation, like the end of the law" (627). Cf. Bynum, *The Resurrection of the Body*, 144; and John A. T. Robinson, *The Body: A Study in Pauline Theology* (London: SCM, 1966): a body requires "more than one member"; the resurrection body of Christ is "articulated in diversity *without ceasing to be a unity*" (59–60); also Dale B. Martin's

view that the Pauline body attains "identity *only* through participation": *The Corinthian Body* (New Haven, Conn.: Yale University Press, 1995), 132 (my emphasis). Cf. Plotinus, *Select Passages Illustrating Neoplatonism*, trans. E. R. Dodds (1923; rpt. Chicago: Ares, 1979), "In the world Yonder the several intelligences [human] will never be lost in unity . . . ; each endures, preserving its proper being in individual difference" (*Enneads*, IV.iii.5).

58. In *Christian Doctrine*, 145–46 (I.ii), Milton rejects the Aristotelian/Thomist exclusion of potentiality (realization *in potentia*) in characterizing God.

59. The Plowman simile is a recurrent thread in John Leonard's *Faithful Labourers: A Reception History of "Paradise Lost," 1667–1970*, 2 vols. (Oxford: Oxford University Press, 2013), 1:327–90, esp. 355–86.

60. See Milton's *A Fuller Course in the Art of Logic*, ed. and trans. Walter J. Ong SJ and Charles J. Ermatinger, in *Complete Prose Works*, 8:139–407: "*Form is the cause through which a thing is what it is*," and "all things which differ in number also differ in essence . . . by reason of [their] proper forms" (232–34, Bk. I.vii); also *Works*, ed. Patterson, 11:58. Cf. Merritt Y. Hughes, "'Myself Am Hell,'" *Modern Philology* 54 (1956): 80–94, esp. 94 (the demonic assumption of "forms after the similitude of their natures"). Hughes relates the demonology of the Florentine Neoplatonists to Milton's hell. Cf. Bynum: in hell the bodies of the damned mirror their "moral disorder and blindness in bodies that are dark, heavy, and infinitely passible" (*The Resurrection of the Body*, 266, on Aquinas).

61. Feisal Mohamed argues that cherubim, such as Ithuriel and Zephon, know only what they experience firsthand; they lack immediate access to ideas (108–10). In sharp contrast, Sugimura suggests that the unfallen angels are locked into a single Grand Possible Intellect; they lack individuation, are "devoid of knowledge of evil," and fail to understand "fallen angelic substance," that is, Satan (*"Matter of Glorious Trial,"* 165, 173, 192–93). I doubt that Milton's cherubs are as limited as either of these views maintain (e.g., "two strong and suttle Spirits," *Paradise Lost*, IV.786).

62. Cassirer welcomes the *logical* subjectivism of Descartes, distinguishing it from the "*religious* subjectivism" of Augustine and Neoplatonists in general (*The Individual and the Cosmos*, 128, 140–41, 184, 186). On the circularity of Descartes's reasoning regarding God's existence, see, e.g., *Meditations on the First Philosophy*: Synopsis of Meditations I–VI, in *Descartes: Selections*, ed. Ralph M. Eaton (New York: Charles Scribner's Sons, 1927), 87: "it is similarly impossible that the idea of God which is in us should not have God himself as its cause." Cf. Descartes's *Discourse on Method*, ibid., 34.

63. See my *Reading the Allegorical Intertext*, 280–320, including notes 43, 103. I do not suggest that *Paradise Lost* as a whole is an allegory: see also *Allegorical Intertext*, 1–15, 21–22.

4. CONNECTING THE CULTURAL DOTS: CLASSICAL TO MODERN TRADITIONS OF ANALOGY

1. While attributing the *theorized* roots of analogy to Aristotle, I recognize that Plato used analogy in his dialogues, most pertinently in the late dialogue *Timaeus*, which entwines mathematical analogy with myth. In "Aristotle on the Imagination," in *Essays on Aristotle's "De Anima,"* ed. Martha C. Nussbaum and Amélie Oksenberg Rorty (Oxford: Clarendon, 1992), 249–77, Malcolm Schofield suggests that Aristotle's presence in the Academy influenced Plato's late dialogues (249). In "Aristotle's Logic of Analogy," revised for *Models and Analogies in Science* (Notre Dame, Ind.: University of Notre Dame Press, 1966), 130–56, Mary B. Hesse shows the applicability of Aristotelian analogy to modern scientific usage and a continuity between Aristotle's scientific and metaphysical analogies (130–31, 155).

2. Fernand Hallyn, *The Poetic Structure of the World: Copernicus and Kepler*, trans. Donald M. Leslie (New York: Zone, 1993), 74, 80–81. On the distinction between Copernicus's heliostatic and Kepler's heliocentric universe, see Owen Gingerich, *The Eye of Heaven: Ptolemy, Copernicus, Kepler* (New York: American Institute of Physics, 1993), 40–41, 379. On metaphor as archtrope and the relation of metaphor to metonymy, see Judith H. Anderson, *Translating Investments: Metaphor and the Dynamic of Cultural Change in Tudor-Stuart England* (New York: Fordham University Press, 2005), 61–62, and chap. 7.

3. Dedre Gentner and Michael Jeziorski, "The Shift from Metaphor to Analogy in Western Science," in *Metaphor and Thought*, ed. Andrew Ortony, 2nd ed. (Cambridge: Cambridge University Press, 1993), 447–80, here 447–49; also Dedre Gentner, "Analogy in Scientific Discovery," in *Model-Based Reasoning: Science, Technology, Value*, ed. Lorenzo Magnani and Nancy J. Nersessian (New York: Kluwer Academic/Plenum, 2001), 21–39, esp. 27–35.

4. E.g., Jacques Derrida, "White Mythology: Metaphor in the Text of Philosophy," trans. F. C. T. Moore, *New Literary History* 6 (1975): 5–74; George Lakoff and Mark Johnson, *Philosophy in the Flesh: The Embodied Mind and Its Challenge to Western Thought* (New York: Basic Books, 1999).

5. Gentner and Jeziorski, "The Shift from Metaphor to Analogy," 475, 478; likewise, Gentner, "Analogy," 36–37: "representational structure," "structured belief representations." Do these (confused?) phrases refer to structural content or to the structure of representation?

6. In *The Machine in the Text: Science and Literature in the Age of Shakespeare and Galileo* (Oxford: Oxford University Press, 2011), Howard Marchitello—citing phrases from Elizabeth Spiller (*Science, Reading, and Renaissance Literature: The Art of Making Knowledge, 1580–1670* [Cambridge: Cambridge University Press, 2004], 2)—exhibits comparable suspicion of

metaphor and imaginative writing (14) and opposes a metaphorical interest in the new science to a methodical one (125); cf. 87, 111, 113, 128. Frédérique Aït-Touati also refers recurrently to *mere* metaphor without theorizing it or its relation to other forms of figuration: *Fictions of the Cosmos: Science and Literature in the Seventeenth Century*, trans. Susan Emanuel (Chicago: University of Chicago Press, 2011), e.g., 44, 73–74, 96, 117, 128, 150. Spiller's study is seminal in its emphasis on the *making* of knowledge (2, 4) and on Aristotelian structures (19–20). Yet she distinguishes the creativity of metaphor from that of literary, as well as scientific, knowledge making (7–8). Hallyn rejects suspicion of "the purely literary," while also distinguishing a metaphor's status by its purpose (*The Poetic Structure of the World*, 28, 128). Cf. Steven Shapin, *The Scientific Revolution* (Chicago: Chicago University Press, 1996), 84; Wolfram Schmidgen, *Exquisite Mixture: The Virtues of Impurity in Early Modern England* (Philadelphia: University of Pennsylvania Press, 2013), xv.

7. David Douglass, "Issues in the Use of I. A. Richards' Tenor-Vehicle Model of Metaphor," *Western Journal of Communication* 64 (2000): 405–24. Max Black, *Models and Metaphors: Studies in Language and Philosophy* (Ithaca, N.Y.: Cornell University Press, 1962).

8. Black belatedly discusses a fifth model, "implicit or submerged," "a conceptual archetype" (*Models and Metaphors*, 239, 241). My subsequent references to Black are directly to his book, unless otherwise indicated.

9. "Bacon, Galileo, and Descartes on Imagination and Analogy," *Isis* 75 (1984): 287–326; Introduction, 287–89; citation, 288; my emphasis. Inclusive page numbers for each of the three grouped articles follows my first citation of the title of each.

10. Thomas Kuhn, "Metaphor in Science," in *Metaphor and Thought*, ed. Andrew Ortony, 2nd ed. (Cambridge: Cambridge University Press, 1993), 533–42, here 540–41.

11. Aristotle, *On the Soul* (*De Anima)*, trans. J. A. Smith, in *The Complete Works of Aristotle*, ed. Jonathan Barnes, rev. Oxford Translation, 2 vols. (Princeton, N.J.: Princeton University Press, 1984), 1:428a11–12; also 427b 15–24, 432a3–9. Cf. "The *De Anima* of Alexander of Aphrodisias" [most reliable Aristotelian commentator of late antiquity], trans. Athanasios P. Fontinis, with commentary by Fontinis, 1978 (Dissertations [1962–2010], Proquest Digital Dissertations, http://epublications.marquette.edu/dissertations/AAI7824337), 83 (2.58, 3–4): "sensation is always true with respect to the proper sensibles, whereas most imaginations are false."

12. Cf. Peter Dear on the contrast between Robert Boyle's Baconian collecting of facts and arguments employed by Galileo and the Jesuits: "Narratives, Anecdotes and Experiments: Turning Experience Into Science in the Seventeenth Century," in *The Literary Structure of Scientific Argument*, ed.

Peter Dear (Philadelphia: University of Pennsylvania Press, 1991), 135–63, here 162–63; according to Dear, the theoretical and epistemological model dominant in Europe for the first half of the seventeenth century was Aristotelian. Similarly, E. J. Dijksterhuis, *The Mechanization of the World Picture: Pythagoras to Newton*, trans. C. Dikshoorn (1961; rpt. Princeton, N.J.: Princeton University Press, 1986), 237–38. On Bacon, cf. Karsten Harries, *Infinity and Perspective* (Cambridge, Mass.: MIT Press, 2001), 285, 289–90; Rhodri Lewis, "Francis Bacon and Ingenuity," *Renaissance Quarterly* 67 (2014): 113–63, here 117, 154.

13. Francis Bacon, *The Proficience and Advancement of Learning*, in *Works*, ed. and trans. James Spedding, Robert Leslie Ellis, and Douglas Denon Heath, 15 vols. (Boston: Brown and Taggard [imprint varies; vols. 6–10: Taggard and Thompson], 1860–64), 6:202–3 (my emphasis); cf. Bacon's *Descriptio Globi Intellectualis*, in *Works*, ed. and trans. James Spedding, Robert Leslie Ellis, and Douglas Denon Heath, 15 vols. (Boston: Brown and Taggard [imprint varies; vols. 6–10: Taggard and Thompson], 1860–64), 10:403–5.

14. L. A. Kosman, "What Does the Maker Mind Make?" in *Essays on Aristotle's "De Anima,"* ed. Martha C. Nussbaum and Amélie Oksenberg Rorty (Oxford: Clarendon, 1992), 343–58. Properly speaking, even the phrase *nous poiētikos* belongs to the Aristotelian tradition rather than to Aristotle per se, as likewise does the very text of *De Anima*: ibid., 343n2; Martha C. Nussbaum, "Introduction A. The Text of Aristotle's *De Anima*," in *Essays on Aristotle's "De Anima,"* ed. Martha C. Nussbaum and Amélie Oksenberg Rorty (Oxford: Clarendon, 1992), 1–6, here 2; cf. Franz Brentano, "*Nous Poiētikos*: Survey of Earlier Interpretations," in *Essays on Aristotle's "De Anima,"* ed. Martha C. Nussbaum and Amélie Oksenberg Rorty (Oxford: Clarendon, 1992), 313–41.

15. Sir Philip Sidney, *An Apology for Poetry or The Defence of Poesy*, ed. Geoffrey Shepherd (Manchester: Manchester University Press, 1973), 99.

16. See Bacon, *The Proficience and Advancement of Learning*, in *Works*, 6:184; the remaining quotations from Bacon in this paragraph are from *The Dignity and Advancement of Learning* [*De Dignitate et Augmentis Scientiarum*], in *Works*, ed. and trans. James Spedding, Robert Leslie Ellis, and Douglas Denon Heath, 15 vols. (Boston: Brown and Taggard [imprint varies; vols. 6–10: Taggard and Thompson], 1860–64), 8:409–11 (Bk. II.ii). Cf. Jonathan Sawday, *Engines of the Imagination: Renaissance Culture and the Rise of the Machine* (London: Routledge, 2007), 210–16.

17. Bacon, *De Dignitate et Augmentis Scientiarum*, in *Works*, 2:189–90, reads, "Sed et illabitur etiam animis hominum aliud subtilius malum; nempe, ut ars censeatur solummodo tanquam additamentum quoddam naturae, cujus scilicet ea sit vis ut naturam (sane) vel inchoatam perficere, vel in deterius vergentem emendare, vel impeditam liberare; minime vero penitus vertere,

transmutare, aut in imis concutere possit." Cf. Bacon's *Descriptio Globi Intellectualis*, in *Works*, ed. and trans. James Spedding, Robert Leslie Ellis, and Douglas Denon Heath, 15 vols. (Boston: Brown and Taggard [imprint varies; vols. 6–10: Taggard and Thompson], 1860–64), 10:407–8 (Latin 7:289); also *Parasceve*, in *Works*, ed. and trans. James Spedding, Robert Leslie Ellis, and Douglas Denon Heath, 15 vols. (Boston: Brown and Taggard [imprint varies; vols. 6–10: Taggard and Thompson], 1860–64), 8:357 (Latin: 2:47).

18. C. P. Snow, *The Two Cultures* (1959; rpt. Cambridge: Cambridge University Press, 1993). On Snow, see Marchitello, *The Machine in the Text*, 185–213; also Harries, *Infinity and Perspective*, 301–17.

19. On the status of mathematics as a science, see Peter Dear, *Mersenne and the Learning of the Schools* (Ithaca, N.Y.: Cornell University Press, 1984), 48–79; on mathematical idealization in Kepler and Galileo, E. J. Dijksterhuis, *The Mechanization of the World Picture*, 404, 407; Shapin, *The Scientific Revolution*, 58–60; Fokko Jan Dijksterhuis, "Constructive Thinking: A Case for Dioptrics," in *The Mindful Hand: Inquiry and Invention from the Late Renaissance to Early Industrialization*, ed. Lissa Roberts, Simon Schaffer, and Peter Dear (Amsterdam: Koninklijke Nederlandse Akademic van Wetenschappen, 2007), 59–82, here 71, 77–80; on Galileo's "uncritical use of analogy," see Harries, *Infinity and Perspective*, 279–85.

20. See Daniel Tiffany, *Toy Medium: Materialism and Modern Lyric* (Berkeley: University of California Press, 2000), 267: from Planck's standpoint, a picture of the atom is "the last picture before the abyss of a purely mathematical and hence *unimaginable* system of material reality."

21. Schofield, "Aristotle on the Imagination," 274; Aristotle, *De Anima*, 2:428a11–12; also 427b16–24, 28–29; 431b2, 432a4–9. On the capacities and reliability of imagination, see also Schofield, "Aristotle on the Imagination," 251, 272–74, 276–77; Dorothea Frede, "The Cognitive Role of *Phantasia* in Aristotle," in *Essays on Aristotle's "De Anima,"* ed. Martha C. Nussbaum and Amélie Oksenberg Rorty (Oxford: Clarendon, 1992), 279–95, here 285, 288–90, 294; and Michael V. Wedin, *Mind and Imagination in Aristotle* (New Haven, Conn.: Yale University Press, 1988), 45–63. Cf. René Descartes's Aristotelian view: *Discourse on Method*, in *Descartes: Selections*, ed. Ralph M. Eaton (New York: Charles Scribner's Sons, 1927), 34: neither "our imagination nor our senses can ever assure us of anything, if our understanding [intellect] does not intervene"; cf. Hallyn, *The Poetic Structure of the World*, 172; Harries, *Infinity and Perspective*, 198.

22. Qtd. and trans. Peter Galison, "Descartes's Comparisons: From the Invisible to the Visible," *Isis* 75, no. 2 (1984): 322 (his brackets), from René Descartes, *Oeuvres de Descartes*, ed. Charles Adam and Paul Tannery, 12 vols. (1897–1910: rpt. Paris: Vrin, 1964–75), 10:439–40; *The Philosophical Works of*

Descartes, trans. Elizabeth S. Haldane and G. R. T. Ross, 2 vols. (1911–12; rpt. with corrections, 1931, 1979), 1:53, 56 (translation modified by Galison). On the term *aequatio*, see Johannes Kepler, *Mysterium Cosmographicum / The Secret of the Universe*, trans. A. M. Duncan, introduction and commentary by E. J. Aiton (New York: Abaris, 1981), 15 (translator's note). Cf. A. I. Sabra, *Theories of Light from Descartes to Newton* (London: Oldbourne, 1967), 28–33.

23. Qtd. and trans. Galison, "Descartes's Comparisons: From the Invisible to the Visible," 312; Descartes, *Oeuvres*, 3:297–98. Cf. Sabra, *Theories of Light*: "Descartes gives no other proof of the laws of optical reflection and refraction" in *De la lumière* except comparison of these phenomena to the mechanical reflection and refraction of a ball (28); "the analogies or comparisons used in the *Dioptrics* . . . indicate the kind of argument" needed "to establish the second deductive step after an intellectual intuition of what constitutes a natural force in general" (32); cf. Bruce Stansfield Eastwood, "Descartes on Refraction: Scientific Versus Rhetorical Method," in *The Scientific Enterprise in Early Modern Europe: Readings from "Isis,"* ed. Peter Dear (Chicago: University of Chicago Press, 1997), 154–75, esp. 158–62. Antoni Malet argues that Descartes found the sine law in Kepler's visual analogies, not in mechanical ones: "Gregorie, Descartes, Kepler, and the Law of Refraction," *Archives Internationales D'Histoire des Sciences* 40, no. 125 (1990): 278–304, here 281, 302. Cf. Aït-Touati, *Fictions of the Cosmos*, 87–88; and Olivier Darrigol, *A History of Optics from Greek Antiquity to the Nineteenth Century* (Oxford: Oxford University Press, 2012), 39–45 (Descartes), esp. 41–42.

24. Cf. Catherine Wilson, *The Invisible World: Early Modern Philosophy and the Invention of the Microscope* (Princeton, N.J.: Princeton University Press, 1995), 61, for early modern examples of analogical thinking; 106, on Descartes's susceptibility; 113, on "scientists' equation of truth with its visualizability [in mechanical terms], or with the analogical similarity of the [explanatory] process to a visualizable process."

25. Joshua P. Hochschild, *The Semantics of Analogy: Rereading Cajetan's "De Nominum Analogia"* (Notre Dame, Ind.: University of Notre Dame Press, 2010), 1, 6.

26. Darrigol refers to "the old Greek principle of the interaction of the like with the like," on which Kepler relied (*A History of Optics*, 32). Descartes's principle was an instance, in effect, of turning back to the future. Cf. Dear, *Mersenne and the Learning of the Schools*, 63–64, on the Aristotelian and Scholastic "law of homogeneity."

27. See Dale Pesmen, "Reasonable and Unreasonable Worlds: Some Expectations of Coherence in Culture Implied by the Prohibition of Mixed Metaphor," in *Beyond Metaphor: The Theory of Tropes in Anthropology*, ed. James W. Fernandez (Stanford, Calif.: Stanford University Press, 1991), 213–43, here 213, 218, 228, 238.

28. Galison, "Descartes's Comparisons: From the Invisible to the Visible," 317n15. Dear, *Mersenne and the Learning of the Schools*, 55, paraphrases a letter from Descartes to Mersenne in which Descartes asserts that "God's laws of mathematics are all 'mentibus nostris ingenitae' (inborn in our minds)." Cf. *Meditations on the First Philosophy*: Synopsis of Meditations I-VI, in *Descartes: Selections*, ed. Ralph M. Eaton (New York: Charles Scribner's Sons, 1927), 87: "*the certainty of geometrical demonstrations is itself dependent on the knowledge of God.*"

29. Brian Vickers, "Analogy Versus Identity: The Rejection of Occult Symbolism, 1580–1680," in *Occult and Scientific Mentalities in the Renaissance*, ed. Brian Vickers (Cambridge: Cambridge University Press, 1984), 95–163.

30. Genter by herself, in her later essay ("Analogy," 2001), views Kepler in a more positive light (31–34, 36–37). In contrast with the Gentner and Jeziorski article, her concern with analogical modeling in 2001 is primarily theoretical, not historical. She does not specify how the chief Keplerian example she cites is structurally an analogy (A:B :: C:D) rather than a simple metaphor or an iconic model (25).

31. Shapin suggests that when new scientists objected to the study of words rather than things, they were objecting "quite specifically to the verbose and wrangling style of natural philosophy in the schools" (*The Scientific Revolution*, 121). E. J. Dijksterhuis identifies the illusion that naming extends knowledge as "the weakest spot in the Aristotelian philosophy of nature" yet promptly adds that names are important because we need concepts to think (*The Mechanization of the World Picture*, 338). Contrastingly, for Claudia Baracchi, Aristotle is "the thinker of unexhausted *aporia*" and "of truth as phenomenal disclosure": "Contributions to the Coming-to-Be of Greek Beginnings: Heidegger's Inceptive Thinking," in *Heidegger and the Greeks: Interpretive Essays*, ed. Drew A. Hyland and John Panteleimon Manoussakis (Bloomington: Indiana University Press, 2006), 23–42, here 36–37.

32. On metaphor and metonymy, see Anderson, *Translating Investments*, esp. 61–62; also chaps. 1–2.

33. According to Hochschild, "Cajetan's treatise [on analogy] is the most influential" in the Aristotelian tradition over many centuries (*The Semantics of Analogy*, xiv, xx). Cardinal Cajetan's career included a debate with Pico della Mirandola and, as papal legate in Germany, negotiations with Martin Luther. Cajetan was Chair of Thomistic Metaphysics at the University of Padua, a hotbed of naturalistic and scientific thought particularly identified with Averroistic Aristotelianism, and later he was master general of the Dominican Order (xv).

34. Paul Ricoeur, *Time and Narrative*, trans. Kathleen McLaughlin and David Pellauer, 3 vols. (Chicago: University of Chicago Press, 1984), 1:32–33, 66. Janet Leslie Knedlick argues that Tasso and Mazzoni (not to mention

Sidney and Milton) understood and applied "the fundamental point of Aristotle's *Poetics*: that in the process of structuring the artistic mimesis, the poetic maker works in a mode intrinsically valid as a way of knowing": "Fancy, Faith, and Generative Mimesis in *Paradise Lost*," *Modern Language Quarterly* 47 (1986): 19–47, here 24–25, 27, 30.

35. Tiffany, *Toy Medium*, 288, cf. 261: the debate among Bohr, Heisenberg, Einstein, and Schrödinger about the viability of intuitive pictures of quantum reality, whose "very existence . . . is to some degree contingent on the event of representation"; cf. 273 (on allegory). Cf. Owen Gingerich on the "leap of faith" from epistemological laws of nature to ontology: "Kepler and the Laws of Nature," *Perspectives on Science and Christian Faith* 63, no. 1 (2011): 17–23. Cf. Marchitello on belief in Galileo's sunspots (*The Machine in the Text*, 115).

36. Aristotle, *Poetics*, trans. I. Bywater, in *The Complete Works of Aristotle*, ed. Jonathan Barnes, rev. Oxford Translation, 2 vols. (Princeton, N.J.: Princeton University Press, 1984), 2:1457b16–30; cf. *Rhetoric*, trans. W. Rhys Roberts, in ibid., 2:1406b30–1407a17.

37. Here Richard Janko translates "so [the poet] will call evening 'old age of the day,' as Empedocles does": Aristotle, *Poetics*, trans. Richard Janko (Indianapolis, Ind.: Hackett, 1987), 29.

38. Aristotle, *Works*, 1:443b6–8, 452b10–13,469b15–20; 2:1093a3–b6, 1240a13–14.

39. Darrigol, *A History of Optics*, 32.

40. On analogy in Aristotle, see Hesse, "Aristotle's Logic of Analogy," 131–40 (biological and poetic analogies); 140–50 (metaphysical analogies). On analogy in Aristotle's *Metaphysics*, cf. Steven A. Long, *Analogia Entis: On the Analogy of Being, Metaphysics, and the Act of Faith* (Notre Dame, Ind.: Notre Dame University Press, 2011), e.g., 73–74.

41. Varro, *De lingua Latina*, trans. Roland G. Kent, 2 vols. (1938; rpt. London: Heinemann, 1977), 2:VIII.ix.21–23.

42. See Judith H. Anderson, *Words That Matter: Linguistic Perception in Renaissance English* (Stanford, Calif.: Stanford University Press, 1996), e.g., 33–40 and 332 (commonplace); and Anderson, *Translating Investments*, chap. 7, here 136, 145.

43. Quintilian, *The Institutio Oratoria of Quintilian*, trans. H. E. Butler, 4 vols. (1920; rpt. London, Heinemann, 1980), 1:I.vi.1.

44. St. Augustine, *On Christian Doctrine*, trans. D. W. Robertson Jr. (Indianapolis, Ind.: Bobbs-Merrill, 1958, 1984), 93 (III.xv.23), 30–31 (I. xxxvi.40), 87–88 (III.x.14). The analogy of faith (*analogia fidei*, also *regula fidei* or *ratio fidei*) originates in Romans 12:6. I have treated this analogy relevantly in "Donne Cooking: Analogy, Proportion, Authority, and Faith," *John Donne Journal* 32 (2013): 1–23, here 15–21.

45. On the familiarity of Varro and Quintilian in the Middle Ages, see Mark Amsler, *Etymology and Grammatical Discourse in Late Antiquity and the Early Middle Ages* (Amsterdam: J. Benjamins, 1989), 24–31 (esp. on Varro); John O. Ward, "Quintilian and the Rhetorical Revolution of the Middle Ages," *Rhetorica* 13 (1995): 231–84.

46. As signaled by quotation and parenthetical reference, my account of analogy in Boethius draws heavily on Hochschild, *The Semantics of Analogy*, 7–9.

47. On focal meaning, see Ralph McInerny, *Aquinas and Analogy* (Washington, D.C.: Catholic University of America Press, 1996), 39–40. McInerny, ibid., 98, 137, expands *ad unum* to *ab uno* and *in uno* ("from one" and "in one"). Cf. also J. F. Ross, *Portraying Analogy* (Cambridge: Cambridge University Press, 1981): 1, 12 (focal meaning); and 17 (equivocation). Ross describes analogy as "differentiation with relatedness" (22).

48. Hochschild, *The Semantics of Analogy*, 9. On Boethius, see also S. K. Heninger Jr., *Touches of Sweet Harmony: Pythagorean Cosmology and Renaissance Poetics* (San Marino, Calif.: Huntington Library, 1974), 29, 33, 53–54, 58.

49. Edmund Spenser, *The Faerie Queene*, ed. A. C. Hamilton, 2nd ed., with text by Hiroshi Yamashita and Toshiyuku Suzuki (Harlow: Pearson, 2001), V.ii.43–49.

50. In *The Rule of Metaphor: Multidisciplinary Studies of the Creation of Meaning in Language*, trans. Robert Czerny, with Kathleen McLaughlin and John Costello (Toronto: University of Toronto Press, 1979), 272–80, Paul Ricoeur treats the *analogia entis*, attempting, like Aquinas and Cajetan, to distinguish analogy from metaphor (261, 277–80).

51. Knowledge of Latin does not ipso facto qualify a translator or interpreter of Scholastic writings: central terms shift in technical and sometimes bewildering ways among them: witness the fortunes of *analogia* itself. On shifts in terminology more generally, see Peter Dear, *Discipline and Experience: The Mathematical Way in the Scientific Revolution* (Chicago: University of Chicago Press, 1995), 152.

52. McInerny's opening chapter in *Aquinas and Analogy* is titled "Where Cajetan Went Wrong"; see esp. 4–6, 28–31. He suggests that Cajetan (con) fuses the real with the logical (47n21). Hochschild argues that Aquinas never adequately addresses the difference between analogous and fallacious equivocation (e.g., *The Semantics of Analogy*, 44, 65–81). Long takes issue with McInerny's logical emphasis on "being" (*Analogia Entis*, 73–79).

53. Cf. Dear's version of the Thomist analogy of being: "Aquinas had described God as an absolutely simple intellect that nonetheless comprehends the diversity of all possible essences, which it can bring into actual being as *analogues* of the divine being (*Mersenne and the Learning of the Schools*, 70; my emphasis). See also Long, *Analogia Entis*, e.g., 1–11 (overview).

54. Hesse, "Aristotle's Logic of Analogy," 135, 141, discusses much the same analogy in Aristotle's *Topics* ("'as sight is in the eye, so is reason in the soul'"). The existence and extent of a separation of body and soul in Aristotle are debated, and the Aristotelian tradition, which attracted Neoplatonic commentaries, further complicates the question.

55. McInerny, *Aquinas and Analogy*, 114, 132. "Raising" refers to the Hegelian concept of *Aufhebung*, which partly raises, partly cancels, and partly continues the original meaning of a word, a description that has more than a little in common with the mean between univocation and equivocation that Aquinas and Cajetan variously sought; on the Ricoeur-Derrida debate about this concept, see my *Translating Investments*, esp. 14–22.

56. McInerny, *Aquinas and Analogy*, 131: citing *Summa Theologica*, Ia, q. 67, a.1, with additional reference to *In I de anima*, lect. 10, n. 157. For the works of Aquinas, McInerny's bibliography indicates the Leonine edition or else the manual editions of Marietti. Aquinas, like Bonaventure, observes the traditional medieval distinction between the term *lux*, primary light, and *lumen*, secondary light (or source and offspring), which is based on Avicenna's *De anima*; the Renaissance Neoplatonist Marsilio Ficino asserts that there is "a single, infinite light, variously participated in," and "no abrupt discontinuity . . . where a linguistic break can be *unequivocably* made": David C. Lindberg, "The Genesis of Kepler's Theory of Light: Light Metaphysics from Plotinus to Kepler," *Osiris* 2, 2nd ser. (1986): 5–42, here 18, 24–25 (my emphasis).

57. Qtd. from McInerny, *Aquinas and Analogy*, 132; the source is Aquinas, *II Sent.*, d. 13, q. 1, a.2. For relevant medieval background, see David C. Lindberg, *Theories of Vision from Al-Kindi to Kepler* (Chicago: University of Chicago Press, 1976), esp. 89–97; David C. Lindberg's introduction to *Roger Bacon's Philosophy of Nature*, ed. and trans. David C. Lindberg (Oxford: Clarendon, 1983); and David C. Lindberg and Geoffrey Cantor, *The Discourse of Light from the Middle Ages to the Enlightenment* (Los Angeles: University of California, 1985).

58. Qtd. from McInerny, *Aquinas and Analogy*, 133; the source is Aquinas, *I Sent.*, d. 45, q. 1, a. 4.

59. E.g., Lindberg, *Theories of Vision*, 80, 161; Sabra, *Theories of Light*, 28–32, 116.

60. Cf. Dear, *Discipline and Experience*, 158–59: to imply that contrived, experimental situations are what happens in nature is in fact to express "a metaphorical identification" that all too readily becomes a metonymic replacement of reality (158n24).

61. Dear, *Discipline and Experience*, 96; cf. 123 (Descartes); also Shapin, *The Scientific Revolution*, 4, 65–66.

62. The Latin text of *De Nominum Analogia* (1498) that Hochschild uses is that edited by N. Zammit (1934) and revised by H. Hering (Rome: Angeli-

cum, 1951). He invokes five other works by Cajetan, mainly in the commentary tradition. The translations he offers are his own.

63. William Shakespeare, *Hamlet*, ed. Ann Thompson and Neil Taylor (London: Thomson Learning, 2006): I.ii.184, cf. I.i.111: "A mote it is to trouble the mind's eye."

64. Cf. Paul Ricoeur on "the as-if": *Time and Narrative*, I:45; also Paul Ricoeur, "Metaphorical Process as Cognition, Imagination, and Feeling," in *On Metaphor*, ed. Sheldon Sacks (Chicago: University of Chicago Press, 1979), 141–57, here 151–52.

65. The origin of the metaphor "the field smiles" or the "meadow laughs" has variously but erroneously been attributed in its exact form (*pratum ridet*) to Quintilian and Vergil. It could easily have been derived from proximate, associated words in descriptions of landscape in Vergil's *Eclogues* or *Georgics*.

66. John Milton, *Paradise Lost*, ed. Barbara K. Lewalski (Oxford: Blackwell, 2007), III.362–64.

67. Giorgio Agamben, *The Signature of All Things: On Method*, trans. Luca D'Isanto, with Kevin Attell (New York: Zone, 2009), 9–32, esp. 18–19; Ricoeur, *Rule of Metaphor*, e.g., 47–48; *Time and Narrative*, 1:31–32.

68. Here Agamben notes his debt to Enzo Melandri, *La linea e il circolo: Studio logico-filosofico sull'analogia* (Macerata: Quodlibet, 2004). Cf. also G. E. R. Lloyd on Aristotle's analogical paradigm: *Polarity and Analogy: Two Types of Argumentation in Early Greek Thought* (Cambridge: Cambridge University Press, 1971), 403–20, 440.

69. Ignoring the reciprocity and partiality of the metaphorical transfer of meaning, Agamben peculiarly distinguishes analogy from metaphor but associates it with allegory, itself a form of continued metaphor (*The Signature of All Things*, 18). His distinction disables neither my assertion of the metaphoricity of analogy nor the broader claims of his discussion.

5. PROPORTIONAL THINKING IN KEPLER'S SCIENCE OF LIGHT

1. See especially Stephen Straker, "The Eye Made 'Other': Dürer, Kepler, and the Mechanisation of Light and Vision," in *Science, Technology, and Culture in Historical Perspective*, ed. Louis A. Knafla, Martin S. Staum, and T. H. E. Travers (Calgary: University of Calgary Press, 1976), 7–25; Steven Straker, "Kepler, Tycho, and the 'Optical Part of Astronomy': The Genesis of Kepler's Theory of Pinhole Images," *Archive for the History of Exact Sciences* 24, no. 4 (1981): 267–93; David C. Lindberg, "The Genesis of Kepler's Theory of Light: Light Metaphysics from Plotinus to Kepler," *Osiris* 2, 2nd ser. (1986): 4–42; David C. Lindberg, *Theories of Vision from Al-Kindi to Kepler* (Chicago: University of Chicago Press, 1976); David C. Lindberg and Geoffrey Cantor, *The Discourse of Light from the Middle Ages to the Enlightenment* (Los Angeles:

University of California, 1985). Also Nick Jardine, "Koyré's Kepler/Kepler's Koyré," *History of Science* 38, no. 4 (2000): 363–76; E. J. Dijksterhuis, *The Mechanization of the World Picture: Pythagoras to Newton*, trans. C. Dikshoorn (1961; rpt. Princeton, N.J.: Princeton University Press, 1986), 303–23; Alexandre Koyré, *From the Closed World to the Infinite Universe* (Baltimore, Md.: Johns Hopkins University Press, 1957); Thomas S. Kuhn, *The Copernican Revolution: Planetary Astronomy in the Development of Western Thought* (Cambridge, Mass.: Harvard University Press, 1957); A. C. Crombie, *Medieval and Early Modern Science*, 2 vols. (Garden City, N.Y.: Doubleday, 1959), 2:180–99; Olivier Darrigol, *A History of Optics* (Oxford: Oxford University Press, 2012), 1–108; A. Mark Smith, *From Sight to Light: The Passage from Ancient to Modern Optics* (Chicago: University of Chicago Press, 2015), 364, 369–71.

2. Johannes Kepler, *Optics: Paralipomena to Witelo and Optical Part of the Astronomy*, trans. William H. Donahue (Santa Fe, N.M.: Green Lion, 2000), 19; unless otherwise indicated, I have not reproduced the extensive italicization (e.g., in propositions) in this edition. For Kepler's Latin, I use *Johannes Kepler Gesammelte Werke*, 22 vols. (Munich: C. H. Beck, 1937—), hereafter *GW*: 1, 8 (*Mysterium cosmographicum*: 1596, 1621); 1 (*De stella nova*: 1606), 2 (*Ad Vitellionem Paralipomena*: 1604), 3 (*Astronomia nova*: 1609), 6 (*Harmonice mundi*: 1619), 7 (*Epitome astronomiae Copernicanae*: 1618–21), 10 (*Tabulae Rudolphinae*: 1627). Where Kepler's Latin is particularly sensitive to translation, Donahue inserts the original word parenthetically or in a footnote, which I include in a bracket. For consistency, my own insertions of Latin within quotations are bracketed in this chapter as well.

3. Johannes Kepler, *Harmonies of the World* [*Harmonice mundi*], *Book Five*, ed. Stephen Hawking, trans. Andrew Motte (Philadelphia: Running Press, 2002), 47, cf. 11, 20, 60, 73; cited hereafter as *Harmonice*, V, or parenthetically, as V, with page number. Kepler's admiration of Plato's *Timaeus* is well recognized: e.g., S. K. Heninger Jr., *Touches of Sweet Harmony: Pythagorean Cosmology and Renaissance Poetics* (San Marino, Calif.: Huntington Library, 1974), 48–49 (*Timaeus*), 107–8 (regular solids), 110–13 (Kepler); also J. V. Field, *Kepler's Geometrical Cosmology* (Chicago: University of Chicago Press, 1988), xv (*Timaeus*), 73–74 (*Mysterium cosmographicum* and *Harmonice mundi*), 98–99 (God as Platonic geometer).

4. On Kepler's conception of the sun, see *Paralipomena*, where the sun is source and preservation of all motion; the densest material body in the world, to which "belongs the pure actuality of [immaterial] light"—a soul or at least a vital faculty, dwelling within the body of the sun, which subjugates the "contumacious matter" of the sun and from whose victory the result is light (237–39); also Kepler's *Mysterium Cosmographicum: The Secret of the Universe*, trans. A. M. Duncan, introduction and commentary by E. J. Aiton

(1596/1621; facsimile rpt. New York: Abaris, 1981), 201: "noble epithets" for the sun, namely, "heart of the universe, king, emperor of the stars, visible God"; also Kepler's *New Astronomy* [*Astronomia nova*], trans. William H. Donahue (1609; Cambridge: Cambridge University Press, 1992), 385: "For it may appear that there lies hidden in the body of the sun a sort of divinity, which may be compared to our soul"; Kepler's *Epitome of Copernican Astronomy* [*Epitome Astronomiae Copernicanae*], Books IV–V, 1620–21, trans. Charles Glenn Wallis (Annapolis, Md.: The St. John's Bookstore, 1939), 1:90–91: as the source of all movement, the sun "has a soul in itself"; Kepler's *Harmonice mundi*: distinction first, then resemblance between Proclus' God-in-the-sun and the son of God (V, 84); analogy, namely, planets:sun :: *dianoia:nous*, i.e., discursive reason to intuitive intellection (V, 87).

5. W. Pauli, "The Influence of Archetypal Ideas on the Scientific Theories of Kepler," trans. Priscilla Silz, in C. G. Jung and W. Pauli, *The Interpretation of Nature and the Psyche* (New York: Pantheon: 1955), 149–240; Gerd Buchdahl, "Methodological Aspects of Kepler's Theory of Refraction," *Studies in the History of the Philosophy of Science* 3, no. 3 (1972): 265–98.

6. See Pauli, "The Influence of Archetypal Ideas," 151–240, here 157, 190–201, 237, and appendices 1 and 2 (213–35) for selections from Fludd; also Raz Chen-Morris, "From Emblems to Diagrams: Kepler's New Pictorial Language of Scientific Representation," *Renaissance Quarterly* 42, no. 1 (2009): 134–65.

7. See Judith V. Field for a refinement of this claim: "Kepler's Rejection of Numerology," in *Occult and Scientific Mentalities in the Renaissance*, ed. Brian Vickers (Cambridge: Cambridge University Press, 1984), 273–96. Having distinguished between abstract, undimensioned numbers and numbers derived from measurement (*numeri numerantes* and *numeri numerati*, respectively), Field ties Kepler's thought to the latter.

8. My introduction treats iconic representation, as does my fourth chapter.

9. See Lindberg, *Theories of Vision*, 57–60; Darrigol, *A History of Optics*, 17–21; A. Mark Smith, "Ptolemy, Alhazen, and Kepler and the Problem of Optical Images," *Arabic Sciences and Philosophy* 8 (1998): 9–44, esp. 21–30.

10. For (immaterial) heat as a property of light and for the temporal action of bleaching by light, see *Paralipomena*, 39–42; cf. Kepler's *Astronomia nova*, 383.

11. In the interest of clarity, I have changed a comma after "points" to a semicolon.

12. Cf. Isabelle Pantin, "Simulachrum, Species, Forma, Imago: What Was Transported by Light in the Camera Obscura," *Early Science and Medicine* 13 (2008): 245–69. Pantin similarly holds that Kepler came to believe

rays and points really to be geometrical lines and points (257). Her argument does not engage issues of representation (256).

13. I would again stress *simply*: e.g., in *Harmonice mundi*, Kepler, citing Proclus, asks impatiently, "And why waste words? Geometry, which before the origin of things was coeternal with the divine mind and is God himself (for what could there be in God which would not be God himself?) supplied God with patterns for the creation of the world": Johannes Kepler, *The Harmony of the World*, trans. and ed. E. J. Aiton, A. M. Duncan, and J. V. Field (Philadelphia: American Philosophical Society, 1997), 304; cited hereafter as *Harmony*. Kepler's statement is not the same as saying that geometry is all God is. His parenthetical question invokes a theological context about the ideas/patterns in the mind of God, the attributes of God, and so on, and it actually (playfully?) qualifies the reduction of God to geometry. Elsewhere in *Harmonice*, Kepler finds geometry inferior to harmony (V, 80). See also Peter Barker and Bernard R. Goldstein, "Theological Foundations of Kepler's Astronomy," *Osiris* 16, 2nd ser. (2001): 88–113: Barker and Goldstein ground Kepler's science in Melanchthon's "doctrine of the natural light" of the intellect that accessed innate knowledge of morals and mathematics (95).

14. See John Milton, *Paradise Lost*, ed. Barbara K. Lewalski (Oxford: Blackwell, 2007), Bk. I, 21: "brooding" means "conceiving" in both its mental and physical senses, i.e., meditating and nesting or hatching; Milton's image derives from the Hebrew and is translated correctly as *incubabat* in the Tremellius Latin Bible (12n17–22).

15. The planets also require "an inherent force [*vis insita*]" by means of which, like ferrymen using rudders, they navigate the current of power flowing out from the sun: *Astronomia nova*, 404–5: the analogy is Kepler's. Kepler introduced the term *inertia*: E. J. Dijksterhuis, *The Mechanization of the World Picture*, 314. On planetary movement, see also *Astronomia nova*, 387–91; *Epitome*, 87–89, 95, 98–99.

16. On this terminological problem, which Donahue considers "supremely important," see his extensive gloss on the words *vis* and *virtus* in his edition of *Astronomia nova*, 25.

17. E. J. Dijksterhuis, *The Mechanization of the World Picture*, 349. Cf. Crombie, *Medieval and Early Modern Science*, 2:182, 194–95. On the Keplerian circle, see also Field, *Kepler's Geometrical Cosmology*, 101–2; and Karsten Harries, *Infinity and Perspective* (Cambridge, Mass.: MIT Press, 2001), 277–81 (comparison of Galileo and Kepler).

18. My paragraph is based on Patrick J. Boner, "Kepler's Living Cosmology: Bridging the Celestial and Terrestrial Realms," *Centaurus* 48 (2006): 32–39, here 36; Boner quotes *GW*, 1:193–94, 6:406–7. The translation and internal brackets are Boner's, the italics mine.

19. In addition to Boner, "Kepler's Living Cosmology," 1, 37, 39, see Crombie, *Medieval and Early Modern Science*, 2:185–86, 191–92, and Pauli, "The Influence of Archetypal Ideas," 177–79, 188. Pauli connects such animistic thinking with late antiquity, Paracelsus, and the like. On the Earth soul as "perceiver of the harmonics of the motions of the heavens (though without discursive thought), form-giver of the marvellous figures in fossils," and much more, see *Paralipomena*, 239–40, including Kepler's marginal note.

20. Cf. Wolfram Schmidgen, *Exquisite Mixture: The Virtues of Impurity in Early Modern England* (Philadelphia: University of Pennsylvania Press, 2013), 24–58, e.g., 53.

21. On *species*, see Lindberg, "Genesis of Light," 19 (history of the term), 35–36 (Kepler), 40 (the species of solar power and those of light as "different species of the same genus"); the basis of Lindberg's view is in my next paragraph. Cf. Donahue's gloss in *Astronomia nova*, 23–24; and the gloss on "*species immateriata* (an invisible and immaterial simulacrum of itself emitted by an object)" in *Harmony of the World*, trans. and ed. E. J. Aiton, A. M. Duncan, and J. V. Field, xli; also Pantin, "Simulachrum, Species, Forma, Imago," 248, 254–56. On Keplerian terminology, cf. also Svetlana Alpers, *The Art of Describing: Dutch Art in the Seventeenth Century* (Chicago: University of Chicago Press, 1983), 26–71, here 36: Alpers invokes a Keplerian distinction between *imago rerum* "(image of the world outside the eye)" and the retinal *pictura* but without specifying its location in Kepler or its relation to Kepler's other uses of *imago/image*. *Species* (< *specio*) most basically means "a seeing" (or "seeings") and thus a "sight, look, view." It is most simply that which is *seen*, whatever it might be, on the eye's screen by its seer, or interpreter. Latin *species* is also a standard translation of Aristotle's *eidos*, which in English means "that which is seen," "form," "shape," "figure," or "species."

22. On play (*Spiel*) in Keplerian science, see Giora Hon, "Putting Error to (Historical) Work: Error as a Tell-Tale in the Studies of Kepler and Galileo," *Centaurus* 46 (2004): 58–81, e.g., 72–73.

23. On significant form in Kepler's work, see Albert Van der Schoot, "Kepler's Search for Form and Proportion," *Renaissance Studies* 15, no. 1 (2001): 59–71; Fernand Hallyn, *The Poetic Structure of the World: Copernicus and Kepler*, trans. Donald M. Leslie (New York: Zone, 1993); Elizabeth Spiller, *Science, Reading, and Renaissance Literature: The Art of Making Knowledge, 1580–1670* (Cambridge: Cambridge University Press, 2004), esp. 116–36; Howard Marchitello, *The Machine in the Text: Science and Literature in the Age of Shakespeare and Galileo* (Oxford: Oxford University Press, 2011); Frédérique Aït-Touati, *Fictions of the Cosmos: Science and Literature in the Seventeenth Century*, trans. Susan Emanuel (Chicago: University of Chicago Press, 2011).

24. The impetus theory, current in Kepler's time but under scrutiny by Galileo, attributed impetus to the projectile after it left the source of its motion. The fourteenth-century nominalist Jean Buridan's formula for it was impetus = weight × velocity: for further explanation, see the *Dictionary of the Middle Ages*, ed. Joseph R. Strayer (New York: Charles Scribner's Sons, 1983), 2:430–32, s.v. *Buridan*; Kuhn, *Copernican Revolution*, 119–23; Harries, *Infinity and Perspective*, 135–38; and Max Caspar, *Kepler*, trans. and ed. C. Doris Hellman, updated by Owen Gingerich and Alain Segonds (New York: Dover, 1993), 136 (Galileo's clinging to "the old Aristotelian distinction between 'natural' and 'violent' motion," which Kepler discards).

25. *Phos* and *phaos*, the latter used in my introduction, are Greek alternatives for "light." *Phos* suggests English *photo*; *phaos* suggests English *phantasy*.

26. A. Mark Smith, "Ptolemy, Alhazen, and Kepler," 40, 42; subsequent reference is to this essay. Smith's more recent book *From Sight to Light* essentially affords the same argument, aiming to separate Kepler from the perspectivists and the physics of light from the psychology of sight (369–70). Ironically, Smith provides a long list of similarities between Keplerian optics and perspectivism (364).

27. E.g., *Paralipomena*, 319–21: "we then finally consider ourselves to have *seen*, i.e. observed, a heavenly body rightly, when we shall have measured the angle of vision accurately" (319; my emphasis); also 337: "Now the speed of the stars has no proportion to sense perception. . . . And so whatever is in our senses concerning the motion of the heavens, we have absorbed thanks to the intervention of reasoning."

28. First citation, *Harmony*, 304; second, Caspar, *Kepler*, 271; in the next sentence, the citation is from Pauli, "The Influence of Archetypal Ideas," 165–66 (Latin and English).

29. See Alan E. Shapiro, "Images: Real and Virtual, Projected and Perceived, from Kepler to Dechales," *Early Science and Medicine* 13 (2008): 270–312, e.g. 271, 287.

30. *Paralipomena*, 77: italics in the text. Donahue glosses "intensional quantities" as "quantitative properties of things that can acquire different degrees of intensity" (77n21).

31. On *intentionale ens* in earlier optical tradition, see Katherine H. Tachau, *Vision and Certitude in the Age of Ockham: Optics, Epistemology, and the Foundations of Semantics* (Leiden: E. J. Brill, 1988), e.g., 15, 21–22 (ambiguity), 43–44 (recognition of the problem), 62 (Scotus's terminological clarification), 90–106 (apparent being). Questions include the nature of the visual species and their relation to knowledge: e.g., whether the species are natural and sensible (extramental) or spiritual and intentional (intramental) and whether they convey reality, truth, or both.

32. See Judith H. Anderson, *Words That Matter: Linguistic Perception in Renaissance English* (Stanford, Calif.: Stanford University Press, 1996), e.g., 2–3, 11–19, 25, 36–40, 43: the recognition that a word is a thing is both ancient Greek (Plato's *Cratylus*) and Augustinian (*Christian Doctrine*).

33. E.g., Steven Shapin, *The Scientific Revolution* (Chicago: University of Chicago Press, 1996); Steven Shapin, *Never Pure* (Baltimore, Md.: Johns Hopkins University Press, 2010); Peter Dear, "Narrative, Anecdotes, and Experiments: Turning Experience Into Science in the Seventeenth Century," in *The Literary Structure of Scientific Argument*, ed. Peter Dear (Philadelphia: University of Pennsylvania Press, 1991), 135–63; and note 38 in this chapter.

34. Ofer Gal and Raz Chen-Morris, "Baroque Optics and the Disappearance of the Observer: From Kepler's Optics to Descartes' Doubt," *Journal of the History of Ideas* 71, no. 2 (2010): 191–217.

35. See Stephen Straker, "The Eye Made 'Other,'" 12 (Alberti), 12–17 (Dürer); Lindberg and Cantor, *Discourse of Light*, v–vi; 42, 42n92 (Dürer).

36. Giora Hon and Yaakov Zik, "Kepler's Optical Part of Astronomy (1604): Introducing the Ecliptic Instrument," *Perspectives on Science* 17, no. 3 (2009): 307–45, esp. 326–32. See also Straker, "Kepler, Tycho, and the 'Optical Part of Astronomy'" and "The Eye Made 'Other.'"

37. Cf. Gérard Simon, "Analogies and Metaphors in Kepler," in *Metaphor and Analogy in the Sciences*, ed. Fernand Hallyn (Dordrecht: Kluwer Academic, 2000), 71–82, esp. 74.

38. Hon, "Putting Error to (Historical) Work," 72–73, 76–78. Giora Hon's earlier essay is "On Kepler's Awareness of the Problem of Experimental Error," *Annals of Science* 44 (1987): 545–91. Cf. Sven Dupré, "Inside the *Camera Obscura*: Kepler's Experiment and Theory of Optical Imagery," *Early Science and Medicine* 13 (2008): 219–44, esp. 232–37, 243.

39. Aristotle, *On the Soul* (*De Anima*), trans. J. A. Smith, in *The Complete Works of Aristotle*, rev. ed. Jonathan Barnes, 2 vols. (Princeton, N.J.: Princeton University Press, 1984), 1:428a11–12; 427b16, 28–29; 431b2, 432a3–10. On the reliability of the images, see Malcolm Schofield, "Aristotle on the Imagination," in *Essays on Aristotle's "De Anima,"* ed. Martha C. Nussbaum and Amélie Oksenberg Rorty (Oxford: Clarendon, 1992), 249–77, here 251; also Dorothea Frede, "The Cognitive Role of *Phantasia* in Aristotle," in *Essays on Aristotle's "De Anima,"* ed. Martha C. Nussbaum and Amélie Oksenberg Rorty (Oxford: Clarendon, 1992), 279–95, here 285, 294.

40. Richard Sorabji, "Intentionality and Physiological Processes: Aristotle's Theory of Sense-Perception," in *Essays on Aristotle's "De Anima,"* ed. Martha C. Nussbaum and Amélie Oksenberg Rorty (Oxford: Clarendon, 1992), 195–225, esp. 198, 208–11. Harries, in *Infinity and Perspective*, setting

the terms infinity and perspective in opposition, treats the distortions of eyesight from Plato through early modernity (e.g., 106).

41. Kepler's Latin reads, "Quale itaque corpus lucis (sine materia), talis et motus (sine tempore) vtrumque analogicum": *GW*, 2:45.

42. On Kepler and Descartes's tennis balls, see Chapter 4, including note 23.

43. Cf. Antoni Malet, "Gregorie, Descartes, Kepler, and the Law of Refraction," *Archives Internationales d'Histoire des Sciences* 40, no. 125 (1990), 278–304, esp. 281. Brian Vickers also treats Kepler's explanation of conics in "Analogy Versus Identity: The Rejection of Occult Symbolism, 1580–1680," in *Occult and Scientific Mentalities in the Renaissance*, ed. Brian Vickers (Cambridge: Cambridge University Press, 1984), 95–163, here 149–50. Cf. Field, *Kepler's Geometrical Cosmology*, 125–26 (on Kepler's objection to Jean Bodin's use of analogy; in contrast, see Vickers, "Analogy Versus Identity," 152).

44. *Paralipomena*, 109, 109nn35, 37: when Donahue translates *proportiones* as "ratios" he is within his rights, but he is skewing the text more exclusively toward mathematics and modern distinctions; he is also obscuring connections with Kepler's use of this word elsewhere and its broader cultural use in this period.

45. Latin *locutio* derives from *loquor*, "to tell, talk about"; "declare, show, indicate, express clearly": Charlton T. Lewis and Charles Short, *A Latin Dictionary* (Oxford: Clarendon, 1966). In Keplerian usage, it could indicate not only verbal expressions but also empirical and numerical ones. It could apply to many of Kepler's own "demonstrations" (so-termed), which include verbal definitions (e.g., "image"), experiments (e.g., aqueous globe), common experience (e.g., reflections through a multipaned window), and even inconclusive or imprecise arithmetical data. My text beneath the inset explains further.

6. ANALOGY, PROPORTION, AND DEATH IN DONNE'S *ANNIVERSARIES*

1. Gérard Simon, "Analogies and Metaphors in Kepler," in *Metaphor and Analogy in the Sciences*, ed. Fernand Hallyn (Dordrecht: Kluwer Academic, 2000), 71–82, here 74.

2. Albert Van der Schoot, "Kepler's Search for Form and Proportion," *Renaissance Studies* 15, no. 1 (2001): 59–78, here 62. See also J. V. Field, *Kepler's Geometrical Cosmology* (Chicago: University of Chicago Press, 1988), 98–99.

3. John Donne, *The Variorum Edition of the Poetry of John Donne*, ed. Gary A. Stringer et al.; commentary, ed. John R. Roberts, et al. (Bloomington: Indiana University Press, 1995), 6:SA 468. This is the text for Donne's *Anniversaries*, *Epicedes*, and *Obsequies*, cited as *Variorum* 6.

4. See Max Caspar, *Kepler*, trans. and ed. C. Doris Hellman, corrected and expanded ed. (New York: Dover, 1993), 83, 213, 217, 336; also Peter

Barker and Bernard R. Goldstein, "Theological Foundations of Kepler's Astronomy," *Osiris* 16, 2nd ser. (2001): 88–113, here 96–97; Field, *Kepler's Geometrical Cosmology*, 49, 127.

5. W. Milgate, ed., *The Epithalamions, Anniversaries, and Epicedes* (Oxford: Clarendon, 1978), xlv; the comparison of Eliot's poem to an epic is I. A. Richard's, in *Principles of Literary Criticism* (London: Keegan Paul, 1926), 291.

6. Aside from typography, the form of the titles I cite is from *The Variorum Edition*. On the relation of *The First Anniversarie* to satire, see Milgate, ed., *The Epithalamions, Anniversaries, and Epicedes*, xxxviii; and on the generic pertinence of Donne's sermons, xl. On the genre(s) of the *Anniversaries*, see also Barbara Kiefer Lewalski, *Donne's "Anniversaries" and the Poetry of Praise: The Creation of a Symbolic Mode* (Princeton, N.J.: Princeton University Press, 1973); also *Variorum* 6:326–34.

7. Cf. Charles Monroe Coffin, *John Donne and the New Philosophy* (New York: Humanities Press, 1958), 134–37; John H. Cartwright and Brian Baker, *Literature and Science: Social Impact and Interaction* (Santa Barbara, Calif.: ABC-CLIO, 2005), 65–66.

8. On the ancient myths, see Edgar Wind, *Pagan Mysteries in the Renaissance*, rev. ed. (Harmondsworth: Penguin, 1967), e.g., 76–92, 209–12 (Spenser).

9. Cf. Marjorie Hope Nicolson, *The Breaking of the Circle: Studies in the Effect of the "New Science" Upon Seventeenth-Century Poetry*, rev. ed. (New York: Columbia University Press, 1960), 32, 114 (topographical irregularities of the earth); 116–21 (Donne's originally relaxed response); 121, 155–56 (disproportion in the old and new astronomy); 145 (Gilbert's magnetism); 147–54 (Kepler, esp. the element of fire).

10. Coffin, *John Donne and the New Philosophy*, 166–72, suggests Donne's knowledge of arguments refuting the existence of the fiery sphere, including his awareness of Kepler's refutation in the *Paralipomena* and *Dioptrice* as a result of his studying optical refraction. Besides Kepler, Tycho Brahe, Johnannes Pena, Christoph Rothmann, and Girolamo Cardano (definitely known to Donne), all discussed by Kepler, rejected the element of fire, as did Kepler's source, the thirteenth-century scholar Witelo. See also Cartwright and Baker, *Literature and Science*, 65–66, who connect the vanishing of old stars lamented by Donne's speaker with Tycho's catalogue of fewer stars than Ptolemy's—777 rather than 1,022; 64, on two new "stars" of 1600 and 1604, discussed by Kepler in *De stella nova*, to which Donne refers.

11. See Francis R. Johnson, *Astronomical Thought in Renaissance England: A Study of the English Scientific Writings from 1500–1645* (Baltimore, Md.: Johns Hopkins University Press, 1937), 23n, 25, 110–11 (equinoxes); 47–50, 101–3 (planetary orbits); Stephen Toulmin and June Goodfield, *The Fabric of*

the Heavens: The Development of Astronomy and Dynamics (New York: Harper and Row, 1961), 27–28, 138–41, 171–72. On the dissemination of scientific information, see note 28 of my Introduction. Donne knew at least one of Kepler's books by the time he wrote *Problem* X and *Biathanatos* (the latter, 1608): see my Introduction and its notes 24–27. Thomas Harriot, a correspondent of Kepler, whose work might have been known to Donne through Henry Percy, was, according to the *DNB*, a Copernican by 1610 and had already conjectured that the planetary orbits were not circular; Harriot "immediately accepted Kepler's theory of elliptical orbits" (accessed online 6/27/14).

12. See R. J. Meyer, "'Fixt in heauens hight': Spenser, Astronomy, and the Date of the *Cantos of Mutabilitie*,'" *Spenser Studies* 4 (1984): 115–29, here 118–19; cf. Sarah Powrie, "Spenser's Mutabilitie and the Indeterminate Universe," *Studies in English Literature* 53 (2013): 75–89, esp. 75 (Donne), 79–85 (radical astronomical implications of Spenserian Mutabilitie). On responses to new bodies in the heavens, cf. also Andrew Fleck, "'None ends where he begun': Astronomical Revolutions in John Donne," in *The Book of Nature and Humanity in the Middle Ages and the Renaissance*, ed. David Hawkes and Richard G. Newhauser (Turnhout, Belgium: Brepols, 2013), 265–85, here 273–76.

13. Victor Harris, *All Coherence Gone: A Study of the Seventeenth-Century Controversy Over Disorder and Decay in the Universe* (London: Cass, 1966); also William J. Bouwsma, *The Waning of the Renaissance, 1550–1640* (New Haven, Conn.: Yale University Press, 2000), e.g., 87. In contrast to such evidence of pessimism, see Julián Jiménez Heffernan, "John Donne and the New Universe: Retaking the Issue," *Sederi: Spanish and Portuguese Society for English Renaissance Studies* 8 (1998): 65–74. The introduction by John Haffenden, ed., to William Empson, *Essays on Renaissance Literature*, vol. 1: *Donne and the New Philosophy*, 2 vols. (Cambridge: Cambridge University Press, 1993), 1:1–61, also provides a substantial review of Donne's knowledge of the new science (1:25–44): the skeptical Donne "was less shaken than stirred by the new science" (1:30).

14. John Donne, *Ignatius His Conclave*, ed. T. S. Healy (Oxford: Clarendon, 1969), xxix, 17: admitting that Copernicus's opinions "may well be true," Donne's Ignatius suggests that Christopher Clavius, who opposed the Copernican "truth," is the one who belongs in hell (ibid.). Cf. R. Chris Hassel Jr., "Donne's *Ignatius His Conclave* and the New Astronomy," *Modern Philology* 68, no. 4 (1972): 329–37: Donne's use of a "naïve, gullible" persona and of "amused ridicule" regarding Copernicus, Galileo, and Kepler (331–32; also 334, 336). Hassel adds that Donne's attitude in *Ignatius* is "too skeptical to suggest any personal disillusionment" (334, 336).

15. Edmund Spenser, *The Faerie Queene*, ed. A. C. Hamilton, with Hiroshi Yamashita, Toshiyuki Suzuki, and Shohachi Fukuda, rev. 2nd ed. (Harlow: Pearson, 2007), V.Pro.4.

16. Variant versions of the macro/microcosm analogy are available in the Renaissance: e.g., one a Platonic/Neoplatonic ladder (or stages) with Man in the middle; another, typically geocentric, with Man as the center, essence, focus, and end of the universe; yet another—to which either heliocentrism (as in Kepler), geocentrism, or the Tychonic compromise between them—could be congenial, with the heavens providing a pattern or model for earth, having an influence on it, and even interacting with it. The common denominator for all these views, which can overlap, is orderly relationship, coherence. Influence is mutual: if the whole affects the part, disruption of a part affects the whole. If God is considered to be in some way atop the sequence, however, the necessity of this last logic is itself disrupted, whether destructively or providentially. Harris concentrates on (originally Pythagorean) Man as microcosm (summarized 198).

17. Allowing for a voice, as well as a persona here, I am not disallowing Judith Scherer Herz's denial of a fictional speaker in Donne's lyrics in favor of a distinctive voice, style, convention (sometimes successively various in a single poem), although I find strong historical precedents for the device of a persona from the Middle Ages to the Renaissance: "'An Excellent Exercise of Wit That Speaks So Well of Ill': Donne and the Poetics of Concealment," in *The Eagle and the Dove: Reassessing John Donne*, ed. Claude J. Summers and Ted-Larry Pebworth (Columbia: University of Missouri Press, 1986), 3–14; see also note 25 in this chapter.

18. Matthew Greenfield recounts a shift from public to more private forms of mourning and locates Donne between these in the *Anniversaries*: "The Cultural Functions of Renaissance Elegy," *English Literary Renaissance* 28, no. 1 (1998): 75–95, esp. 76–77, 86; see also Philippe Ariès, *The Hour of Our Death*, trans. Helen Weaver (1981; rpt. Oxford: Oxford University Press, 1991), 227–30.

19. Louis L. Martz, *The Poetry of Meditation* (New Haven, Conn.: Yale University Press, 1954), 233; for more complaints about the syntax, logic, and coherence of *The First Anniversarie*, see James Andrew Clark, "The Plot of Donne's *Anniversaries*," *Studies in English Literature* 30 (1990): 63–77, esp. 68–69. Citing Martz, Hugh Grady argues that Donne's *First Anniversarie* is an example of Walter Benjamin's *Trauerspiel* (mourning play): "Donne's *First Anniversary* as Baroque Allegory: Fragmentation, Idealization, and the Resistance to Unity," *John Donne Journal* 32 (2013): 107–29, here 110, 115–16. Martz later modified his judgment: see note 23 in this chapter.

20. Frank Manley, ed., *John Donne: The Anniversaries* (Baltimore, Md.: Johns Hopkins University Press, 1963), e.g., 125n12, 139n145.

21. See the movement of the gods' chariots to the rim of heaven [i.e., the heavens], where they stand to gaze on the reality of the Forms beyond it:

Plato, *Phaedrus*, trans. Alexander Nehamas and Paul Woodruff (Indianapolis, Ind.: Hackett, 1995), 247a–e (pp. 32–33).

22. "Ben Jonson's Conversations with William Drummond of Hawthornden," in *Ben Jonson: The Complete Poems*, ed. George Parfitt (New Haven, Conn.: Yale University Press, 1975), 461–80, here 462. On decorum and proportion more generally, see Gavin Alexander, "Sidney, Scott, and the Proportions of Poetics," *Sidney Journal* 33(2015): 7–28.

23. For detailed outlines of the structures of the *Anniversaries*, see esp. Martz, *The Poetry of Meditation*, 222–23, 236–37; Lewalski, *Donne's "Anniversaries" and the Poetry of Praise*, 228, 248–62, 285–302; and Edward W. Tayler, *Donne's Idea of a Woman: Structure and Meaning in "The Anniversaries"* (New York: Columbia University Press, 1991), 76–90; also, *Variorum* 6:335–54. Subsequently Martz changed his conception of the structure of *The Second Anniversarie*, on which see Tayler, *Donne's Idea of a Woman*, 80–81, for a sympathetic summary and response; cf. also Louis L. Martz's later essays "Metaphysical and Meditative: Donne's *Anniversaries* and Eliot's *Quartets*," *Literature and Belief* 19 (1999): 25–42; and "Donne's *Anniversaries*: The Powers of the Soul," in *Reading the Renaissance: Ideas and Idioms from Shakespeare to Milton*, ed. Marc Berley (Pittsburgh, Penn.: Duquesne University Press, 2003), 78–89. More recently, Catherine Gimelli Martin has argued that *The First Anniversarie* is "a thinly disguised assault" on Francis Bacon's *Advancement of Learning* (1605): "*The Advancement of Learning* and the Decay of the World: A New Reading of Donne's *First Anniversarie*," *John Donne Journal* 19 (2000): 163–203, esp. 174, 186.

24. On alternative arguments for a persona, see *Variorum* 6:317–25: "The Poet and His Audience," esp. Stanwood (1971), Elliott (1972), Lewalski (1973, 1976), Parrish (1973, 1977), Shami (1984), Marotti (1986).

25. For the purpose of this essay, I consider the terms *persona*, *speaker*, *voice*, and the like to be on a continuum; of the three here specified, roughly in order of degree dramatized, one can shift or merely slide into another; *role*, like *persona*, is dramatized to a strongly marked degree. The date suggested for *Essays in Divinity* is 1614, but Evelyn Simpson remarks that the evidence would fit any date from 1611 to early 1615: Evelyn Simpson, *Essays in Divinity*, ed. Evelyn M. Simpson (Oxford: Clarendon, 1952), ix–x; cf. R. C. Bald, *John Donne: A Life* (Oxford: Oxford University Press, 1970), 281–82, 298–300.

26. Unless otherwise indicated, texts of Donne's *Songs and Sonnets* are from *The Complete English Poems*, ed. A. J. Smith (1971; rpt. London: Penguin, 1996).

27. On the chronology of composition of "A Fvnerall Elegie" and the *Anniversaries*, about which there has been some disagreement, see *Variorum* 6:281–84.

28. George Puttenham, *The Arte of English Poesie* (1589; rpt. Kent, Ohio: Kent State University Press, 1970), 61–62.

29. Marshall Grossman, *The Story of All Things: Writing the Self in English Narrative Poetry* (Durham, N.C.: Duke University Press, 1998), 154–55, 193.

30. Elizabeth D. Harvey and Timothy M. Harrison, "Embodied Resonances: Early Modern Science and Tropologies of Connection in Donne's *Anniversaries*," *ELH* 80 (2013): 981–1008, here 982.

31. Helkiah Crooke, *Mikrokosmographia* (London: William Jaggard, 1615), 699: I have removed a superfluous comma after "is"; the spacing suggests that it is an error.

32. Aristotle, *Poetics*, trans. I. Bywater, in *The Complete Works of Aristotle*, ed. Jonathan Barnes, 2 vols. (Princeton, N.J.: Princeton University Press, 1984), 2:1457b7–9.

33. Harvey and Harrison, "Embodied Resonances," 986–87, take the one contiguous example of metonymy in Puttenham (container for contained), apart from Puttenham's other examples (191–92), and combine it with a quotation from Kaja Silverman (*The Subject of Semiotics* [Oxford: Oxford University Press, 1983], 111) that identifies metonymy as the trope of contiguity. For Silverman, as for Lacan, this identification is Jakobson's. Jean-François Lyotard, for one, has ably questioned Jakobson's interpretation of metonymy: "The Dream-Work Does Not Think," in *Discourse, Figure*, trans. Antony Hudek and Mary Lydon (Minneapolis: University of Minnesota Press, 2011), 233–67, esp. 245–49. George Lakoff and Mark Johnson, *Philosophy in the Flesh: The Embodied Mind and Its Challenge to Western Thought* (New York: Basic Books, 1999), identify the container/contained trope as a basic metaphor within the "Classical Theory of Categorization" (341; and index: Container Metaphors).

34. This is what the U.S. Supreme Court did when it decided that a corporation (< Latin *corpus*, "body") is really a person and thus denied centuries of linguistic change and civil intellection.

35. "Literally" most basically means "letter-ally" (from Latin *littera*, "letter). To equate the literal with the physical (or actual = real) is to ignore the mediation of letters—that is, to make language transparent. If *literal* is used only to mean "nonfigurative," its letterality is not compromised, but it becomes merely a negation.

36. Materials in this paragraph are based on my *Translating Investments: Metaphor and the Dynamic of Cultural Change in Tudor-Stuart England* (New York: Fordham University Press, 2005), esp. 129–65.

37. See Carol M. Sicherman, "Donne's Timeless Anniversaries," in *Essential Articles for the Study of John Donne's Poetry*, ed. John R. Roberts (Hamden, Conn.: Archon, 1975), 374–86, here 384–85.

38. This is the perception that a deity is merely the projection and conception of human desire. The old joke that God made Man in his image, and Man returned the compliment comes to mind. Cf. Grossman's suggestive comment, which refers to both *Anniversaries*, that Donne moves "from metonymy to metaphor, along the established path of the analogy of macrocosm and microcosm" (*The Story of All Things*, 193). Grossman's rhetorical terminology, hence its application, is specialized. His assumptions indicate an underlying tautology here, yet one that issues in "an excess of signification" and "an ironic questioning of mimetic representation itself" (193–94).

39. Although the omission of the deceased person's name is conventional, it is not universal, as Ben Jonson's Cary-Morison Ode, as well as his elegy on his first son, memorably attests. Any poet worth his/her salt, moreover, can use or modify a convention in a newly significant way. Greenfield, "The Cultural Functions of Renaissance Elegy," 85, notes that personal names became "increasingly prominent" in Renaissance funerary inscriptions of the sixteenth and seventeenth centuries.

40. See A. C. Hamilton, "Our New Poet: Spenser, 'well of English undefyld,'" in *A Theatre for Spenserians*, ed. Judith M. Kennedy and James A. Reither (Toronto: University of Toronto Press, 1973), 101–23, here 110. I am only noting a suggestive contemporary (1605) translation, not connecting Elizabeth Drury with the Queen.

41. *The Vision of Pierce Plowman* (1550; rpt. Richmond: Kingprint for Paradine, 1976), Bi^{v}. There were two editions and two reprints of Langland's poem in the sixteenth century. Since the *OED* cites only earliest instances of use in a given text—late fourteenth century for Langland—it lacks these timely occurrences of *drury*. Donne's pun on *drury* gains added resonance in relation to his verse apology ("Though I be dead, and buried") to the Countess of Bedford: "I have to others lent / Your stock, and over prodigally spent / Your *treasure*": *The Satires, Epigrams, and Verse Letters of John Donne*, ed. W. Milgate (Oxford: Clarendon, 1967), 104 (my emphasis). Since my next sentence cites the confusion of *drury* with *dowery* (i.e., *dowry*), I add as antidote to silliness that by 1606, George More was paying with interest the dowry early denied Donne (Bald, *John Donne*, 153). For the Chaucerian references (and other medieval ones), see *MED*, s.v., *druerie*.

42. I cite the King James version; the Geneva Bible is essentially the same. Interestingly, the Vulgate uses *thesaurus* for "treasure" here, thus enlarging the scope of the word to memory and words themselves, given a little poetical meditation: on *thesaurus*, see the indexical entry in my *Words That Matter: Linguistic Perception in Renaissance English* (Stanford, Calif.: Stanford University Press, 1996). On naming as knowing and Donne's interest in biblical names, see Raymond-Jean Frontaine, "'The name of shee': The Biblical

Logocentrism of Donne's *Anniversaries,*" *Publications of the Missouri Philological Association* 22 (1997): 28–39. On the heart, see Robert A. Erickson, *The Language of the Heart: 1600–1750* (Philadelphia: University of Pennsylvania Press, 1997): by the early modern period, *heart* is "the single most important word referring both to the body and to the mind" (11).

43. Citations from Manley, ed., 2; Yamen Liu, "The Making of Elizabeth Drury: The Voice of God in 'An Anatomy of the World,'" in *John Donne Journal* 8 (1989): 89–102, here 95. Cf. Peter L. Rudnytsky, "'Sight of God': Donne's Poetics of Transcendence," *Texas Studies in Literature and Language* 24, no. 2 (1982): 185–207, here 195–96. Howard Marchitello considers Mistress Drury "a screen onto which the narrator projects his own (poetic) identity": *The Machine in the Text: Science and Literature in the Age of Shakespeare and Galileo* (Oxford: Oxford University Press, 2011), 124. Grady concludes that *She* represents "the advent of a modernity of fragments" (128–29).

44. Making this generalization, I am not simply accepting Lewalski's identification of *She* (*Donne's "Anniversaries" and the Poetry of Praise*, 219–20, 244, 247), but I am reducing a number of major variants to their common denominator, the Augustinian Word within, the Trinity, the image of God. One example is Manley's Wisdom, another Martz's Trinity, and still another Tayler's Christianized intelligible species (notably deriving from Bryskett's *Discourse of Civill Life*). To name is to know, and Donne's speaker in *The First Anniversarie* emphatically does not know.

45. On aether, see Robert Fludd (1574–1637), *Mosaicall Philosophy* (London: Humphrey Moseley, 1659), 221. On *pneuma*, see Giorgio Agamben, *Stanzas: Word and Phantasm in Western Culture*, trans. Ronald L. Martinez (Minneapolis: University of Minnesota Press, 1993), 91–95: the Christian-Stoic-Galenic-Neoplatonic *pneuma* has become "the breath that animates the universe, circulates in the arteries, and fertilizes the sperm; . . . that, in the brain and . . . heart, receives and forms the phantasms of the things we see, imagine, dream, and love. Insofar as it is the subtle body of the soul, it is in addition the intermediary between the soul and matter, the divine and the human . . . [enabling] influxes between corporeal and incorporeal" (94). As Ramie Targoff, *John Donne: Body and Soul* (Chicago: University of Chicago Press, 2008) observes, Donne's lines would be a problem for an Aristotelian because immateriality lacks (material) substance and therefore subtlety; that is, there are no degrees of immateriality (98–99). But there is no need to restrict a reading here to Aristotelian principles. To require that a juncture or interface between matter and spirit be expressed in purely material terms is, ipso facto, to doom Donne's imaginative exercise to failure. On the vexed meaning of *substance* in the Reformation period, see my *Translating Investments*, 48–50; also David C. Lindberg, "The Genesis of Kepler's Theory of

Light: Light Metaphysics from Plotinus to Kepler," *Osiris* 2, 2nd ser. (1986): 4–42.

46. The printing of *The First Anniversarie* is considered to have been careful and perhaps even reviewed by Donne himself: see *Variorum* 6:282–84.

47. See the notes on these lines in *Variorum* 6:442–43; also Theresa M. DiPasquale, "'To good ends': The Final Cause of Sacramental Womanhood in *The First Anniversarie*," *John Donne Journal* 20 (2001): 141–50, esp. 146.

48. For discussion of theories about the Supper and their implications, see my *Translating Investments*, 36–77. Luther insisted on real presence while retaining the substances of bread and wine (not just their appearances, as in Catholicism).

49. See *Variorum*, 6:449, note on line 469. I share with Gary Kuchar the recognition that "Incomprehensibleness" constitutes an enabling condition at the end of *The First Anniversarie*: *The Poetry of Religious Sorrow in Early Modern England* (Cambridge: Cambridge University Press, 2008), 208. Otherwise, my emphasis on a mediating persona and my skepticism regarding unmediated biography or autobiography in *The First Anniversarie* differ considerably from Kuchar's analysis of what he takes to be trauma in the poem—Donne on the couch, as it were.

50. I follow Donne in using "her" for the speaker's soul, in accordance with the feminine gender of Latin *anima*, "soul."

51. Cf. Marchitello's description of one of Galileo's many thought experiments, which were imagined or thought rather than actually conducted (*The Machine in the Text*, 87).

52. While noting George Saintsbury's observation that the last day is a time of judgment as well as of salvation, a *Dies Irae* (God's wrath) as well as a *Venite* (welcome; literally, "come"), I do not consider the former active at this point: see *Variorum* 6:466–67. The Last Judgment might be described as subliminally present here without actual or deliberate irony; it is sufficiently taken into account later.

53. My brackets in the citation. A modern, alliterative translation of the same lines reads,

> For I am Lord of Life, love is my drink
> And for that drink today I died upon earth.
> I struggled so I'm thirsty still for man's sake.
> No drink may moisten me or slake my thirst
> Till vintage time befall in the Vale of Jehoshaphat,
> When I shall drink really ripe wine, *Resurrectio mortuorum*.
> ....
> And all that are both in blood and in baptism my whole brothers
> Shall not be damned to the death that endures without end.

Qtd. from William Langland, *Will's Vision of Piers Plowman*, trans. E. Talbot Donaldson, ed. Elizabeth D. Kirk and Judith H. Anderson (New York: Norton, 1990), XVIII.365–78.

54. My phrasing echoes Donne's in *Deaths Duell* in *The Sermons of John Donne*, ed. George R. Potter and Evelyn M. Simpson, 10 vols. (1959: rpt. Berkeley: University of California Press, 1984), 10:229–48, here 248.

55. Like Kepler's, the position of the English church on the Supper was closer to Calvin's than to Luther's by the early Jacobean period—with significant individual variations, however. Provocatively, Kuchar identifies Donne's position with Luther's and takes it in *The First Anniversarie* to include Luther's skepticism regarding analogical thought (204, 207–8, 211). But the more humanistic thinking of Philipp Melanchthon (1497–1560), Luther's eventual successor (and author of a rhetoric), affords relevant complication: Melanchthon, like Kepler and Donne, considered the will more active in belief than did Luther and was more open to compromise with Rome and Geneva on the doctrine of the Eucharist, which involved controversies regarding metaphor: e.g., see Jaroslav Pelikan, *The Christian Tradition: A History of the Development of Doctrine* (Chicago: University of Chicago Press, 1984), 4:143–44, 176–77, 185–86. Donne refers to Melanchthon in his *Sermons*.

56. See *Variorum* 6:470; Targoff, *John Donne: Body and Soul*, 92. Although the passage in question has attracted commentary, Targoff, focusing on the relation of body and soul, is apparently original in finding it shocking and confused in this regard (92). Her observant reading has called my attention to the passage, although I offer a different interpretation of it.

57. See Katharine Park, "The Organic Soul," in *The Cambridge History of Renaissance Philosophy*, ed. Charles B. Schmitt et al. (Cambridge: Cambridge University Press, 1988), 464–84: Park draws on Melanchthon's *Liber de anima* and additional sources to establish that the *spiritus*, vaporized or exhaled from blood and "disseminated throughout the body by the arteries and nerves . . . [is] The source of all activity in the living body" and is often called "the 'first instrument' of the soul" (469). The distinction between soul and spirit is common in the period. Robert E. Stillman establishes the nature and breadth of Melanchthon's influence definitively in *Philip Sidney and the Poetics of Renaissance Cosmopolitanism* (Aldershot: Ashgate, 2008).

58. Manley's gloss (*John Donne: The Anniversaries*, 183), is still the best on Donne's treatment of the soul's origin. He considers Donne's view in lines 157–68 "ambivalent," in lines 157–58 to be creation *ex traduce* (the soul's origin the result of human inheritance), and in lines 159–68 to be creation *ex nihilo* (direct infusion by the Creator of an individual soul into the body). Manley cites the relevant passages in Donne's earlier *Letters* and later *Sermons*.

See also *Variorum* 6:484–85; for a different view, cf. Targoff, *John Donne: Body and Soul*, 93–94, 99–100. The closest parallel remains one of Donne's letters from Mitcham, which *calmly* suggests the need of a better explanation since *ex traduce* fails to guarantee the immortality of the soul or to prove that more than one soul exists in the body while *ex nihilo* fails to uphold the doctrine of original sin or to ensure the immortality of the intellective soul: John Donne, *Letters to Severall Persons of Honour (1651)* (Delmar, N.Y.: Scholars' Facsimiles, 1977), 11–19, here 16–17.

59. Wind, *Pagan Mysteries in the Renaissance*, 122.

60. The existence and number of the crystalline spheres varied with the theorist. Coffin remarks that in the ascent of the soul in *The Second Anniversarie*, "the *Primum Mobile* does not exist at all" (190). The point appears instead to be that it is invisible alike to naked eye or telescope. The planetary order that Donne observes (Venus before Mercury) is evidently Tycho Brahe's, as noted by John Shawcross, ed., *The Complete Poetry of John Donne* (Garden City, N.Y.: Doubleday, 1967), 296n189–206. Tycho's system was to a limited extent a compromise, mainly geocentric but with some degree of heliocentrism.

61. The meaning of *firmament* in the seventeenth century is flexible. Here, context leads me to take it as *Online OED* 2. "In Old Astronomy: the sphere containing the fixed stars; the eighth heaven of the Ptolemaic system." 2b. "Hence, applied sometimes to the other celestial spheres." In *Online OED* 1 the meaning is looser: "The arch or vault of heaven overhead, in which the clouds and the stars appear; the sky or heavens" (accessed 2/5/15). Powrie describes the Copernican rendering of "the sphere of fixed stars into a stationary and functionless feature" as its "most explosive threat to the traditional cosmos" (81); cf. Harris, *All Coherence Gone*, 178.

62. On Augustinian time, see Paul Ricoeur, *Time and Narrative*, trans. Kathleen McLaughlin and David Pellauer, 3 vols. (Chicago: University of Chicago Press, 1984), 1:5–30, esp. 16–22.

63. Pertinently, Jean Aitchison indicates that "where more than one meaning [of a word] is plausible," subjects in experiments "are likely to activate all of them," and for an ambiguous word, they "simultaneously activate more than one meaning . . . even when the context biases them in one direction": *Words in the Mind*, 2nd ed. (Oxford: Blackwell, 1994), 214–21. Aitchison's explanation could be extended to a syntactical construction.

64. On the beads as a rosary or necklace, see Harvey and Harrison, "Embodied Resonances," 981; on the analogies more generally, see *Variorum* 6:493–96.

65. On the word *string* in this passage, see Harvey and Harrison, "Embodied Resonances," 1001–2. Here, as elsewhere, their skillful concordancing attends to fascinating connections. They have little interest, however, in the

difference that inheres in metaphor/analogy, in the constraints of context, or in the movement, or *progress*, of *The Second Anniversarie*. In my reading, the broken lute strings at the outset of the poem connect by significant contrast, not simply by static recurrence, to the strings of beads, bones, and other relationships in the transcendence and reflective descent of the speaker's soul later on. In the passage with analogical strings, Harvey and Harrison also read the word *cord* (spinal cord), which is another word not in the poem, as an etymological pun on *chord* that connects with their theory of *resonance*, and they find in this wordplay the embodiment of cosmic order. Their view differs from my own, in which the body is at once canceled, continued, and raised, becoming simultaneously a connector to death and to heaven. Harvey and Harrison stop with embodiment and finally, it would appear, with the body per se. This is a dead stop. It is a rewriting of, and an alternative to, Donne's poem. For the roots, associations, and historical usages of *pith* and *cord*, see *Online OED* (accessed 8/30/2014) and note 66 in this chapter, which supplements the *OED*.

66. I have skipped the further translation of spinal cord to pith both because it is less likely that Donne would have recognized it as such, although it is possible that he did: *pith* is an Anglo-Saxon word in origin, and by 1612 it was a term used for the spinal cord. A digital search for *spinal cord* between 1550 and 1612 found no record. In *Mikrokosmographia* (1615), Crooke has a chapter "Of The spinall Marrow or pith of the Backe," which notes that scripture calls this "the siluer cord" (15, 480); his may be the first instance of English *cord* in a context that is specifically anatomical. *The Geneva Bible* needs to gloss the same "siluer corde," which occurs in a metaphorical reference to age, as the "marrow of the backe bone *and* the sinews" (1560; rpt. Madison: University of Wisconsin Press, 1969), Ecclesiastes 12:6; my emphasis. In the Vulgate, the same cord is *funiculus* (rope, cord, string) *argenteus* (silver).

67. William Shakespeare, *The Tempest*, I.ii.399, in *The Riverside Shakespeare*, ed. G. Blakemore Evans et al., 2nd ed. (Boston: Houghton Mifflin, 1997).

68. See Martz, *Poetry of Meditation*, 241–42; Donne, *Sermons*, 10:230–31.

69. Kepler follows tradition in affirming the inaudibility of heavenly music on earth. Harvey and Harrison refer to the speaker's "new ear" at this point of *The Second Anniversarie* as instancing the "enhanced organs of hearing" his soul is "to grow" ("Embodied Resonances," 995); their emphasis on embodiment and literalism seems incongruous here. Intellective souls lack physical organs, and metaphorical language is not flesh. Language itself has a material dimension as sound and script, however, and is sensed as such. It also has an affective dimension.

70. See Chapter 6. Readings that take the center of the circle to be earth go backward from SA 439 to the geometric image and take "thoughts" in this

line to be employed on earthly matters (e.g., Grierson, *Variorum* 6:525, Manley, ed. 197n439), whereas I see in the abstract image itself a withdrawal from earth to center on God and take the latter half of line 439 to mean "thy thoughts while on earth."

71. In SA 141–46, the "chain," while potentially promising as an allusion, refers only to the body while living and is further compromised by its origin in "Fate."

72. See, for example, George Herbert's "Affliction (I)," 63; "The Collar," 31; "Love (III)," 16; in *The English Poems of George Herbert*, ed. Helen Wilcox (Cambridge: Cambridge University Press, 2007). On the provisional liberation available to the soul while in the body, cf. Suzanne Smith, "The Enfranchisement of the 'In-mate soule': Self-Knowledge and Death in Donne's *Anniversaries*," *Literature and Theology* 24 (2010): 313–30, esp. 314, 326.

73. On dual souls, see notes 57–58 in this chapter and my discussion of Bryskett in Chapter 2.

74. Rolls, although an antiquated technology, were still in official use in Donne's time (Robin Robbins, ed., *The Complete Poems of John Donne*, rev. ed. [Harlow: Pearson, 2010], 919n504), and as Milgate, ed., observes, Dean Donne's name appears on one of these in the P.R.O., which lists members of the Court of High Commission (175n504). Interpreters of the Bible have suggested an allusion to the Hebrew Torah scroll in the double-sided scroll of Revelation 5:1. The image of a double-sided roll (i.e., scroll) also occurs in Ezekiel 2:9 and Zechariah 5:1–3: see "sealed scroll" in http://answersintheendtimes.com and http://oncedelivered.net and further reflected in what follows. Two-sided scrolls are unusual and the three in the Bible are associated with vision and prophecy. The roll in Revelation also suggests an ancient Hebrew deed or marriage contract, which overlaps with the genealogical concerns of the Pentateuch, or Torah, not to mention the Marriage Supper of the Lamb. Torahs themselves are one-sided, however, and so only a Torah-type scroll with two lateral rollers, rather than precisely a Torah, qualifies for Donne's graphic image.

75. Following Grossman (*The Story of All Things*, 189–91), Harvey and Harrison ("Embodied Resonances," 996–97) identify Donne's "written Rols" with Augustine's depiction of the firmament as a skin, a two-sided parchment roll, in his meditation on the creation in Genesis at the conclusion of the *Confessions*. Proleptically, Augustine's depiction also includes references to the Fall, to the deaths of the ancient biblical writers, and to the abiding w/Word of God. But it further glances at Rev. 6:14, a back reference to the scroll of Rev. 5:1 and related biblical passages: see Augustine, *Confessions*, trans. F. J. Sheed, ed. Michael P. Foley, 2nd ed. (Indianapolis, Ind.: Hackett, 2006), 299; cf. 300 (XIII.xv.16, 18): cf. note 74 in this chapter. Whatever the allusive

range of Donne's "written Rols," the double-sidedness of this recto-verso image is at odds with Harvey and Harrison's suture.

76. *Online OED*, s.v. *For*, II.4, III.7, V.16, 17; IX.26, 27 (accessed 2/5/15).

77. *Online OED*, s.v. *Piece v.*, 1.a. "To mend, make whole, or complete" by addition of a piece(s); 2.b. *fig.* "To join oneself *unto*, *to*, unite *with* [suggestive, although the prepositions are lacking in Donne's use]; 3.*intr.* "To unite, come together" [associatively suggestive, although Donne's use is transitive and the first record is 1622, but cf. s.v. *Peece v.*] (accessed 4/8/15). *Variorum* 6:532 cites only Kermode's "augments"; Manley, ed., suggests "adds to," noting the paradox of supplementing a symbol of perfection (200). Robbins, ed., suggests "repairs." The importance of the word *growth* and its implications go unremarked by all.

78. Cf. Elizabeth D. Harvey, "'Mutuall elements': Irigaray's Donne," in *Luce Irigaray and Premodern Culture: Thresholds of History*, ed. Theresa Krier and Elizabeth D. Harvey (London: Routledge, 2004), 66–87, esp. 83; also on Christian-Neoplatonic eros, the longing of the soul, see the final section of Chapter 3 in this volume.

7. MILTON'S TWILIGHT ZONE: ANALOGY, LIGHT, AND DARKNESS IN *PARADISE LOST*

1. John Milton, *Paradise Lost*, ed. Barbara K. Lewalski (Oxford: Blackwell, 2007), IX.1053–55. Reference is to this edition unless otherwise specified.

2. Cf. Edmund Spenser, *The Faerie Queene*, ed. A. C. Hamilton, with Hiroshi Yamashita and Toshiyuki Suzuki, rev. 2nd ed. (Harlow: Pearson, 2007), II.Pro.v:

> The which O pardon me thus to enfold
> In couert vele and wrap in shadowes light,
> That feeble eyes your glory may behold
> Which ells could not endure those beames bright
> But would bee dazled with exceeding light.

Dazzling brightness is enfolded in shadow in order to enlighten feeble, human eyes, enabling them to behold true glory. This veil reveals truth rather than obscuring it.

3. *Paradise Lost*, Book I.768–88; John Milton, *Complete Shorter Poems*, ed. Stella Revard, with Latin poems trans. Lawrence Revard (Chichester: Wiley-Blackwell, 2009), *Paradise Regained*, IV.40; cf. IV.57: "Aerie Microscope." See note 5 below on the term (epic) *simile*.

4. For a nonsatanic, telescopic instance, see *Paradise Lost*, V.261–63.

5. Dominance of the current, narrow meaning of *simile* as a comparison using *like* or *as* is quite modern and can be misleading; I use *epic analogy*, as

applicable. *Online OED* (accessed 9/15/14), s.v. *simile*, 1, gives Langland (1393) as the first use in English, but Langland does not intend the narrow, modern meaning. The *OED*'s next citations are 1589 and 1602, still without the modern meaning. The last example prior to Milton's death the *OED* offers is dated 1646: "Playing much upon the simile or illustrative argumentation," another testimony to the broad meaning of the term, evidently a substitute for *similitude* (Latin *similitudo*). In John Leonard's chapter on epic similes in *Faithful Labourers: A Reception History of Paradise Lost, 1667–1970*, 2 vols. (Oxford: Oxford University Press, 2013), *simile* for "epic simile" (short for "similitude"?) is common from the early eighteenth century (1:328–42). On *simile/similitudo* (and the like) in classical Latin and Renaissance rhetorics, see Judith H. Anderson, *Translating Investments: Metaphor and the Dynamic of Cultural Change in Tudor-Stuart England* (New York: Fordham University Press, 2005), 131, 135, 143–44, 162, 153–56. Julia Staykova offers a recent study of Milton's "*topos* of specular instability," specifically including his telescopic epic similes: "Structures of Perception in the Similes of *Paradise Lost*," *SEL* 53 (2013): 157–78, here 164.

6. On Milton and Kepler, see, for example, *Paradise Lost*, III.582–83 (magnetism), IV.32–34 (cf. note 4 of Chapter 5 of this volume: Kepler on the sun), and III.723–24, VIII.140–44, 150 (cf. *Paralipomena*, 242–43, 263–68: communicated light); see Chapter 5 in this volume for other connections. On Milton and Kepler, see also Harinder Singh Marjara, *Contemplation of Created Things: Science in "Paradise Lost"* (Toronto: University of Toronto Press, 1992),65–67 and "Kepler," index; Leonard, *Faithful Labourers*, 2:814: *Paradise Lost*, V.620–24, suggests Kepler; Anita Lawson, "'The Golden Sun in Splendor likest Heaven': Johannes Kepler's *Epitome* and *Paradise Lost*, Book 3," *Milton Quarterly* 21, no. 2 (1987): 46–51; Dennis Danielson, *Paradise Lost and the Cosmological Revolution* (Cambridge: Cambridge University Press, 2014), chap. 6 and 185–87 (John Wilkins's use of Kepler in *The Discovery of a World in the Moone* [1638, 1640]).

7. John Milton, *Areopagitica*, vol. 2, ed. Ernest Sirluck, in *Complete Prose Works of John Milton*, gen. ed. Don M. Wolfe, 8 vols. (New Haven, Conn.: Yale University Press, 1953–82), 480–570, here 514. In early modern Europe, Craig Koslofsky finds endemic a binary logic of inversions with respect to night and day, good and evil, religious worship and its diabolical perversions, and so on: *Evening's Empire: A History of the Night in Early Modern Europe* (Cambridge: Cambridge University Press, 2011), 39–40.

8. Cf. Karen L. Edwards, *Milton and the Natural World* (Cambridge: Cambridge University Press, 1999), 69: "for a brief period in the middle of the seventeenth century, experimental philosophy and experimental devotion coincided"; also Norman T. Burns, *Christian Mortalism from Tyndale to Milton* (Cambridge, Mass.: Harvard University Press, 1972), e.g., 57, 68.

9. Qtd. from Anne Ferry, *The Inward Turn of Language: Sonnets of Wyatt, Sidney, Shakespeare, Donne* (Chicago: University of Chicago Press, 1983), 12; cf. Spenser, *The Faerie Queene*, III.xi.1. Ferry observes the common overlap of affective and mental operations.

10. *Paradise Lost*, II.269–70; cf. IV.110–11 (Satan's explicit voicing of the dualist's claim).

11. Kepler, *Paralipomena*, 51; see note 2 in Chapter 5 of this volume for bibliographical detail.

12. Other terms for Milton's philosophic monism include animist materialism, scientific vitalism, equivocal monism, and their variants, which I have discussed in note 12 of Chapter 3. With reference *exclusively* to Satan and hell, cf. A. D. Nuttall's suggestive argument that Milton subscribes to the Gnostic heresy: *The Alternative Trinity: Gnostic Heresy in Marlowe, Milton, and Blake* (Oxford: Clarendon, 1998).

13. Introduction, by E. R. Dodds, ed. and trans., *Select Passages Illustrating Neoplatonism* (1923; rpt. Chicago: Ares, 1979), 19. Dennis Richard Danielson separates moral/ethical evil, which requires rational agents, both from natural evil, such as pain, disease, volcanoes, and (presumably) death, and from metaphysical evil, namely the theory that evil is a privation of good: *Milton's Good God: A Study in Literary Theodicy* (Cambridge: Cambridge University Press, 1982), 4–7. As Danielson acknowledges, such distinctions get "blurred" in practice, life, and Milton's epic (5, 7).

14. For the citation from Milton's *De Doctrina Christiana*, see note 5 in Chapter 3 of this volume.

15. John Milton, *Christian Doctrine*, vol. 6, ed. Maurice Kelley, trans. John Carey, in *Complete Prose Works of John Milton*, 307–9 (Bk. I.vii). Unless otherwise indicated, quotations in English of Milton's *Christian Doctrine* in this chapter are from this edition. For Milton's original Latin (and, if needed, for the translation), in this chapter I have used *De Doctrina Christiana*, vol. 8, ed. and trans. John K. Hale and J. Donald Cullington, with additional material by Gordon Campbell and Thomas N. Corns, in *The Complete Works of John Milton*, ed. Thomas N. Corns and Gordon Campbell, 11 vols. (Oxford: Oxford University Press, 2012), here 8:290–94. This Latin/English edition is cited as *De Doctrina*. I provide the page reference for Milton's Latin after each reference to the Kelley-Carey edition.

16. *Christian Doctrine*, 309 (Bk. I.vii); *De Doctrina*, 292–94.

17. *Paradise Lost*, III.475, X.702, 847; XI.738.

18. *Paradise Lost*, IX.180, VI.511, 515; VII.238.

19. *Paradise Lost*, XII.41, I.405.

20. *Paradise Lost*, IX.463–64. From this point forward, I capitalize Night and Light, as well as Death, either in accordance with critical tradition or with personified status.

21. I have benefitted greatly from essays on Milton's Night by John Leonard, "Milton, Lucretius, and 'the Void Profound of Unessential Night,'" and John Rumrich, "Of Chaos and Nightingales," in *Living Texts: Interpreting Milton*, ed. Kristin A. Pruitt and Charles W. Durham (Selingsgrove, Penn.: Susquehanna University Press, 2000), 198–217 and 218–27, respectively.

22. John Keats, "Ode on a Grecian Urn," l. 44, in *The Norton Anthology of English Literature*, ed. M. H. Abrams et al., rev. ed. (New York: Norton, 1968), 2:530–32.

23. *Excess* derives from Latin *excedere/excessus*, "to surpass, exceed, tower above"; cf. I.589–91. On the relation of Miltonic to *essential* form, see Neil D. Graves, "'The whole fulness of the Godhead dwells in him bodily': The Materiality of Milton's God," *Christianity and Literature* 52 (2003): 1–20, esp. 14.

24. See Alastair Fowler, ed., *Milton: Paradise Lost*, rev. 2nd ed. (Harlow: Pearson, 2007), 97n596–9. Cf. Leonard, *Faithful Labourers*, 1:345: "the matching of diabolical malevolence with pastoral beauty is *in*congruous." Leonard reviews responses to this solar simile, most of them admiring Satan's sublimity, nobility, and grandeur: *Faithful Labourers*, e.g., 2:398, 401, 406, 402, 413, 442; he analyzes it himself in *Naming in Paradise: Milton and the Language of Adam and Eve* (Oxford: Clarendon, 1990), 111–12. Cf. William Poole, *Milton and the Idea of the Fall* (Cambridge: Cambridge University Press, 2005), 149–50; also Peter C. Herman, "'Warring Chains of Signifiers': Metaphoric Ambivalence and the Politics of *Paradise Lost*," *TSLL* 40 (1998): 268–92, here 274–75.

25. *Paradise Lost*, I.244: "mournful gloom." Cathleen M. Swaim perceptively treats solar imagery in Milton's epic, although mainly in passages other than those I discuss: *Before and After the Fall: Contrasting Modes in "Paradise Lost"* (Amherst: University of Massachusetts Press, 1986), 67–69, 76–77, 82–83, 88, 167; she connects the sun to the Son throughout, excepting hell. Marjara, *Contemplation of Created Things*, devotes a section of his study to "The Pre-eminence of Milton's Sun" (118–23).

26. Cf. *Paradise Lost*, VI.783–84: when God restores the hills to their places, "Heav'n his wonted face renewed, / And with fresh Flourets Hill and Valley smil'd."

27. On various readings of the present solar analogy, see Leonard, *Faithful Labourers*, 1:351. See also note 28, just below.

28. Stanley Fish's emphasis on the fallen reader, caught repeatedly in his (or her) sinful responses, is not my point, although I am concerned with the responses of readers: see his *Surprised by Sin: The Reader in Paradise Lost* (London: Macmillan, 1967).

29. *Paradise Lost*, II.917, 932, 941, 960, 969–70. In Book II.941, "crude consistence," drawing on Latin *crudus*, *cruditas*, suggests "raw, rough, unfin-

ished, undigested uncultivated," and drawing on Latin *consistere*, suggests something "set, firm, cohering." Book I.50–51 affords an indication of the timelessness of hell itself in *earthly* terms. "The wilde Abyss" of Night and Chaos that Satan and Sin see is where time and place are lost" (II.890–894, 910). In *De Doctrina Christiana*, Milton observes that "nothing compels agreement to the common belief that motion and time (which is the measure of motion) could not, according to the ideas of 'earlier' and 'later', have existed before this world was established; since Aristotle taught that in this world, which he judged to be eternal, motion and time were nonetheless givens": translation from *De Doctrina*, 299 (Bk. I.vii). Leonard, *Faithful Labourers*, 815: Milton's saying that time and place are lost in Chaos is not the same as saying that they are timeless (isn't the opposite equally true?); cf. note 35 in this chapter.

30. My reading assumes a pause (the equivalent of a comma) after "Eternal" in line 2. Barbara Kiefer Lewalski considers the whole proem to Book III "an invocatory hymn": *"Paradise Lost" and the Rhetoric of Literary Forms* (Princeton, N.J.: Princeton University Press, 1985), 29–33, here 29.

31. *Christian Doctrine*, 206 (Bk. I.v); *De Doctrina*, 128; *Paradise Lost*, III.63, 140–41, 170–72. In the gospel of John, the Son is identified with and as light: e.g., 1:1–8. Possible readings of Milton's Invocation are legion: e.g., Fowler, ed., 165–67; Merritt Y. Hughes, ed., *John Milton: Complete Poems and Major Prose* (New York: Odyssey, 1957), 199, section 56. Michael Lieb surveys contexts bearing on light as "a sacral phenomenon in Milton's thought": *Poetics of the Holy: A Reading of "Paradise Lost"* (Chapel Hill: University of North Carolina Press, 1982), 185–211, here 194. See also Noam Reisner, *Milton and the Ineffable* (Oxford: Oxford University Press, 2009), esp. 173 ("negative theology's bright cloud of darkness"), 180–89, 198–99. Cf. Stephen M. Fallon, *Milton's Peculiar Grace: Self-Representation and Authority* (Ithaca, N.Y.: Cornell University Press, 2007), 219, 254. Citing John 4:24 among other texts in *De Doctrina*, 34–35 (Bk. I.ii), Milton asserts that "God in his simplest nature is a spirit (*simplicissima Spiritus* [*sic*])," but what he means by "spirit" (e.g., a gas, breath?!) is ambiguous in his writings, as in those of others. Relevant to the present Invocation is Pseudo-Dionysius's belief that the Good "By itself . . . generously reveals a firm, transcendent beam [of light], granting enlightenments *proportionate* to each being" (50; my emphasis: *The Divine Names*, in *The Complete Works*, trans. Colm Luibhéid [New York: Paulist, 1987]); Pseudo-Dionysius alludes to Romans 12:6, touchstone for the analogy of faith.

32. David Quint, *Inside "Paradise Lost": Reading the Designs of Milton's Epic* (Princeton, N.J.: Princeton University Press, 2014), chap. 4, argues against the continuity of light in this Invocation, positing a Baconian gap between

God's works and their "Fountain." For Quint, the poet's inspiration is quite separate from nature's works (101), and so analogy is invalid (115). He overlooks the poet's nightingale as well as Milton's having enjoyed sight for over forty years. Quint couples his dismissal of analogy with recognition of its presence (e.g., 113, 119).

33. Readings of Milton's Chaos differ widely: Regina M. Schwartz, uncovering references to the cosmogonic myth of the Chaos monster, finds Chaos preeminently evil: *Remembering and Repeating: On Milton's Theology and Poetics* (Chicago: University of Chicago Press, 1993), esp. 11, also 22–31; Michael Lieb considers Chaos neutral, a potent source for good or evil: *The Dialects of Creation: Patterns of Birth and Regeneration in "Paradise Lost"* (Amherst: University of Massachusetts Press, 1970), 16–17; John Peter Rumrich proposes that Chaos "represent[s] the material dimension of God's own being" and therefore participates "in the fullness of divinity as neither the Son nor the Holy Spirit do": *Matter of Glory: A New Preface to "Paradise Lost"* (Pittsburgh, Penn.: University of Pittsburgh Press, 1987), 7–8; more recently, Rumrich has described Chaos as representing "a principle of indeterminacy and randomness essential to divine power" and as "God's womb": *Milton Unbound: Controversy and Reinterpretation* (Cambridge: Cambridge University Press, 1996), 141; cf. vi. Leonard holds that Night "is in some ways the more sinister of the two [Night and Chaos]," his comparative implying that Chaos is suspect too: "Void Profound," 206, cf. 198, 209; John Rogers calls Chaos "specifically Hobbesian": *The Matter of Revolution: Science, Poetry, and Politics in the Age of Milton* (Ithaca, N.Y.: Cornell University Press, 1996), 219; David Norbrook finds in Chaos "the cosmos's default mode" and "a cosmic civil war," although Chaos is nonetheless "not inherently evil": *Writing the English Republic: Poetry, Rhetoric, and Politics, 1627–1660* (Cambridge: Cambridge University Press, 1999), 472; Gordon Teskey concludes that Chaos is "everything in the objective world, in history as well as in nature, that remains undecided, untouched by the creative impulse to change": *Delirious Milton: The Fate of the Poet in Modernity* (Cambridge, Mass.: Harvard University Press, 2006), 85. William B. Hunter, like Lieb, considers Chaos neutral: *Visitation Unimplor'd: Milton and the Authorship of "De Doctrina Christiana"* (Pittsburgh, Penn.: Duquesne University Press, 1998), chap. 8, here 25. (I do not agree with Hunter's rejection of Milton's authorship of *De Doctrina*, but some of his learned arguments have other provocative applications.)

34. The quotation, like the proposal, is Leonard's, "Void Profound," 212. My reservations about Leonard's otherwise illuminating essay on the Lucretian void and Milton pertain to the identification of the abyss with Night alone (e.g., 207), to the speakers on whom he bases some of his critical evidence regarding Night as primal (beyond references to "eldest Night"), and

to his insistence on the separation of Night and Chaos (e.g., 206). Evidence spoken by devils involved in rhetorical situations (e.g., 207) may have an element of exaggeration or distortion in it. Its situation asks to be considered with care. Leonard cites and rebuts Danielson's dismissal of the possibility of an eternal Chaos (Leonard, "Void Profound," 212; Danielson, *Milton's Good God*, 51; cf. Danielson, *Cosmological Revolution*, 28, 42–46). N. K. Sugimura, building on Leonard's views, suggests that the void may be God: *"Matter of Glorious Trial": Spiritual and Material Substance in "Paradise Lost"* (New Haven, Conn.: Yale University Press, 2009), 276, cf. 268, 273–75; she also indicates that Night threatens creation with preexistent darkness and death, citing Isaiah 45:7 and Milton's *Christian Doctrine* (300: Bk I.vii; *De Doctrina*, 282) to the effect that darkness is something, not nothing (Sugimura, *"Matter of Glorious Trial,"* 266); notably, the passage from Isaiah refers to created darkness, however. In relation to Milton's speaking "in his own voice as a poet" in the Invocation to Light, cf. Victoria Silver: Milton's "proems are an ironic expression of the author's self-understanding, with the speaker of *Paradise Lost* a persona, an actor in the individual drama of creature and creator," as in his previous poetry: *Imperfect Sense: The Predicament of Milton's Irony* (Princeton, N.J.: Princeton University Press, 2001), 189; cf. 191, 195, 206. Cf. Anne Davidson Ferry: *Milton's Epic Voice: The Narrator in "Paradise Lost"* (Cambridge, Mass.: Harvard University Press, 1963), e.g., 20, 87.

35. As I suggested in *Reading the Allegorical Intertext: Chaucer, Spenser, Shakespeare, Milton* (New York: Fordham University Press, 2008), 298–302, "his sea," Satan's sea, the medium through which he has traveled though Chaos, signals that, like hell, it is an interior condition as well as an exterior one, and the *personification* of Chaos is Satan's projection (II.1011); cf. Danielson: Chaos is "Satan's view in more ways than one" (*Milton's Good God*, 50–51; cf. Danielson's *Cosmological Revolution*, 47–48). On the relative subjectivity of Chaos, see also note 17 in Chapter 3 of this volume. Leonard, *Faithfull Labourers*, 2:785–86, observes that *eternal*, according to the *OED*, does not have to mean "without beginning or end" theologically but inclines to this meaning in Milton's phrase "eternal Anarchie," referring to Night and Chaos, insofar as *anarchy* etymologically signifies "'without beginning.'" But see *Online OED*, s.v. *eternal*, 1.b, exemplified by Hobbes (1651): pertaining "merely to things as viewed by finite intelligence . . . : Not conditioned by time; not subject to time relations" (accessed 12/11/14); cf. *Paradise Lost*, II.893–94: "where . . . time and place are lost"; and note 29 in this chapter.

36. E.g., Paul Ricoeur, "The Metaphorical Process as Cognition, Imagination, and Feeling," in *On Metaphor*, ed. Sheldon Sacks (Chicago: University of Chicago Press, 1978), 141–57; Aristotle, *On the Soul* (*De Anima*), trans. J. A. Smith, in *The Complete Works of Aristotle*, rev., ed. Jonathan Barnes (Princeton,

N.J.: Princeton University Press, 1984), 642–92: the *phantasmata* (images) basic to thought are sensorily derived and emotionally laden; necessarily, they are perceptual—mediated and filtered by the senses of the perceiving subject.

37. On Zion, see Lieb, *Poetics of the Holy*, 163–68. On the privation of light that is blindness, Annabel Patterson's examination of negativity in Milton is helpful: *Milton's Words* (Oxford: Oxford University Press, 2009), 178–79. On the negative suffix *un-*, see as well Poole, *Milton and the Idea of the Fall*, 154; and Thomas N. Corns, *Milton's Language* (Oxford: Basil Blackwell, 1990), 84–86.

38. In *Christian Doctrine*, hell is said to be both an internal state of loss and a "place (*locus*)" that is "situated outside this world" (628–30, Bk. I.33; *De Doctrina*, 890–93).

39. Cf. Koslofsky, *Evening's Empire*, 46–47: the "intense chiaroscuro" of baroque art, or "active darkness" made darkness "inseparable from light." Qtd. from Maria Rzepinska, "Tenebrism in Baroque Painting and Its Ideological Background," *Artibus et Historiae* 13, no. 7 (1986): 91–112.

40. Schwartz, *Remembering and Repeating*: "Milton's synonyms for Chaos are used interchangeably for hell" (23). See Poole, *Milton and the Idea of the Fall*, 153, on the language of "abortion, male pregnancy, and uncreation" in Chaos and hell.

41. Even if the parenthesis around Night were the product of a Miltonic scribe or of the print shop, the clause would remain a subordinate one, having the quality of a conspicuous afterthought.

42. *Paradise Lost*, V.685–86; VI.478, 511, 554.

43. On God's mount, see Lieb, *Poetics of the Holy*, 140–70, esp. 140–44 (cultic and cosmic, biblical and nonbiblical traditions).

44. Discussing Jakob Böhme, whom Charles I and Cromwell's chaplain, Peter Sterrey, admired, Koslofsky recounts Sterrey's preaching "in Behemist terms that 'Darkness, and light, are both in God; not only representatively but really; not in their ideas only, but their Identities'" (*Evening's Empire*, 64, cf. 65–66).

45. Fredric Jameson, *The Political Unconscious: Narrative as a Socially Symbolic Act* (1981; rpt. Ithaca, N.Y.: Cornell University Press, 1988), 41–42.

46. *The Faerie Queene*, IV.x.32. My point is the precedence of Hate over Love, also a reflection of the father's superiority to the mother's. Concord here has her being in the relationship between the brethren. She is Coming Together and Consanguinity with respect to them.

47. *Paradise Lost*, VII.525–26; *Christian Doctrine*, 317 (Bk. I.vii); *De Doctrina*, 300; Genesis 2:7. Kester Svendsen, *Milton and Science* (1956; rpt. New York: Greenwood, 1969), 87, remarks the pervasive association of the word *exhalation* with ominous meteorological phenomena. Satan's palace in hell rises "like an Exhalation" (I.711).

48. My view also opposes that of James S. Baumlin, for whom such Miltonic double voicing is the "demonization" of multiple, competing truths (xl, 199): *Theologies of Language in English Renaissance Literature: Reading Shakespeare, Donne, and Milton* (Lanham, Md.: Lexington, 2012).

49. Annotating *Paradise Lost*, III.108, "Reason is also choice," Lewalski, ed., suggests that reason is meaningless unless it is free; the sentence more obviously indicates that reason itself is informed by choice. Such a basic description of right reason as that of Douglas Bush remains useful: "the candle of the Lord is the *recta ratio* of the humanistic tradition, and *recta ratio* is found . . . where *fides* is found. . . . [For] Milton and others, reason signifies not the mere logical and critical faculty but the Platonic capacity for attaining divine truth, the whole unified personality, the philosophic conscience, of the well-disposed man": *English Literature in the Earlier Seventeenth Century, 1600–1660*, 2nd ed. (Oxford: Oxford University Press, 1962), 360–61. Cf. Michael Lieb, *Theological Milton: Deity, Discourse, and Heresy in the Miltonic Canon* (Pittsburgh, Penn.: Duquesne University Press, 2006), 65–67; also *Christian Doctrine*, 132 (Bk. 1:2): "No one would try to be virtuous . . . if the voice of Conscience or right reason did not speak from time to time [*subinde*: 'constantly, repeatedly'] in the heart of man, reminding him . . . that a God [*Deum aliquem*: 'some God'] does exist." Poole notes that Alexander Gil, head of St. Paul's when Milton attended, "espoused a rationalistic Christianity" in which, although reason was subordinate to faith, faith was "under constant interrogation by reason" (*Milton and the Idea of the Fall*, 125); Poole cites Augustine: "'hence let no one seek from me to know what I know that I do not know, except it be in order to learn how not to know what we should know cannot be known'" (Poole, *Milton and the Idea of the Fall*, 29: from *City of God*, XII.7). With respect to double-voicing, cf. Virginia R. Mollenkott, "Milton's Technique of Multiple Choice," *Milton Studies* 6 (1974): 101–11, esp. 101–5.

50. Danielson finds a threefold creation in *Paradise Lost*: first, a stage "in which prime matter is *in some way* 'alienated' from God [its origin], rendered external to him" (i.e., Chaos); second, a stage in which some of this matter is selected as "the stuff of this visible world"; third, a "stage in which that stuff receives its actual forms" (*Milton's Good God*, 45–46; my emphasis); cf. Danielson, *Cosmological Revolution*, 42–43. The alienation is an obvious pressure point for interpretation.

51. Respectively, Rogers, *The Matter of Revolution*, 135; Walter Benjamin, *The Origin of German Tragic Drama*, trans. John Osborne (1998; rpt. London: Verso, 2003), 227. Cf. Leonard, *Faithful Labourers*, 1:375–76. Satan's Hebrew name signifies "Adversary," aligning him with "dregs / Adverse to life."

52. In *Christian Doctrine*, 145–46 (I.ii), *De Doctrina*, 38, Milton rejects the Aristotelian/Thomist exclusion of potentiality (realization in *potentia*) in characterizing God. Simply put, Milton's God can will the begetting of the Son when he chooses to do so, rather than from eternity (cf. Kelley, ed., 145–46n47). Cf. Michael Lieb, *Theological Milton: Deity and Heresy in the Miltonic Canon* (Pittsburgh, Penn.: Duquesne University Press, 2006), 99–100.

53. Leonard, "Void Profound," 206–7. "Void" translates both the Hebrew substantives *tohu*, "formlessness, confusion, unreality, emptiness, [wasteness]," and *vavohu*, "emptiness," and the Latin adjectives *inanis* and *vacua*, "empty" and "void, vacuous," in Genesis 1:2, as found in the Geneva, King James, and Duay-Rheims Bibles, for example. In the same verse, the "deep" over the face of which darkness hovers, in Hebrew is *tehovm*, "deep, sea, abyss," *vechoshech*, "darkness, obscurity," and *penei*, "face, faces"; the Latin is *tenebrae super faciem abyssi*, "darkness on" or "upon, over," "the face of the abyss." *Tenebrae* is especially associated with the darkness of Night, but also notably with death and the infernal region, and rarely and poetically with blindness: Charlton T. Lewis and Charles Short, *A Latin Dictionary* (Oxford: Clarendon, 1966), s.v. *tenebrae*. Again, the three English Bibles essentially agree in translating these words, "darkness was upon the face of the deep," although Geneva omits "the face of": biblehub.com/lexicon/genesis/1–2.htm; also *The Geneva Bible*: facsimile of the 1560 edition (Madison: University of Wisconsin Press, 1969).

54. Joseph H. Summers, *The Muse's Method: An Introduction to "Paradise Lost"* (Cambridge, Mass.: Harvard University Press, 1962), 39; Stephen M. Fallon endorses Summers's characterization of Sin and Death, yet he takes Summers's term "real" only to mean abstract whereas in context it more clearly means "effectual" or "actual": *Milton Among the Philosophers: Poetry and Materialism in Seventeenth-Century England* (Ithaca, N.Y.: Cornell University Press, 1991), 184. See my discussion of Milton's Death in Chapter 1.

55. On paradox as "a mode of explanation" in Milton's writing, see William Kolbrener, *Milton's Warring Angels: A Study of Critical Engagements* (Cambridge: Cambridge University Press, 1997), 137; relatedly, *discordia concors* (6–7, 134), "competing narrative perspectives" (108: monism and dualism).

56. Milton's God himself refers to the passage of time in heaven (VI.685), as does Raphael (V.580–82). With the primacy (or not) of age, the story of Esau and Jacob comes to mind: Genesis 25:27.

57. E.g., *Paradise Lost*, III.636 ("stripling"); IV.552 ("Youth"), 845 ("youthful"); XI.246 ("Manhood where Youth ended"). Usually the reference is to a cherub: in the first instance to Satan disguised; in the next two, to the angels guarding Eden under Gabriel's command; in the last, to the Archangel Michael expelling the human pair from Eden.

58. *Ayenbite of Inwyt*, 104: http://archive.org/stream/DanMichelsAyenbiteOfInwyt/DanMichelsAyenbite_Michel_484PgsFO.7052899036_djvu.txt. Kentish is not the (most accessible) dialect that develops into modern English, namely, East Midland. See also *Online OED*, *s.v. Eldest*, 1–3 (accessed 3/7/15). The closest linguistic relatives of Old English *eldest* are in Old Frisian, Old High German, and Gothic.

59. See *OED*, *s.v. Court. n.* III.9; *Hold*, *v.* 8.a; an EEBO search for *hold court* produces multiple examples of the expression, mainly legal, although its origin is political. See also Jonathan Goldberg's perceptive discussion of the antiquity of Night: *The Seeds of Things: Theorizing Sexuality and Materiality in Renaissance Representations* (New York: Fordham University Press, 2009), 207–8. Goldberg, however, discounts the figuration of Night on the grounds that other personifications are present in the realm of Chaos and Night. Since this realm is courtly, it seems right that it should be populated, if only with abstractions. The Lucretian void, after all, as a place is theoretical—that is, theorized to exist. It is required to enable the movement of the "embryon Atoms" (II.900). Teskey, *Delirious Milton*, 88, postulates a tension between concept and metaphor to be "at the heart of the system of *Paradise Lost*"; his example is God as substance and God as a body, but it could as easily be Night as void and Chaos as atomic matter, as distinct from the two as personified Anarch and Consort.

60. Qtd. from Michel Serres, *The Birth of Physics*, ed., David Webb, trans. Jack Hawkes (Manchester: Clinamen, 2000), 173. For Lucretius, I have used *De Rerum Natura* (*On the Nature of Things*), trans. W. H. D. Rouse, rev. Martin Ferguson Smith (Cambridge, Mass.: Harvard University Press, 1992), II.1144–74.

61. Lucretius, I.146–635, V.416–508. If not joined at the hip, Chaos and Night as Sovereign and Consort are joined in the marriage bed.

62. Use of the word "sinister" recalls Leonard's description of Night ("Void Profound," 206). In Latin, *sinister* refers to the left (hand, side), thus implying a binary, the right (hand, side); similarly binary are Night and Light—or Night and Chaos, for that matter. On the genealogy of Night, see especially Spenser, *The Faerie Queene*, I.v.22. Night's nephew in Spenser is Sans Joy, the Redcrosse Knight's morally diseased double, and Night calls Duessa daughter and child when this duplicitous figure seeks assistance. See also Henry Gibbons Lotspeich, *Classical Mythology in the Poetry of Edmund Spenser* (New York: Gordian, 1965), 91.

63. Leonard, "Void Profound," 207; Edward Le Comte, *Milton and Sex* (London: Macmillan, 1987), 69: in this instance, Le Comte's puns are historical and so familiar that he does not document them, but see *secrets* and immediately proximate expressions in Eric Partridge, *Shakespeare's Bawdy*, rev. ed.

(New York: Dutton, 1969); also *thing* in Partridge. Leonard limits sexuality to Night, personified as a female. Identifying Night with the Lucretian void, he finds Night in itself/herself barren. Granted, but as Leonard knows, embryon atoms populate Milton's Lucretian void, over which the fertilizing Spirit of Milton's God moves, activating their potentiality. On the sexuality and gender of Night and Chaos, see also Goldberg, *The Seeds of Things*, 204–9, who offers a comprehensive, critical review of earlier opinions and their relationships to one another, including that of Catherine Gimelli Martin, *The Ruins of Allegory: "Paradise Lost" and the Metamorphosis of Epic Convention* (Durham, N.C.: Duke University Press, 1998); Martin interprets Chaos as entropy and Night as negentropy, making Night the more constructive figure (162–79). On God's secrecy, see also notes 65–66 in this chapter. More generally, see the perceptive article by Karma deGruy, "Desiring Angels: The Angelic Body in *Paradise Lost*," *Criticism* 54 (2012): 117–49.

64. See *Online OED*, s.v. *Divulge. v.* 2. "To declare or tell openly (something private or secret); to disclose, reveal." *Online OED*, *s.v. Scan*, *v.* 3. *trans.* "To examine, consider, or discuss minutely." 4. "To interpret, assign a meaning to" (both accessed 1/12/15).

65. Cf. *Christian Doctrine*, 424 (Bk. I.xiv): referring to Christ's hypostatic union, Milton observes, "We do not know how it is so, and it is best for us to be ignorant of things which God wishes to remain secret." *De Doctrina*, 480–81, reads, "Modus ignoratur et ignorari certè praestat quod Deus ignotum est (how it is so, is not known, and it is assuredly best that what God wills to be unknown should not be known)." The word "*mysteriis* [mysteries]" appears in the next sentence, as earlier does "*arcana* [secrets]": e.g., 478–79.

66. On *privete(s)* and its cognates, the full entries in the *MED* are fascinating: examples come from a variety of texts, including Chaucer and the Wycliffe Bible. James Dougal Fleming explains that "Created *secretio* [separation] explicitly denotes the *secretum* [secret, secret place, retreat] of the Creator: *Milton's Secrecy and Philosophical Hermeneutics* (Aldershot: Ashgate, 2008), 7–8. The secrets of deity are the *arcanum/arcana dei* to be found in every major religion.

67. *Paradise Lost*, I.21, VII.233–37. Danielson perceptively analyzes the creation of the sun and its echoes of the Psalms in Raphael's narrative in Book VII.354–86 as "a marriage celebration" and "a cosmic epithalamion" (*Cosmological Revolution*, 149–50).

68. Saint Bonaventure's influential formulation of the relation of the Father, Son, and Spirit seems made for Milton's image of the brooding Dove as well as for Satan's parodic perversion of it: "Just as the Father engenders the eternal knowledge of the Word who expresses Him, and as the Word is in turn united with the Father by the Holy Spirit, so memory or thought, big

with the ideas which it encloses, engenders knowledge of the intellect or word, and love is born from both by the bond that unites them. . . . The structure of the creative trinity [thus] conditions and therefore explains the structure of the human soul": Etienne Gilson, *The Philosophy of St. Bonaventure*, trans. Illtyd Trethowan and F. J. Sheed (New York: Sheed and Ward, 1948), 224. Milton, no Trinitarian himself, modified this traditional formulation to suit his own purposes, eliminating the third person of the trinity, retaining the relation of love between the Father and Son that the Spirit traditionally expresses and its internalization: the Spirit of God that broods on the deep, dovelike, and conceives by the creative Word analogically also broods, poetlike, in thought. See Milton's *Christian Doctrine*, 281–86, esp. 283 (I.vi); *De Doctrina*, 244–53, esp. 246–51.

69. With respect to *semen* as well as to the bisexuality of Milton's brooding dove, see James W. Broaddus's discussion (based on Helkiah Crooke) of the seed present in the reproductive organs of males and females: *Spenser's Allegory of Love: Social Vision in Books III, IV, and V of "The Faerie Queene"* (Cranbury, N.J.: Associated University Presses, 1995), e.g., 26; also John Erskine Hankins on the mixture of male and female coital fluids in the Garden of Adonis, as well as his discussion of the physical allegory there: *Source and Meaning in Spenser's Allegory* (Oxford: Clarendon, 1971), 234–86, esp. 246, 255; and my "Flowers and Boars: Surmounting Sexual Binarism in Spenser's Garden of Adonis," in *Reading the Allegorical Intertext*, 214–23. Rogers finds an argument for the ontological equality of male and female in Raphael's creation narrative (*The Matter of Revolution*, 118–22): *Paradise Lost*, VII.276–84; cf. Rumrich, *Milton Unbound*, 105–8.

70. *Paradise Lost*, VI.478–79, 511; Marjara, *Contemplation of Created Things*, 163–77, esp. 173–74.

71. Rumrich, *Milton Unbound*, approaches this argument without making it: e.g., 94, 141–42, 145. He affirms male and female attributes of God but locates them rather loosely in the poem (excepting Chaos as womb) rather than in the specifics of the cave within God's mount and the deep of both Night and Chaos. He offers an informative review of opinions earlier than his own regarding the femininity of chaos, or chaotic matter, as inherent in God (142–44).

72. On Keplerian *inertia*, see E. J. Dijksterhuis, *The Mechanization of the World Picture: Pythagoras to Newton*, trans. C. Dikshoorn (1961; rpt. Princeton, N.J.: Princeton University Press, 1986), 314: although Kepler invented the term *inertia*, for him it meant only "absence of activity, the desire to remain at rest." On privation, which I do not use here in the Augustinian sense, cf. Danielson's distinction between metaphysical inferiority and its transformation in Augustinian theology to "actual-evil-as privation" (*Milton's Good God*, 6–7, 124).

73. *Paradise Lost*, VII.168–69: my emphasis; II.892 . Cf. Danielson, *Milton's Good God*, 46, 48.

74. I am struck again by the importance of keeping in mind and, when pertinent, distinguishing the chronology of events in the poem and the order of their telling in the narrative.

75. Spenser, *The Faerie Queene*, III.vi.36. Leonard, "Void Profound," 198. Danielson, *Cosmological Revolution*, 44–45, considers Milton's "presentation of Chaos and creation consciously mythic." Spenser's Garden of Adonis, a mythic tour de force, is the inescapably relevant precedent.

76. Milton recalls Spenser's Garden of Adonis in the Spirit's epilogue in *Comus* and in the initial approach to the Edenic mount, atop a steep, "hairie" wilderness, in *Paradise Lost*, IV.133–36. Milton's cave within the godly mount also alludes to Hesiod, who locates the home of Day and Night in alteration in the abyss: Fowler, ed., 339–40nn2–4. In the same note, Fowler records the allusion of Milton's cave to Spenser's.

77. On the relation of the various tropes to metaphor and to one another, see Chapter 6 of this volume.

78. On Auerbach's and Paul Ricoeur's theories of history and narrative, see my *Reading the Allegorical Intertext*, 9–15.

79. Marjara, *Contemplation of Created Things*, 31–37, has a section on scientific analogy and metaphor yet discounts the extent to which these practices have always been under pressure, as indeed, rhetoric more generally has been. Harinder Singh Marjara's "Analogy in the Scientific Imagery of *Paradise Lost*," *Milton Studies* 26 (1990), 81–99, affords a survey of Milton's use of analogy in the epic. Swaim, contrasting Raphael's narrative in Milton's Books V–VIII with Michael's in Books XI–XII, identifies epistemological analogy with the former and typology with the latter (e.g., 158–59). Momentarily, even she acknowledges that "typology is analogy of a special kind" (199). The problem is one met earlier in my fourth chapter, the confusion of analogical content with analogical structure. For Milton's own statements about analogy, see *A Fuller Course in the Art of Logic Conformed to the Method of Peter Ramus*, ed. and trans. Walter J. Ong and Charles J. Ermatinger, in *Complete Prose Works of John Milton*, 8:139–395, here 285–88 (Bk. I.21: "On the Similar"; for the Latin, *The Works of John Milton*, ed. Frank Allen Patterson et al. [New York: Columbia University Press, 1935], 11:192–203). Milton, having identified similarity as English *proportion* or Greek *analogia*, appropriately emphasizes that "*nothing similar is identical.*" He attends at length to a proper analogy involving four terms, which he describes as "a disjunctive similarity" whose relations can be exchanged in whole or in part, as in Aristotle (Ong and Ermatinger, ed., 284–87; Patterson, ed., 192–202).

80. In this paragraph, I draw on my *Reading the Allegorical Intertext*, 318–19.

81. Cf. Regina Mara Schwartz, *Sacramental Poetics: At the Dawn of Secularism* (Stanford, Calif.: Stanford University Press, 2008), 59–70; and Fallon, *Milton Among the Philosophers*, 98–107, here 106–7. Fallon's discussion also bears on what follows regarding Raphael's scale of being, as does Sugimura's, *"Matter of Glorious Trial,"* 42–55. More recently, see Sophie Read, *Eucharist and the Poetic Imagination in Early Modern England* (Cambridge: Cambridge University Press, 2013): Read offers a critical review of a wide range of interpretations of the Miltonic narrator's word "transubstantiate" (189–95), arguing that Milton's use of the word is metaphorical and opposes its use in Roman Catholicism (193–94). What could be lost in her argument is the root meaning of *transubstantiate*. It is this etymological kind of meaning, typically material or physical, which Milton engages here, as do I: he *re*materializes the transubstantive process.

82. E.g., see the indexical entries for both *matter* and *substance* (verbal, lexical, rhetorical, logical, material, metaphysical) in my *Words That Matter: Linguistic Perception in Renaissance English* (Stanford, Calif.: Stanford University Press, 1996), and for *substance* in *Translating Investments*; for *matter*, see also Daniel Tiffany, *Toy Medium: Materialism and Modern Lyric* (Berkeley: University of California Press, 2000), 1–33; likewise, Jonathan Gil Harris, *Untimely Matter in the Time of Shakespeare* (Philadelphia: University of Pennsylvania Press, 2009), e.g., 1–19. The title of Sugimura's book on *Paradise Lost*, *"Matter of Glorious Trial,"* puns on *Matter*, and its subtitle is *Spiritual and Material Substance*. Stephen Toulmin and June Goodfield's *The Architecture of Matter* (1962; rpt. Chicago: University of Chicago Press, 1982) still affords a useful introduction to this subject.

83. *Paradise Lost*, VI.347: Satan and the other angels, Raphael explains, "Cannot but by annihilating die," a statement that seems at least to raise the possibility of annihilation. But see *Christian Doctrine*, 310–11 (Bk I.vii): "no created thing can be utterly annihilated. . . . God neither wishes to nor, properly speaking, can altogether annihilate anything . . . because by making nothing he would both make and not make at the same time, which involves a [logical] contradiction": my bracket. In part, *De Doctrina*, 295, reads, "non posse quicquam rerum creatarum in nihilum interire (no created thing can perish into nothing)"; *annihilationis*, "annihilation," and *annihilari*, "to be annihilated," occur in the next two sentences. By this reasoning, however, Satan's self-making, if true, would enable his annihilation as a logical consequence. On Satan's prying, cf. Fleming, *Milton's Secrecy and Philosophical Hermeneutics*, 11: "endless secrecy is the choice that God leaves open to him [Satan]," and he takes that choice early in Book IV: "in the lowest deep a lower deep / Still threatening to devour me opens wide" (76–77); also cf. Stanley Fish's Freudian reading of the end of the War in Heaven, when the frightened rebels recoil from the "wastful Deep" that the breach in heaven's

wall discloses to them, just before they cast themselves "headlong" into its "bottomless pit" (VI.863–64, 865–66): *How Milton Works* (Cambridge, Mass.: Harvard University Press, 2001), 102–3. Reisner observes that "The devils' roving, inexpressible thoughts—inexpressible because meaningless in the poem's logocentric materialist sense—fizzle out into nothing in the 'womb of uncreated night'" (231). Such fizzling might be taken as an omen of their eventual fate. Quint, like Yaakov Mascetti ("Satan and the 'Incomposed' Visage of Chaos: Milton's Hermeneutic of Indeterminacy," *Milton Studies* 50 [2009]: 35–63), considers Sin and Death unqualifiedly internalized in Book II; Quint refers conclusively to "Satan's own spiritual death" (47, 56–57, 152). What "spiritual death" means within Milton's monism is elusive.

84. Lucretius, 114–19: II.251–93: Lucretius, who believes in the free will (or at least in the unpredictability, not to say the willfulness) of human beings, locates its origin in the random swerving of the atoms.

85. On the spirits in the human body, see Noga Arkha, *Passions and Tempers: A History of the Humours* (New York: Harper, 2007): in 1628, William Harvey's views regarding "the spirits that *putatively* flowed within the blood," as Arkha describes them, ranged from Harvey's noncommittal ("'there are many and opposing views'") to his conciliatory ("'blood and spirit, like a generous wine and its bouquet, mean one and the same thing, . . . so also . . . blood without spirit is no longer blood but the equivocal gore'" (189, my emphasis on *putative*); note Harvey's recourse to analogy, wine:bouquet :: blood:spirit. Harvey further suggests that since the nature of the spirits remains in doubt, they "'serve as a common subterfuge of ignorance'" (189); citation of Harvey from *The Circulation of the Blood*, ed. and trans. Kenneth J. Franklin (London: Dent, 1963), 149–50; to Arkha's, I add this citation: "if for driving factors spirits are to be understood, whatever in living bodies has force and drive would be styled spirit. And not all spirits are airy in substance, power, and quality: similarly, they are not all incorporeal" (150). Helkiah Crooke explains bodily spirit as an "exhalation" of the blood: "A subtle and thinne body alwayes moouable . . . and the vehicle or carriage of the Faculties of the soule." The spirit, wind-like, "passeth and repasseth at his pleasure, vnseene, but not vnfelt": *Μικροκοσμογραφια* [*Mikrokosmographia*]*: A Description of the Body of Man, together with the Controversies and Figures thereto belonging, Collected and Translated out of all the Best Authors of Anatomy, especially out of Gasper Bauhinus and Andreas Laurentius* ([London:] W. Iaggard, 1616), 173–74; thus described, such a spirit sounds as mysterious (not to say mystifying) as the Holy Spirit. Cf. Marjara, *Contemplation of Created Things*: "Spirits are qualitatively superior to corporeal substances, but they are the same substances in their essentials" (233–34). Cf. Harvey, *The Circulation of the Blood*, 155, where his point is not to prove that spirit per se is visible but that blood provides vital warmth.

86. Fallon, *Milton Among the Philosophers*, 104; Sugimura, *"Matter of Glorious Trial,"* 49–50; cf. 54. Cf. Marjara, *Contemplation of Created Things*, 229–39. Marjara's treatment is important, but he does not discuss (see?) the strains or gaps in the text that Fallon and Sugimura find or those that I do.

87. *Christian Doctrine*, 309–10 (Bk. I.vii): "For not even God's virtue and efficiency could have produced bodies out of nothing . . . unless there had been some bodily force in his own substance [*nisi vis corporea quaedam in substantia Dei*], for no one can give something he has not got"; *De Doctrina* translates the Latin, "unless there had been some corporeal power in his own substance" (294–95). Milton then cites Col. 2:9 ("*the whole fullness of the Godhead dwells in him* [*Christ*] *bodily*"), a text, as Kelley, ed., notes, that Milton twice interprets otherwise elsewhere (310n37). Milton goes on to observe the compatibility of the belief that "a bodily force should . . . issue from a spiritual substance [God]" with the belief that "something spiritual should be able to arise from a body." Graves, "'The whole fulness of the Godhead dwells in him bodily,'" 8, observes that Milton's argument is analogical; he also traces Milton's view to a precedent in Tertullian (6–7).

88. Joad Raymond, *Milton's Angels: The Early-Modern Imagination* (Oxford: Oxford University Press, 2010), 7. Cf. Brian Cummings, *Mortal Thoughts: Religion, Secularity, and Identity in Shakespeare and Early Modern Culture* (Oxford: Oxford University Press, 2013), 324: "The problem of metaphor lurks behind every seventeenth-century attempt to create a new reconciliation of the claims of mind and body." Aquinas and Cajetan come back to mind.

89. *Art of Logic*, 284, 287; Latin, 192, 202. *Online OED*, s.v. *Force*, sb.[1]: 7c (accessed 12/28/14). Pertinently, the first example cited comes from Thomas Wilson's *Logic*: "This [argument] that followeth is of good force."

90. When Raphael first adds that the soul receives reason from the aspiring (subliming) process, he is endorsing a psychological process traditional since Aristotle, Galen, and their various interpreters. In his next addition, "reason is her [the soul's] being, / Discursive or Intuitive" (V.485–88), his meaning has proved less clear. That discursive reason should be fed by the same process is unproblematical, as long as it works with data originating in the senses, as Eve does when she "reasons" about the reptilian tempter's words. That such reason should be the essence of the soul is less easily parsed—at least if we stop with that, as Raphael does not. Intuition, a higher reason, hints at another option that is less normative (at present) for human beings. It further suggests *recta ratio*, "not the mere logical and critical faculty but the Platonic capacity for attaining divine truth," which is inseparable from faith and therefore also from the will (see note 49 in this chapter).

91. Michael C. Schoenfeldt notes that Raphael's word "aspire" is the same term used in the epic to depict sinful ambition: *Bodies and Selves in Early*

Modern England: Physiology and Inwardness in Spenser, Shakespeare, Herbert, and Milton (Cambridge: Cambridge University Press, 1999), 146.

92. *Christian Doctrine*, 319 (Bk. I.vii); *De Doctrina*, 302–4. Milton refers here to this life, but in unfallen Eden, this life is continuous with another, higher one, and for Milton, on fallen earth what is beyond this life entails belief in mortalism until the end of earthly time.

93. *Christian Doctrine*, 309 (Bk. I.vii: my emphasis); *De Doctrina*, 294.

94. On phrases like "they say," "some say," and the like in Spenser and Milton, see my *Reading the Allegorical Intertext*, 308–310; in *Paradise Lost*, see the metamorphosis of the demons into snakes (X.575). On the ontological status of the intellect, cf. the discussion of Bryskett in Chapter 2 of this volume.

95. On the Scholastic terminology, see Kelley, ed., *Christian Doctrine*, 309n34; Marjara, *Contemplation of Created Things*, 232; Anderson, *Translating Investments*, 54–55 (Ridley, Cranmer, Eucharistic controversy).

96. Raymond, *Milton's Angels*, 83–84, approaches this question in a section titled "Are Humans Superior to Angels?" but, while recurrently stressing the similarities between the two in *Paradise Lost*, he does not settle the issue of their ultimate relationship.

97. Raymond, *Milton's Angels*, 272, similarly gives as one example of an angel's mistake, Raphael's misunderstanding "Adam's account of his need of Eve."

98. Corns, *Milton's Language*, 100–1, notes the "mysterious insubstantiality" of this passage without relating it to Chaos and Night; he aligns the presentation of the prayers to the Father's throne with the way "suitors may be presented at a royal court."

99. *Paradise Lost*, VI.781, 784, 828, 832.

100. John Donne, *The Sermons of John Donne*, ed. Evelyn M. Simpson and George R. Potter (1962; rpt. Berkeley: University of California Press, 1984), 10:229–48.

101. See Brian Rotman, *Signifying Nothing: The Semiotics of Zero* (Stanford, Calif.: Stanford University Press, 1987) on the history of this mathematical sign-concept, zero.

INDEX

Distinctions in entries between author and work generally refer to chapters: e.g., Milton (chapter 3) but *Paradise Lost* (chapter 7). Terms that recur continually, such as *form* and *figure/figuration*, are indexed mainly for definitions and defining uses.